Environmental Movements around the World

Volume 2 Europe, Asia, and Oceania

Environmental Movements around the World

Shades of Green in Politics and Culture

Timothy Doyle and
Sherilyn MacGregor, Editors

PRAEGER

AN IMPRINT OF ABC-CLIO, LLC
Santa Barbara, California • Denver, Colorado • Oxford, England

Library of Congress Cataloging-in-Publication Data

Environmental movements around the world : shades of green in politics and culture / [edited by] Timothy Doyle and Sherilyn MacGregor.
p. cm.
Includes index.
Volume 1—North America, South and Central America ; Volume 2—Europe, Asia and Oceania
ISBN 978-0-313-39353-2 (hardback) — ISBN 978-0-313-39354-9 (ebook)
1. Environmentalism. 2. Environmental sciences. 3. Environmental policy.
I. Doyle, Timothy, 1960–
GE195.E584 2014
363.7—dc23 2013020879

ISBN: 978-0-313-39353-2
EISBN: 978-0-313-39354-9

18 17 16 15 14 1 2 3 4 5

This book is also available on the World Wide Web as an eBook.
Visit www.abc-clio.com for details.

Praeger
An Imprint of ABC-CLIO, LLC

ABC-CLIO, LLC
130 Cremona Drive, P.O. Box 1911
Santa Barbara, California 93116-1911

This book is printed on acid-free paper ∞

Manufactured in the United States of America

Contents

Acknowledgments vii

1. Through the Green Kaléidoscope: Environmental Movements around the World 1
   Timothy Doyle and Sherilyn MacGregor

Part I: Europe

2. The Institutionalization of Environmentalism in Germany 23
   Jochen Roose

3. The Environmental Movement in Great Britain 45
   Christopher Rootes

4. The Framing Processes and Outcomes of Environment and Conservation Campaigns in Ireland 69
   Liam Leonard

5. The Green Movement in France: History, Ideology, and Action Repertoires 95
   Sylvie Ollitrault

6. The Russian Environmental Movement between 1985 and 2010: Changing Trends, Shifting Values 117
   Oleg Yanitsky

7. Pragmatic Ethics of Environmentalist Movements in Turkey 139
   Gürcan Koçan and Ahmet Öncü

Part II: Asia

8. Movement against Protected Areas in India: Social Networks and the Role of External Actors 165
Ashok Swain

9. The Environmental Movement in Sri Lanka 185
Sarath W. Kotagama

10. An Overview of the Green Movement in China 207
Da Mao

11. The Environmental Movement in South Korea: The Discourses of Anti-developmentalism and of Saengmyung (Life) 235
Woojoo Chang

12. Development as a Double-Edged Sword: Managing an Environmental Movement in Singapore's Urban Environment 259
Sofiah Jamil

Part III: Oceania

13. The Australian Environment Movement: Between Conflict and Cooperation 287
Cassandra Star

14. The Land Has Eyes and Ears 303
Mesake Koroi and Robert Wolfgramm

15. Environmentalism in New Zealand 331
Ton Bührs

Conclusion: Context Is All: The Many Environmentalisms 359
Adela Alfonsi, Sherilyn MacGregor, and Timothy Doyle

Index 375

About the Editors and Contributors 403

Acknowledgments

This book is designed for students, researchers, activists, and others who are interested in the many shades and shapes of environmentalism around the world. Coming from different countries and cultures, and presenting a range of perspectives, the contributors share a common commitment to placing environmental movements in the context of larger struggles for eco-political change, globally and locally. We dedicate this book to them and to all those engaged in these necessary struggles.

We owe tremendous thanks to Dr. Adela Alfonsi, whom we relied on heavily as assistant editor, and without whose expert editorial assistance, diplomacy, sound judgment, patience, and kindness, this project simply would not have been possible. Without her willingness and ability to work closely with authors whose first language is not English in order to improve their chapters, we would have a very different kind of book.

We would also like to thank the team at Praeger/ABC-CLIO who believed in this project from the start and who saw it through to production. Emily Birch, Beth Ptalis, Anthony Chiffolo, and Valavil Lydia Shinoj (Apex CoVantage) deserve extra special mention. Our colleagues at Keele University and the University of Adelaide provided support in a number of ways, including making it possible for Sherilyn to spend time in Adelaide to providing funding for Adela's work. We would also like to acknowledge that significant parts of this two volume title draw upon the research of a much larger three-year project entitled "Building an Indian Ocean Region DP120101166," which is funded by the Australian Research Council (ARC) Discovery Projects Scheme for funding in 2012/2015.

Finally, we thank our family members, especially Fiona Lawrence and Simon Pardoe, for their understanding and encouragement throughout the project.

Chapter 1

Through the Green Kaléidoscope: Environmental Movements around the World

Timothy Doyle and Sherilyn MacGregor

ka·lei·do·scope [kuh-lahy-duh-skohp] noun. 1. A continually changing pattern of shapes and colors.

The study of environmental movements brings together a vast range of theoretical and ideological material. Ideas that gather under the green rubric are thrown together in a myriad of ways such that any one form of organizing principle does not do justice to their diversity. Environmentalisms can be analyzed using numerous forms of academic categorization. That some of these are more useful than others is an argument that runs through, and is illustrated by, this collection. As both volumes vividly demonstrate, the study of environmentalisms is enhanced by learning about what people do and how they think about environmental questions on the ground in different parts of the world. Their stories help the "worlding"[1] process by challenging traditions and frames that have developed in what has always been a Western-centric field of scholarship. Their stories enable a richer and more colorful understanding of green politics than we have had up to now. When we look through the "green kaleidoscope," we see that the shapes and shades of environmentalisms are constantly changing.

Imperfect but Useful Categories

From the outset of this project, we have been careful not to project, a priori, our own ideological worldviews as Northern scholars—our assumptions, categories, and organizing principles—onto the empirical information provided by the contributing authors to the collection. We wanted alternative categories of ideas and actions to the ones we take as disciplinary standards to emerge from the case studies of our contributors. The fact is that different authors do use diverse scales, continuums, and taxonomies to organize movement ideas and repertoires. As our first volume clearly shows, and as this second volume will confirm, there is no doubt that opening space for different nomenclatures and using systems for ordering ideas and actions make good sense. In any given context, selecting one ordering principle will be useful for describing a national experience at a specific time in its movement's history; in another, students of environmental movements will do well to choose elements of each. Attempts at categorization are simply analytical tools to help understand the maelstrom of ideas, rather than boxes into which we must force reality as we experience and understand it.

In the first volume, we introduced the categories of global North and global South to depict the fundamental dichotomy between environmentalisms operating in different parts of the globe: those movements largely operating in the poorer, majority worlds (what people used to call "the third world") and those located in the more affluent, minority worlds. One reason for beginning with this simple and imperfect dichotomy is that it provides a clear geopolitical map for understanding the power relations and asymmetries that sometimes get lost amid talk of global civil society, global threats, and "our common future." We think this distinction between global North and global South still holds weight when understanding ideas and actions of diverse environmentalisms. But two other categorizations (or frameworks) have emerged from our reading of these movement stories: Left and Right and what we call the "three posts"—post-materialism, post-industrialism, and post-colonialism.[2] These are imperfect categories, but it is worth visiting them briefly.

Left and Right

Students of politics will have met the Left-Right spectrum before. It is an ideological continuum that emerged after Europe's Industrial Revolution and sometimes seems a little out-of-date today. Still, many of the contributors to this collection have used the categories from this form of traditional politics. Using the Left-Right spectrum of ideology and praxis as an organizing principle, we can observe that nearly all environmental movements in the global South still seem to cling to ideas from the Left. This is hardly surprising since many of these countries have not yet undergone the kind of industrialization seen in the global North, and, in those who are currently undergoing that process, the Left-Right ideological battlefield seems very apposite. In most parts of the world, capitalism has not yet won. In fact, in many parts of the global South, environmental struggles tend to be framed as "the people" versus either a large transnational corporation or the state. The Latin American cases in the first volume illustrate this framing nicely. In these struggles, the language of Marxism and neo-Marxism are never far from the surface: the owners of the modes of production are in conflict with the labor force.

Environmental movements have traditionally been movements of the Left, but in the global North there has been a blurring of the political continuum in recent years. Meanwhile, when ideologies on the radical left are invoked in many countries in the majority world, the language of unreconstructed Marxism is never far away. The state is often constructed as the enemy, as the major shareholder in the capitalist enterprise. In addition, transnational corporations are seen as partners to the bourgeois-led state in stripping away ownership of resources from the people. Many environmental campaign repertoires, when seen in this light, are not about generating more green science through environmental education to allow better environmental management or about stopping an industrial practice that is environmentally degrading. Rather, many issues are seen as being resolved when the people have control over the basic elements that sustain life: water, air, fire (energy), earth. In strict Marxist terms, through their campaign framings and repertoires, the people are taking control of the "modes and means of production" and are seeking to regain access to key resources. They are seeking to reconnect their lives to their land and their livelihoods.

There is nothing "post-political" here. As we discuss in the concluding chapter to Volume one, it is possible to observe a depoliticization of environmentalism in places where neoliberalism has been normalized and environmentalists pursue consensus within dominant framings of the crisis. We noted that climate change has played a major role in enclosing the green public sphere, leaving environmental groups with few options but to engage in symbolic politics (performing for the media) and call for individualized "lifestyle choices." But politics in most parts of the majority world is not just a piece of street theatre or a game with lots of choices: it is the bread and butter of survival. There can be no fence-sitters. Engaging with the powers-that-be, whether they be the state or big business, is tantamount to failure. Usually, the only source of an environmental movement's power is *solidarity*: the coming together of the people in a solid block of resistance. Dealing with a large corporation or the state means being "picked off" and losing the only source of power the powerless have—their togetherness. In the global North, however, working with the state, big business, and other powerful interests has become *de rigueur*. Here, environmental struggle often assumes the language of the Right: environmental reforms are pursued within existing institutional and corporate structures. It is imagined that tweaking the knobs and dials of capitalism and advanced industrialization may be enough to save the planet. In the global South, by contrast, more radical pathways and more profound changes are imagined as necessary.

Of course, these are generalizations. The chapters in both volumes in our collection suggest that environmental ideology and activism have moved away from simple Marxist political equations on many occasions. In the North, there are instances where eco-activists are pursuing a more radical line. Still, in the North, the more collectivist and structuralist responses of the radical Left are replaced with more atomized and piecemeal repertoires based on liberal, anarchist, and post-political worldviews. In the North, there are even those environmentalists articulating neoliberal—market-driven—solutions to environmental issues and problems. While there is no doubt that many Northern environmental groups are predominantly liberal and reformist in both the framing of their campaigns and their repertoires of action for achieving their goals, other national movements (and networks amongst and between these groups) within the more affluent world are not. Indeed, some share the same type of radical goals and repertoires as their less-affluent movement compatriots. What strikes us, though, is that even when both North and South repertoires are radical, they often constitute a different *type* of radicalism that manifests itself most plainly in strategic and tactical spheres.

The Three Posts

So, the Left-Right spectrum is still useful when one understands a world that is not necessarily *after* the Industrial Revolution; the Industrial Revolution has simply moved its major sites of production and consumption. This issue of *after*, of course, depends on what you are talking *after* about. It is here that we can usefully talk about reorganizing environmental ideas and actions into another form of categorization: the three posts of post-materialism, post-industrialism, and post-colonialism. The idea that environmentalism is a product of post-materialism (a condition where people are affluent enough to be able to think about issues beyond immediate survival) has been around since the 1970s.[3] It is an idea that is symptomatic of the Western/Northern framing of environmentalisms and, as a result, it has lost much of its explanatory power as green movements in the global South and in the-South-within-the-North emerge and are examined by environmental scholars. It has become increasingly important to analyze green movements in relation to their proximity to industrialization. In the North, we largely talk about industrialization in the past tense: we are now post-industrial societies. In the South, industrialization is still largely a future process or is something that is currently happening. Let us remember that the majority of Southern economies are still dominated by primary production. Through a post-industrial lens, then, environmentalism challenges the excesses of the industrialist project; the rights of corporations to pollute and degrade and the dwindling of the earth's resources as they are fed into the advanced industrial machines. As we discuss in the concluding chapter of Volume one, climate change is a classic post-industrial issue and it is no wonder that, first, it has dominated minority world green agendas and, second, that until very recently, it has been largely absent from the campaigns of the global South.

We can see when comparing the North American studies of Volume one with the European studies of Volume two that climate change is not just a North-South issue, or a conflict between the post-industrialists and the yet-to-be-industrialized. Moreover, there is a distinct difference between the environmentalisms that have longer histories in the "old world" and those emerging in "new world" societies such as the United States, Canada, Australia, and New Zealand. In these societies, the ravages of industrialization are more recent and so large tracts of wilderness and populations of wildlife ("other nature") still stand. In these countries, the green tradition has been more about *preserving* this wilderness and these threatened species than it has been about closing down the fossil-fueled "dark Satanic mills."[4] Using a post-materialist frame, all of a sudden other issues define environmental movement ideas and praxis: the protection of wilderness areas and the saving of threatened species. In this vein, we understand that climate change is not only a Northern issue but has also emerged from specifically Western European movement centers. It lacks the same currency in the majority worlds as well as new worlds of North America and Oceania.

Of course, there are still enormous similarities between the Anglosphere in parts of Europe, North America, and Oceania. And there still remain stark differences between these parts of the globe and the experiences of the majority. Using elements of the North-South dichotomy, as well as the Left-Right spectrums, we can view another *after*: after colonization. Using post-colonialism as a narrative frame, green concerns are cast in light of the colonizer versus the colonized; the dichotomous world of affluence and poverty. In different parts of the world, these frames or storylines are often used to explain the causes and effects of environmental issues and problems. In the South, the frames of post-colonialism and structuralism usually dominate. In many parts of Europe, post-industrialism is deep-seated, whereas in new-world cultures in North America and Australia, post-materialism can usually be employed to interpret environmental politics. But no sooner do we arrange any form of categorization, then environmental movements, with their almost pathological need to cross borders (whether these be conceptual, cultural, or geographical borders), make nonsense out of feeble positivist attempts to understand what is a diverse and continually changing phenomenon. There is no doubt that there is increasing pressure to homogenize the diverse interpretations of green identity into one unified global movement. But it is a central purpose of this collection to show that, despite the recent trend toward a grand green narrative (thanks in large part to the post-politics of climate change in a neoliberal world), the actual environmental issues on the ground are profoundly different in the South when compared to the North, and across various regional blocs.

Overview of Volume Two

Like the first volume in the collection, this volume is divided into three sections according to major geopolitical regions: Europe, Asia, and Oceania. We begin with six case studies from post-industrialist Europe. At first glance, it could be argued that environmental movements in Europe are overrepresented. But the fact remains that since the mid- to late-1960s, European environmental movements have been critical players in the world's greening. We must also recognize that Europe, despite current geopolitical imperatives, is a very diverse continent, consisting of regions and countries that speak different languages, celebrate different cultural and religious forms, and have very distinct histories than, say, the continent of North America. At the subregional level, movements in Western Europe, including the United Kingdom, Ireland, France, and Germany, have been major contributors to the post-industrial green tradition. Russia is included because we sorely needed to review a movement that first emerged on the other side of the "iron curtain" before its fall in the late 1980s and early 1990s. Eastern Europe (again too large a subregion to be overly confident about generalizations) has its own unique movement traditions, many of which were forged under state-led development and state-centric command-and-control governance structures. As with the case of Iraq in the previous volume, and of China in this volume, the kind of environmentalism and the types of repertoires people use to pursue their ends under authoritarian regimes are often very distinct from those we can observe in liberal democracies. Finally, the case of Turkey is investigated, a country historically depicted as the bridge between civilizations. In this collection, we adhere to this time-worn convention: Turkey provides a link between the global West and the global East, and leads the way on our journey south to the Asian and Oceanian regions.

Part One: Europe

We begin in Germany. Ever since Die Grünen (the West German Green Party) walked into their parliament back in 1983, dressed in jeans and casual shirts, demanding more participative forms of democracy, more support for grassroots environmental initiatives, and fairer representation of women, the German green experience has wrestled with the wins and trade-offs of power politics. Very early on in the party's experience, there were sustained battles between the fundamentalists ("Fundis") and the realists ("Realos"), and there is no doubt that the latter's pragmatic approach to politics usually won out. In more recent times, as with other powerful, mainstream movements, the debate has shifted to hand-wringing over the pros and cons of further institutionalization, as environmentalism steadily becomes part of the state. Indeed, Jochen Roose's discussion of environmentalism in contemporary Germany begins with the premise that the green movement has become deeply embedded in dominant German institutions. He argues, however, that what is actually institutionalized is the idea, not its realization. Furthermore, he provides a short history of German environmental movements in which their particularities are highlighted. Drawing upon a number of recent studies, including analyses of national newspaper reports, he plots the sorts of issues that have inspired protests over the years. Although Roose highlights the relative weakness of environmentalism in the Eastern German Democratic Republic (GDR), there can be no doubt that green movements and parties were a major force for unification both before and after the dismantling of the Berlin Wall. One other such particularity is the German movement's exceptional sensitivity to the issue of nuclear energy. All German environmental movements have converged on this issue and there has always been strong public opposition to nuclear development in Germany. In his final section, Roose focuses on environmentalism in three domains: politics, knowledge and science, and everyday behavior.

Next we move to the United Kingdom (U.K.). Christopher Rootes, a leading expert on environmental politics in the U.K. describes an extremely diverse and rich movement with a long history. He offers us a historical overview of the environmental movements from their inception in the last century, with a special focus on the strategies and frames since the 1950s. For most of this time, British environmentalisms have been able to gain access to the halls of power and have often played an insider role as opposed to an openly confrontational one. Unlike in Germany, the U.K. movement no longer holds a uniformly strong antinuclear position even though there was opposition in the 1960s and 1970s. No doubt, with the dominance of the all-consuming climate change debate in the U.K. and in other parts of Western Europe such as France, nuclear energy is often portrayed as "cleaner and greener" than fossil fuel–based energy sources.

Additionally, Rootes describes the development of more radical, direct action movements during the 1980s, the time of Margaret Thatcher, and beyond. These direct action groups created what he calls a "radical flank" effect whereby more "modern" environmental nongovernmental organizations (ENGOs) were more likely to be heard when there were more radical and angry voices shouting from outside. Since the 1990s, increased radicalization and grassroots participation in less formalized groups has occurred in British environmentalism. Another key point is that after the Rio Earth Summit in 1992, despite tensions that still flare between networks with contrasting aims, there has been a clear shift toward the major actors broadening their issues, agendas, and tactical repertoires toward a greater focus on social justice issues. But climate change, as discussed in the conclusion to Volume one, has created a new fault line. It has often occluded other environmental issues, and in the search for answers to the problems, it has exposed divisions between environmental movements. Rootes concludes with the idea that the dominant "insider approach" within the U.K. movement may have become a bit passé, as "a mature environmental movement is confronted by the limits of its power." He argues that U.K. greens may need to consider adopting a more adversarial "outsider" strategy, thereby becoming more distanced from government and corporations.

From the U.K. we move across the Irish Sea, where Liam Leonard presents an engaging analysis of the mobilization, framing processes, and outcomes of environmental campaigns in Ireland. By locating contemporary disputes about public space in Ireland within an understanding of the transition from a traditional agrarian society into that of a modern industrialized and urbanized European state, a wider understanding of the circumstances surrounding such contests emerges. This form of "issue history" provides scope for an analysis that goes beyond a social narrative, and facilitates an in-depth understanding of the consequences emanating from the framing and reframing of heritage within the context of a society where an obsession with history has been displaced by a zealous predilection for economic development. While this transition from traditionalist concern to modernist utility has placed great strains on the environment and heritage of the nation, it has also created opportunities for widespread political mobilization in successive decades during periods of economic growth or decline. These outbreaks of collective action have provoked widespread debate in the media as several thousand people have taken part in heritage-based protests in Ireland over this period. In his conclusion, Leonard argues that a shift in policy direction undertaken by successive neoliberal Irish governments has led to a move toward a primarily ecological modernization-orientated policy framework as set out by the European Union. The emphasis on economic growth over environmental protection, or concerns for local communities, has subsequently become a priority. This has been exacerbated by the Eurozone crisis and the bailout of the Irish economy by the International Monetary Fund (IMF) since 2008. Ultimately, the costs of economic growth have led to greater environmental degradation and an undermining of significant aspects of the nation's heritage. In particular, rural communities have borne the cost of Ireland's transition from an agrarian society into an industrialized and globalized economy. While communities have demonstrated a commitment to retaining some aspect of the traditional values and heritage of the Emerald Isle, Leonard argues that the state's policy of prioritizing rapid infrastructural development in order to underpin private consumption and multinational investment over local community services may prove to be ill-advised as fluctuations on the global markets continue. As the cycle of the "Celtic Tiger" gave way to the era of IMF-sponsored austerity in the wake of the post-2008 economic crisis, the values and concerns of these local communities may provide a new direction for the post-boom era.

Back over the Irish Sea and across the English Channel lies France, sandwiched geographically between Germany and Britain. Sylvie Ollitrault's chapter on environmentalisms in France provides an excellent overview of its history, ideology, and action repertoires. She examines the mix of global forces and local factors influencing the modern French environmental movements. In the 1970s, the French greens fought state and government decisions that would harm the environment, whereas now more criticism is directed at the private sector and economic neoliberalism. The contemporary green movement encapsulates concerns for other progressive causes besides traditionally green ones (this may be similar to what Salih has referred to as "hybrids" in the context of Iraq). Ollitrault sees French citizens as acting on their concerns as individual consumers via consumption practices and private citizens using forms of civil disobedience to exercise "a new form of everyday resistance." She says green activists are being "attached to a form of anarchism" and describes France as having a long history of local struggles and civil disobedience. She presents a picture of a population that is media-savvy, attached to anarchic concepts of freedom, and wary of globalization. The French appear to be exercising market-based choices through consumption that resists global agricultural practices threatening to local and traditional forms of agriculture (which they also associate with health). Repertoires of pacifist direct action, civil disobedience, and a general challenging of the authority of the state are discussed. The French green movement is connected with wider social movements, although some on the Right are more preservationist and less critical of capitalism broadly. Ollitrault concludes that this is mostly a middle-class movement—an urban phenomenon—and the challenge now is how to expand its reach more broadly throughout French society.

The Russian case, a *cause célèbre* of Eastern Europe, obviously has a history of political ideas and actions that is very different from that found in the West. Drawing on his vast research and knowledge of Russian environmentalism, Oleg Yanitsky begins his analysis at a time when Russia was still under the communist control of the USSR, and then traces these early days in 1985 up to and including 2010. He describes an environmental movement defined by "changing trends and shifting values." In its early communist days, it started off very rooted in the circles of cultural and scientific elites but, nevertheless, it was able to mobilize mass popular support. Despite the fall of the Iron Curtain, one hostile political opportunity structure was replaced with another. Under the increasingly hostile conditions of the

1990s, environmental movements have become highly bureaucratized and internationalized, compete with each other for resources, and are supervised by a government that regards them with suspicion.

This brings us to an interesting question about post-authoritarian regimes. Once upon a time in the Russian context, activists struggled under state-led forms of oppression. But what has followed is not necessarily emancipation. Instead, as we have seen in other case studies in this collection, previously one-party states may now have multiple parties, but these parties are still ruled by powerfully entrenched elites. In post-authoritarian states such as Russia, the state is now often seen as working in partnership with large corporations, and opposition to these state-corporate partnerships are resisted almost as completely as those dissidents who dared to utter opposition under the previous regime. Post-authoritarianism does not necessarily lead to the emergence of genuinely liberal-democratic states (which are increasingly rare) but may now be open to a new form of authoritarianism, one that is more closely disciplined by global market imperatives. Within this context, Yanitsky plots the major stages in the evolution of Russian environmentalisms, their value shifts and their political opportunities—or lack thereof.

From Eastern Europe we go next to Eurasia. The Turkish case study provides a story of pragmatic environmental ethics and practice. Gürcan Koçan and Ahmet Öncü begin their very informative work with a taxonomy of ethical responses to the environment, arguing that Turkey's environmental ethics are of a pragmatic, consequentialist type. This can be a problem, they say, because when environmental movements try to critique an issue or project that is deemed important to the nation's economic development, they are unable to counter these state- and market-led arguments. Furthermore, the authors examine the history of industrialization and modernization since the leadership of Ataturk. They argue that from the 1980s, in the pursuit of "etatist" or state-led processes of industrialization (where wealth is harnessed for the good of the nation), Turkey has opened up to foreign capital and to the global market. Against this political backdrop, they then select several themes through which to examine the emergent ethical frames, strategies, and actors. The picture that emerges is of increasing citizen mobilization and successful use of judicial and legal strategies—both nationally and internationally. This emerges from three case studies: conservation and parks; pollution and energy (including antinuclear resistance and opposition to hydroelectric dams); and animal rights. Koçan and Öncü argue that pragmatic ethics are developed when people respond to particular issues that come up and mobilize responses. The three cases show that institutional successes come on the legal or judicial front, but not in the policy-making arenas. Because the movement is largely issue-based, it is not founded on common, consistent moral understanding. Environmentalism in Turkey, they contend, does not have a deep theoretical basis. It is therefore particularistic and strategic in its responses to local constituencies. What emerges then are consequentialist and anthropocentric green ethics. As such, the movement is unable to articulate a strong opposition to capitalism; rather it reads positive consequences in capitalist "development" and the interests of the market always come first.

Part Two: Asia

From Eurasia to Asia, and the first country we visit in Part two is India. India calls itself the biggest democracy in the world and with over 1.3 billion people it certainly has a claim to this title. As Germany is to Europe, so must the status of iconic green case go to India in the Asian regional context. Ever since 1974 when a small band of women hugged trees to stop them going under in the path of bulldozers, the Gandhian tradition of Indian nonviolent resistance—known as *satyagraha* (firmness in truth)—has defined many forms of environmentalist direct action around the world. These earlier forest conflicts of the Chipko Movement are very different from those that Ashok Swain describes today. In this chapter, he concerns himself with mobilizations against unjust forestry management in southeast Orissa. Swain asks that age-old question in relation to classical social movement theory: what causes social movements to succeed or fail? He then takes his theoretical analysis further, arguing that classical social-movement theories, forged in the Industrial Revolutions of the global North, and based specifically around the concept of class, do not account for the recent mobilizations in the hinterland of Orissa. He examines the nature of group ties and social bonds that allow for collective defiance and argues that the tribal peoples—with preexisting strong tribal alliances and hostile relationships—had actually been incapable of wider-scale collective action. With the entrance of ultra-Left and ultra-Right groups into the area, new collective identities have now been forged. The Maoists, on the Left, have provided a narrative of collective resistance against the repressive state—a "dream of revolution"—and from the Right, competition between Hindus and Christians has forged a new politics of identity, facilitating new alliances and a previously unheard-of collective defiance. Another quite fascinating thing about Indian environmentalisms is that it is still the state that is most often cast in the role of the adversary, whereas in many parts of a globalized world of Northern elites, the state has often been sidelined in many conflicts, with the role of villain (in the eyes of many environmentalists) being now played by large corporations.

On the subcontinent, Sri Lankan environmental politics is also largely oriented around the role of the state. Sarath Wimalabadara Kotagama provides an overview of the development of an environmental movement in Sri Lanka since the 1960s. He examines government conservation and sustainable development policies and outlines the challenges for his country: land degradation, water pollution, air pollution, solid and hazardous wastes, loss of biological diversity, unsustainable use of natural resources, and environmental health and social well-being. He then selects a number of key campaigns to demonstrate a defining trend in the development of Sri Lankan environmentalisms. Beginning with the very early establishment of the Sinharaja World Heritage Reserves from 1965, he records the early days that saw the creation of a top-down environmental approach involving the passing of legislation and the involvement of technical and scientific elites in researching and advising government policy. In the 1980s increased popular concern emerges, with the mobilization of religious leaders. By the 1990s, there is a large flow of environmental reporting and an increased organizing force within environmental movements in response to governmental development plans affecting marshes and forests. It is in these more recent times that we start to see popular resistance move beyond passive resistance to forms of resistance based on coercion, violence, and bloodshed. During this more recent period, other more "legitimate" forms of activism are also pursued, with a landmark success experienced by green activists in the courts around a mining issue. Over time, Kotagama writes, Sri Lankan environmentalists have increasingly run into determined opposition from the state when they contradict developmental policies and projects. Occasional successes through the courts continue, but there are also notable losses, including serious breaches of human environmental security. He reflects that there was a moment of glory in the 1990s when public awareness was high and the media did a professional job. Since the civil war, however, NGOs have been tamed—endorsing state activities—and the media coverage of public affairs has turned vicious and partisan.

Of course, the other great social and economic engine of Asia, and the world, is currently China, the most populous country on the planet. No work on environmental movements would be complete without reference to this country in the early years of the 21st century. This issue of regime-type is incredibly important here. The kind of regime green movements are operating within will often determine the style of politics that can be played, including what kind of ideas can be expressed, and what types of political repertoires can be pursued. In China, a one-party authoritarian state, ENGOs, until very recently, were illegal. As is so often the case in many authoritarian and hybrid-authoritarian states, it was environmental activists and their organizations that were among the first to occupy a civil-society sector in China. Da Mao provides us with an excellent overview of Chinese environmentalisms by choosing four campaign case studies. As with the Sri Lankan case, environmental change began via a top-down approach emanating from the government in the 1970s. Da Mao shows how over time civil society starts to take a greater role, not surprisingly being more successful in those issues that do not challenge government developmental policies. These early ENGOs were dominated by intellectuals and professionals of the humanities and social sciences rather than the natural sciences. From the late 1990s, more grassroots mobilizations emerged around a myriad of themes, from the post-material to the post-industrial. In a very traditional post-materialist campaign based around protecting the Tibetan antelope, we see the emergence of international NGOs moving into the Chinese domestic polity from the outside, with the intervention of such global celebrities like former British prime minister Tony Blair. In this light, sometimes domestic green actors within authoritarian regimes are tarred with the brush of international provocateur by their own governments, depicted as "helping outsiders" with dubious—in this case, "non-Chinese"—agendas. In other cases, Da Mao talks of some big, iconic anti-dam campaigns where ENGOs mobilizing opposition learn that they can counter only these state-led projects by providing technical and scientific counterarguments.

The case against Apple demands special attention in the chapter on China. The campaign begins as local NGOs draw attention to Apple Corporation's role in contributing to water pollution. Indirectly, this attracts additional attention for Apple, which is vehemently criticized for not adequately supervising its Chinese production chains that are polluting. Coincidentally, a campaign by a U.S. NGO draws attention to Foxconn's bad labor protection measures, thus dovetailing these two issues and NGOs, forging an international campaign to curb the negative impacts of globalized production processes. In sum, he finds that NGOs are increasingly mature, organized, and internationally connected, but that domestic politics still plays the dominant role in determining the topic selection and direction of environmental movements in China. No doubt, as the Apple case shows us, it is still easier to critique "outsider" practices of transnational corporations than it is to question development projects more directly linked with the state as the major stakeholder.

Woojoo Chang introduces us to South Korea, with a chapter on the history of her country's environmental movement. She provides an overview of Korean environmentalism, with a specific focus on its two main discourses: global anti-developmentalism and a more locally specific *Saengmyung* (life). The latter is a holistic philosophy that resonates culturally with Koreans, rooted as it is in familiar spiritual concepts. These two discourses, in turn, are discussed with reference to two case studies that are also an opportunity for Chang to further consider the evolving role of women in environmental activism; a constant theme in green politics in the global South. She demonstrates how the discourse of *Saengmyung* has inspired the *hansalim*—small local organic cooperatives—that facilitate the participation of women because they respond to women's roles as caregivers within families and managers of family economies. She concludes, however, that neither set of discourses has been able to effect mainstream social and political structural change and, as a consequence, she prefers to see them more as subcultures.

Part Three: Oceania

Like all regional constructions, as was made clear at the outset of this collection, map-making is sometimes an arbitrary process, but oftentimes it is a fiercely political and strategic one. In recent times, we have witnessed the rise of the Indo-Pacific, the regional substance of the final part of this work, a region occupying the connecting seas between the Indian and Pacific Oceans— called Oceania. The green tides of environmental movements have also had their own distinct substance and character in this part of the globe.

To introduce the final region in the collection, we have Sofiah Jamil's chapter on Singapore, the Southeast Asian city-state at the end of the Malay Peninsula, a major gateway between the Pacific and the Indian oceans. It has its own unique history, and shares characteristics of past European occupation with its own Asian heritage. Singapore is another case of an Asian authoritarian regime, often referred to by its leaders (and political scholars) as an example of a hybrid regime. Additionally, Singapore is predominantly a heavily urbanized society, so environmental issues are usually interpreted within these heavily populated domains. Jamil begins the chapter by providing a brief history of Singapore's urban development strategies from independence in 1965 to the late 1980s, and then illustrates how existing urbanization policies provide support for the cultivation of the two main currents in Singapore's environmental movement: conservation and eco-efficiency. She then discusses how the processes of urbanization have shaped the development of these currents of environmentalism and suggests potential avenues through which a third and perhaps more radical current of "environmental justice" might emerge in Singapore in the future. An interesting question that is raised by this discussion is how might concepts such as environmental justice be viewed within a hybrid authoritarian society like Singapore? This is the kind of question that informs Jamil's conclusions. Importantly, she argues that initiatives leaning toward environmental justice have the potential to be, and should be, pursued further by eco-activists in Singapore. She contends that they help to address the challenges that conservation and eco-efficiency groups face by emphasizing the socioeconomic needs of a broader range of stakeholders in society, including poor and elderly people. Only in this manner will it be possible to avert ever-present deadlocks with government bodies and for the movement to overcome the feeling in some quarters that it is elite and disconnected from large sections of mainstream society.

Moving south from Singapore, we reach Australia next, with its own rich history of green politics and activism. Cassandra Star sets her sights squarely on the current state of Australian environmentalism, arguing (in a way that resonates with Rootes's chapter on the British case) that there is an overreliance on "insider politics." She investigates the perspective of the changing relationship between the environmental movement and the state—its internal dynamics of conflict as opposed to cooperation, and its lack of ability to forge alliances with other social movements. She provides a brief history of the movement, arguing that earlier Labour governments were able to secure electoral leverage from green issues, whereas under Conservative governments the relationship was one of overt hostility. Unlike in the U.K. adverse circumstances have not produced much cooperation with other key movements. If Australian environmental movements want to remain relevant, she argues, they must move beyond the "preservation-of-wilderness approach" to embrace other progressive social movements (including those of indigenous groups). Instead of focusing on "where people are *not*," a focus that impedes further collaboration with other groups engaged in social justice issues, the greens in Australia need to reframe their understanding of environment.

The penultimate stop on our journey around the green globe is the small island nation of Fiji. In their chapter, Mesake Koroi and Robert Wolfgramm remind us that there is no one way of knowing social movements and political experiences. As we pointed out in the introductory chapter to Volume one, the theoretical structures through which we understand environmental and social movements have largely been constructed from a Western perspective. In their account, they present a very different way of understanding environmentalism, which they describe with the phrase "the land has eyes and ears" as well as by introducing the activists who bear witness. Although the chapter begins in the manner of most chapters in this collection, with a brief overview of environmentalism in Fiji, the bulk of the writing presents Koroi's extremely personal description and vivid memories of a people who once lived in an environmentally respectful and sustainable way. The role of Western economic rapaciousness is discussed, but, alongside this, there is reference to some benefits that came along with contact with Western culture, principally via Christianity, which did away with practices such as cannibalism. The authors describe a holistic way of living and understanding the environment: it is sacred, a living spirit. They draw on local knowledge of the land, sea, plants, and animal life to show how humans were integrated into patterns of plant and animal cycles and behaviors, inside the laws of seasonality. The chapter finishes with a discussion of some of the challenges facing Fijian youth (brought by increasing urbanization and some unfortunate impacts of Westernization) and a critique of the relentless hunger for money that is causing indiscriminate logging, the depletion of tropical fish, and the destruction of Fiji's coral reefs.

In the final chapter of this volume, Ton Bűhrs completes our maritime journey with an extremely thorough investigation of environmentalism in New Zealand. Ever since the days in the early 1980s when French intelligence agencies blew up Greenpeace's flagship in Auckland Harbour for its opposition to nuclear testing in the Pacific, New Zealand has taken center stage in the minority-world green movement. Bűhrs argues that environmentalism in New Zealand needs to be understood in its broader political-economic and social-cultural context. He identifies the major streams of environmentalisms in his country, outlines the evolution of their different ideas, reminding us nicely at the conclusion of this book that there is not one environmentalism but many, even within one national context. The ideas change over time, some assuming dominance, some dropping off the green agenda but continuing to be embraced by the more radical voices. If Fiji is only recently wrestling with the ravages of Western markets, New Zealand is often presented as an ideal type of the global North's free-market model. The extreme neoliberal context of New Zealand politics has led to the deregulation, marketization, and decentralization of environmental management. In this manner, there is no longer a national locus for the public discussion and policy development of environmental concerns. Instead, issues have been devolved regionally to poor effect, with environmental movements being depoliticized as a result of these dynamics. For example, environmentalists now tend to work with government in an overly consultative and collaborative way rather than maintaining adversarial or critical stances.

In the final pages of his chapter, Bűhrs offers reasons to be cheerful about the future potential of green parties, and this is a good way to tie up the threads of Volume two. In the first chapter of this volume, we begin by celebrating those young greens who first marched into the West German Parliament in the early 1980s to argue for a new politics that challenged the traditions of how politics was structured in the institutions of the state. Die Grünen's four pillars—ecology, participatory democracy, nonviolence, and social and environmental justice—remain the catch-cries of new green parties, emerging from green movements across all the regions that we have investigated in this two-volume set. No other political party of the late-20th and early-21st centuries, in whatever country one observes, has this kind of global reach. But it must be remembered that green parties emerge from the movements that spawn them; parties do not replace movements, they are simply an important *part* of green movements. As green politics becomes part of the global mainstream, it is important to celebrate and to continue to support those parts of green movements that remain outside the state. While positive environmental change can be made by talking power-to-power in short-term, three-to-four-year cycles, it remains just as critical as it ever was to continue to listen to those voices outside the room. It is only in pursuing diverse cultural, social, and political pathways that environmental movements will continue to be instigators of change in the medium- to long-term, playing a critical role in shaping the way people live and die on this beautiful blue planet into the next century, millennium, and beyond.

Notes

1 Waever and Tickner, *International Relations around the World.*

2 Doyle and McEachern, *Environment and Politics.*

3 Inglehart, *The Silent Revolution: Changing Values and Political Styles among Western Publics.*

4 Blake, "And Did Those Feet in Ancient Time."

Bibliography

Blake, W. "And Did Those Feet in Ancient Time." In *The Norton Anthology of English Literature,* Fifth Edition, edited by M.H. Abrahams, Vol. 2, 77. New York and London: W.W. Norton & Company, 1986.

Doyle, T., and D. McEachern. *Environment and Politics,* Third Edition. New York and London: Routledge, 2007.

Inglehart, R. *The Silent Revolution: Changing Values and Political Styles among Western Publics.* Princeton: Princeton University Press, 1977.

Waever, O., and A. B. Tickner. "Geocultural Epistemologies." In *International Relations Scholarship around the World: Worlding beyond the West,* edited by A. B. Tickner and O. Waever, 1–31. London and New York: Routledge, 2009.

PART I

Europe

Chapter 2

The Institutionalization of Environmentalism in Germany

Jochen Roose

Introduction

Looking at German culture today, environmentalism can be found everywhere. Regardless whether one walks through the streets and notices advertisements for organic or environmentally friendly products, or whether one looks at the political discourse, the basic idea of considering the impact of human behavior on the natural surroundings has permeated thoroughly into German culture. Even Pope Benedict XVI, who is not seen as being the most progressive power, praised the environmental movement during his visit to Germany in 2011. Environmentalism has become a fundamental norm that one can refer to in almost all situations to substantiate an argument. In the social sciences, this phenomenon of a norm becoming self-evident is described as institutionalization. This of course does not imply that in Germany behavior is factually oriented toward avoiding harm to the environment. German society is by no means sustainable in the sense that its current activities and resource consumption do not harm the life chances of future generations.[1] What the idea of the institutionalization of environmentalism refers to is the self-evident *idea,* not its realization.

Environmental impact has not always been a legitimate argument in public debate. Going back only four decades, the picture would have looked totally different. Environmentalism was discredited as an irrelevant or even damaging, dangerous idea, only propagated by left-wing extremists and loonies. And going back two more decades, this idea was by and large unknown. Such a fundamental cultural change in such a short time is an astonishing and very unusual development. In this chapter, I explore how this change came to gradually disperse into society and to what extent it can be found in German culture today. The concept of institutionalization is used as a guideline.

Of course, other countries have witnessed similar cultural changes.[2] My aim here, however, is to focus on the German case. Though the approach is not explicitly comparative, I will try to underline German particularities that tend to be unusual (though often not unparalleled) in international comparison. In the following sections, I will first outline shortly the idea of institutionalization and its measurement, before I turn to the German environmental movement as a primary driving force of this cultural change. I will then explore the question of where the institutionalization of environmentalism has led to.

Institutionalizing Environmentalism as a Concept

The concept of institutions has been the basis of a whole paradigm in social sciences, with different currents in sociology, organizational studies, and political science. Though in itself highly complex, the main ideas of the approach have been summarized in very valuable work, on which I will primarily rely here.[3]

While in everyday language institutions are often associated with formal organizations that have a kind of official status, the term is understood differently in the social sciences. Scott defines institutions as "regulative, normative and cultural-cognitive elements that, together with associated activities and resources, provide stability and meaning to social life."[4] Institutions are the basic elements of our cultural understanding of the world. While compliance with regulative institutions comes out of expedience, normative institutions are followed out of social obligation, and cultural cognitive institutions are taken for granted and, therefore, compliance is regarded as inevitable. We are only partly aware of them and especially, we only partly realize their socially constructed character. Social rules and even facts reveal their social effect only when they are broadly accepted. It is to this social acceptance that the idea of institutions refers.

The empirical measurement of institutions is a problem of its own.[5] While research has focused for some time on formal rules in organizational settings, this approach remains unsatisfactory, especially as our interest concerns primarily normative and cultural-cognitive elements. A possible indication for this can be attitudes and even more so, practices. Both approaches have their own specific problems. Attitude research is always in danger of measuring pseudo attitudes. Research attempts to find out what people think about certain problems, but sometimes people have never thought about these issues. This problem is by no means new.[6] However, in the case of analyzing institutionalization, it becomes even more virulent as the relevance hierarchy and cognitive awareness are part of the phenomenon itself. Still, surveys can be of interest if the question wording is open enough to mirror the importance credited to a problem by the respondents.[7]

Using practices as an indication for institutions is the most common approach.[8] However, environmentalism is a preference, an idea that is transposed into action. In which cases environmental concerns have been influential is not as self-evident, as other preferences, opportunities, and resources can be influential as well. Literature on the relationship between attitudes and behavior addresses exactly this problematic link.[9] At the same time, providing opportunities for environmentalism and the position of environmentalism in the hierarchy of preferences are part of the phenomenon itself. Therefore, the environmental aspect of behavior is a complex indicator of environmentalism. Nevertheless, behavior must be part of a fuller picture of environmentalism as an institutionalized part of culture.

Environmentalism can be relevant in many domains of life. Some aspects of environmentalism in thinking and behavior are tied to specialized roles, professional behavior, or organizational rules, while in other respects environmentalism is consequential for everyday behavior. To cover the whole breadth would require an enormous spectrum of research. In this chapter, I focus on three fields in which environmentalism can be institutionalized: everyday behavior, politics, and knowledge.[10] This is a somewhat arbitrary choice, but the chosen fields can be regarded as basic and influential on other domains, as they are either influenced by opportunities produced in other domains or they are a result of environmental preferences in other domains.

Before I take a closer look at the institutionalization of environmentalism, I first want to turn to the way in which environmentalism was introduced into German culture. This takes us to the history and development of the environmental movement.

The Development of the German Environmental Movement

The development of the German environmental movement can be described from many angles. Here I provide two perspectives. A narration of main developments and events is supplemented with a more systematic description of protest event analysis.

History of the German Environmental Movement

Precursors of environmentalism as an idea can be dated back to Romanticism in the late 18th century.[11] In these early sources we find the idea of valuing and partly also protecting the natural environment. However, the first wave of a social movement that was concerned rather with protecting nature and not with environmentalism in a modern sense can be found at the beginning of the 20th century. In these years some major organizations were founded in Germany (and elsewhere in Europe) that still exist today.

The National Socialist totalitarian system and World War II led to a sharp break in the environmental movement. Civil society was completely destroyed, and in the new democracy of the Federal Republic of (West) Germany, organizations for the protection of nature had to be refounded. However, in the decades after the war, recovery from destruction and economic prosperity dominated the political and private agendas, by and large leaving aside concerns of nature protection. Political activism in social movements or other arenas beyond institutionalized politics was also discredited by the National Socialist movement.

During the early 1970s, environmental problems in the sense of substantial damages to the basis of human life on Earth entered the political agenda in the Western world. In the late 1960s and early 1970s, a number of local initiatives started to discuss local environmental problems. Parallel and not directly connected to these early initiatives, the first programs of environmental policy were implemented in the late 1960s, mainly targeting air pollution by end-of-pipe technology. The local and at first isolated groups made fertile ground to take up the broader global debate on environmental problems and the "limits of growth." In 1972, a network of local initiatives was formed, the Bundesnetzwerk Bürgerinitiativen Umweltschutz (BBU, Federal Network of Environmental Citizen Initiatives), which became a de facto center of the environmental movement for the next 10 years. Beyond local initiatives, the more traditional nature protection organizations took up the new issue sooner or later. During the 1980s additional influential organizations were formed, such as the national chapter of Greenpeace and the Bund für Umwelt und Naturschutz Deutschland (BUND, Federation for Environment and Nature Protection Germany), which later became the German chapter of Friends of the Earth International. Other traditional organizations became part of the movement network and added broader environmental issues to their nature protection policies, like the German chapter of WWF or the former Deutscher Bund für Vogelschutz (DBV, German Association for the Protection of Birds), which changed its name to Naturschutzbund Deutschland (NABU, German Nature Protection Federation) to mark its broader approach including environmental issues.[12] In the 1990s, these larger organizations began to dominate the field while the smaller local groups remained vital, at least to some extent. At the beginning of the new millennium it seems that the vitality of local groups has somewhat diminished, leaving the issue to the large environmental interest groups.

So far this same story could have been told for the environmental movements in many Western European countries. However, there is one particularity of the German movement: the role of the struggle over nuclear energy.[13] Right from the beginning the environmental movement incorporated the antinuclear energy movement. "Resistance against the nuclear program was the common denominator of all movement organizations and initiatives."[14] At all building places for nuclear plants of any kind, local resistance groups formed and cooperated with national organizations. While the protest was at first moderately confrontational with place occupations, for example, at the first building site in Wyhl, the conflict escalated as soon as the government pursued its initiatives more actively. The 1981 battle in Brockdorf, the building site for a nuclear power station in northern Germany, became a symbolic event for a highly radicalized movement. Though other projects also met radical resistance by the environmental movement, such as the extension of the Frankfurt airport, it was predominantly nuclear energy plants that led to violent protest. With the nuclear issue and its aftermath, the whole environmental movement polarized West German society in the 1980s.

Due to the prominence of the nuclear issue, the Chernobyl accident became a turning point not only for the antinuclear movement but also for the environmental movement as a whole. As predictions of an upcoming catastrophe seemed to be proven to the sensitized German public, the whole environmental movement gained broader acceptance.[15] Though polarization of society did not disappear overnight, the interpretation of environmental pollution as a fundamental problem and nuclear energy as a risky technology gained acceptance in broad parts of society. This change also became visible with the establishment of the federal ministry of the environment.[16]

Though in the 1990s the radicalism of the environmental movement diminished and local groups became weaker, the antinuclear movement remained lively. The transport of nuclear waste and its storage became the focal point of large mobilizations time and again. While in other European countries nuclear movements disappeared due to the respective governments' decisions to phase out nuclear energy (e.g., Austria) or due to the movement's failure to make any substantial impact on government policy (e.g., France), in Germany the movement survived.[17] Even the decision of the federal government in 2000 to phase out nuclear energy stepwise did not discourage the antinuclear protests. Further transports of nuclear waste were again a reason for large-scale mobilizations. The nuclear accident in Fukushima, Japan, had similar fundamental effects as the Chernobyl accident. German public opinion shifted immediately to opposing nuclear energy. While the Conservative government had partly reversed the decision of 2000 and allowed the nuclear power stations a longer period of energy production, only days after the Fukushima accident the decision was reversed to an even faster phasing out. In 2022, the last nuclear power station should shut down according to the current legislation.

In the Eastern GDR the situation was completely different.[18] An independent civil society could not evolve. Accordingly, nature protection and environmentalism were limited mainly to the state-controlled Gesellschaft für Natur und Umweltschutz (Society for Nature and Environment Protection). Here, beyond practical nature protection initiatives, some discussions of environmental problems also arose under constant threat of suppression. In the late 1980s, some small clandestine groups formed along environmental issues and opposition to nuclear energy. The Chernobyl accident was also influential here. The Umweltbibliothek (Environmental Library) in East Berlin, which collected information on environmental problems independent of state authorities, became a focal point of opposition and repression.

Due to the prominence of environmental issues in the oppositional movement after the fall of the Berlin Wall, environmental policies ranked quite high in the few months of the freely elected government of East Germany before the two countries reunited in 1990. During the turmoil of transition from state economy to market economy, environmental issues and civic engagement lost prominence. Left-libertarians concentrated their engagement on the struggle against right-wing radicalism. Environmental groups lost power and are still even now much weaker in East Germany than in West Germany.

Activities and Issues of the Movement

Beyond a historical narrative, social movement research provides us with more systematic descriptions through protest event analysis. Based on newspaper reporting, protests are analyzed according to the claim, the form of protest, the location, the participants, and so forth. The most comprehensive source for protest data in Germany, primarily West Germany, is the PRODAT project.[19] Using two German national newspapers, the *Frankfurter Rundschau* and the *Süddeutsche Zeitung*, for all Monday issues and in every fourth week for all issues, protests in West Germany for 1950 to 1990 and for the whole of Germany from 1991 to 2002 have been collected.[20] We refer here to all protests on environmental issues and against nuclear energy projects.[21]

Though they appear infrequently from the beginning of the time series, environmental protests were not reported to a considerable extent until the mid-1970s (see Figure 2.1). The reason for this take-off is the antinuclear movement. The conflict over nuclear energy intensified in these years. Particularly at the various building sites, local groups in cooperation with national organizations mobilized strong opposition. In the early 1980s, the number of protests against nuclear energy declined. The number of potential building sites became smaller as some planned projects entered the

Figure 2.1 PRODAT study, own calculation

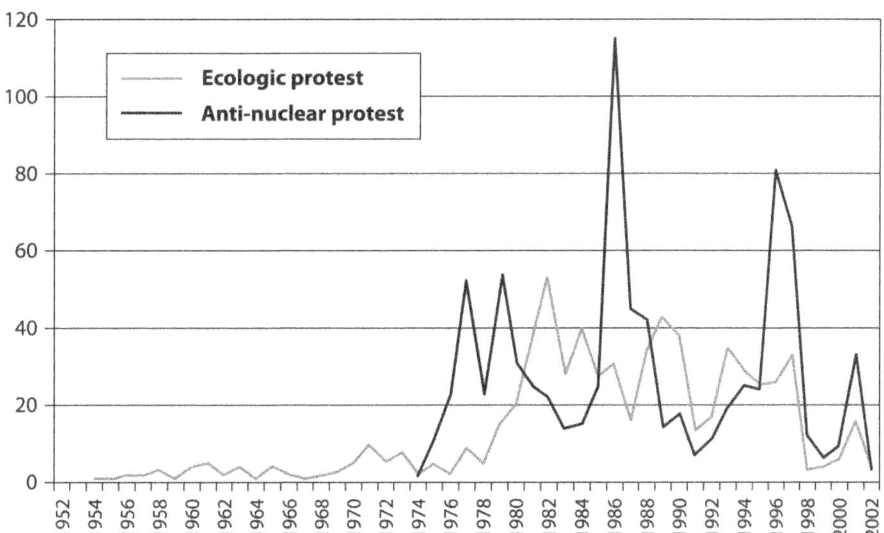

building phase, while others were abandoned. At the building sites, protesters remained active throughout.

In the 1980s, ecological problems became more salient. The most prominent issue was the (assumed) danger to woods. The belief that air pollution would lead to hyperacidity resulting in the death of all forests (*Waldsterben*) within one or two decades was first published in 1979 and became a leading assumption in the following years.[22] Broad programs for reducing air pollution were put in place and emissions declined considerably. The *Waldsterben* issue was not only a driving force behind protests, but was also a key instigator of policies and, not the least, the acceptance of the idea of environmental pollution as a relevant modern problem.[23] Parallel to this, protest resistance against other infrastructure projects, especially the extension of the Frankfurt airport, mobilized protest, which was often radical and violent.[24]

In 1986, the agenda shifted dramatically with the accident in Chernobyl. Again, nuclear energy was at the center of the movement, motivating many protests. The protest wave decreased only after decisions were made to build no further nuclear power stations and to withdraw from the planned construction of a reprocessing plant in Wackersdorf in 1989.

Though environmental protests diminished directly after the fall of the Wall when the problems of German unification dominated, they recovered again during the mid-1990s. Biodiversity, genetically modified organisms (GMOs), and particularly climate protection were among the important issues. In 1996 and 1997, the transport of nuclear waste and its disposal were highly controversial, motivating large and sometimes radical protest. The intensity of these protests came as a surprise to many spectators, as the conflict about nuclear energy had calmed.[25] The protests documented that the antinuclear movement had survived in latency, but was still able to mobilize. The decentralized character of this part of the environmental movement, with local but highly active and connected groups making up the backbone of the movement, proved to be very stable.

In 1998, the Green Party became part of the national government for the first time. For the environmental movement this had immediate effects. Germany's corporatist tradition was extended to the domain of environmental policy. Critical voices with a movement background were invited to various consultation committees. This can be regarded as the last step in the process of acceptance of the environmental movement and its representatives. Due to better opportunities for lobbying with (fairly) open addressees in the national polity, the activities of environmental organizations shifted from protest to lobbying.[26]

A short revival of protest intensity can be seen in 2001, when another transport of nuclear waste took place. It is interesting to see that in the context of intensified antinuclear energy protests, ecological protests also gained momentum. This applies to the protest wave in 2001, and also applies to earlier developments. The close connection of antinuclear protest and wider ecological issues as integral parts of the German environmental movement is also documented by this mutual support.

The time series of the PRODAT data set ends in 2002. Rucht refers to another analysis, based on the left-alternative newspaper *taz*, which additionally includes the years 2003 and 2004 and supports the finding of less protest in the new millennium, with the exception of 2002.[27] At the same time, this source documents that environmental protest did not disappear completely. Environmental protest as such has become institutionalized in the sense that it is regarded as an issue worth protesting about, and protest is regarded as a relevant option for citizens reacting to environmental problems.

Institutionalizing Environmentalism in Germany

The environmental movement has been both the carrier and the result of the blossoming of environmentalism in Germany. In the following part of the chapter I will take a closer look at environmentalism in three domains: politics, knowledge and science, and everyday behavior.

Environmentalism in Politics

In the polity of the Federal Republic of Germany, environmentalism found its place first in 1974 when the Federal Environment Agency (Umweltbundesamt) was founded. The agency's tasks are to consult the government on environmental issues, assist with the transposition of environmental law, and inform the public about environmental measures. The Federal Environment Agency was an early regulative reaction to the upcoming discussion about environmentalism.[28] Though some policy measures were taken, environmental policy remained marginal at first. A more influential regulative institution was the ministry for the environment, as mentioned earlier. The ministry became an important addressee and political coalition partner for environmental interest groups. The receptivity of the federal government and parliament for environmental concerns increased following the establishment of the ministry for the environment.[29] This also documents the normative institutionalization of environmental matters as relevant issues in government and parliament.

This development in the political system is backed by political concerns in the general public. Since the mid-1970s, the majority in Germany regards environmental problems as very important.[30] However, as all political problems can be regarded as "very important," this result is not too telling. Forced to choose the two most important political problems, the environment ranked still quite high during the second half of the 1980s.[31] Between 1986 and 1990, more than a quarter of all Germans regarded environmental problems as one of the two most important political problems. However, in the aftermath of German unification and the difficulties of transformation in East Germany, the prominence of environmental problems decreased. By the end of the 1990s, only about 6 percent of Germans regarded environmental problems as one of the two most important problems in the country. This share remained on a comparable level, with some variation.

The Politbarometer survey, on which this information is based, contains questions on general political issues. However, if we refer to another study focused on environmental issues,[32] results change dramatically. In these surveys, in the years after 2000, consistently, about 20 percent of the population regards environmental pollution and climate protection as one of the two most pressing political problems. This indicates first the unreliability of such survey questions. However, it also indicates that a small change in the frame of reference—in this case, the announcement of a survey focusing on environmental problems—pushes the issue relatively high on the agenda. A possible interpretation would be that environmentalism is among those problems that are regarded as very relevant and whether it ranks first, second, or lower in the current hierarchy depends on the mentioning of issues in the context of the survey.

Statements in surveys do not bear any costs and are therefore only a weak indicator of actual relevance and the depth of institutionalization. Stronger indicators would be practices, though they are of course affected by other influences. One indication for practices documenting the institutionalization of environmental problems as a relevant political concern is the protest activity documented under "Activities and Issues of the Movement." Another indicator is the support of environmental interest organizations. National organizations with close links to the environmental movement witnessed more or less continuous growth in membership and in received donations from the mid-1980s to the present.[33] About 8 percent of the population consists of members of an environmental organization and nearly a fifth donated money to them.[34] These results indicate both the institutionalization of the environmental movement as such,[35] and also the normative and cultural-cognitive institutionalization of environmentalism in the population that results time and again in donations or participation. However, environmental organizations report a decrease in active long-term participation in their activities.[36] People are hesitant to commit themselves for longer periods of time.

Environmental Knowledge

Environmentalism also entered the sciences. As in other countries,[37] the effects of various kinds of pollutants on nature and human health became an important issue. The exchange of scientific expertise across national borders has been common for decades. Accordingly, developments on research issues have also been internationally connected. Remarkable for the development of environmental knowledge in Germany is another issue: distrust.

The sciences are generally seen as having high prestige and high trust by the broader public. Scientific results are regarded as an authoritative source for objective knowledge. Scandals about fraud or misconduct have not harmed this trust in fundamental ways. However, for environmental knowledge and particularly for knowledge on nuclear energy, this is only partially true at best. Particularly during the 1980s, distrust of findings and expertise distributed by official sources was widespread.

This distrust was the base of a specific development in research on environmental problems and the security of nuclear energy. Assumptions about harmful effects are at first assumptions about causal influences in nature that are subject to scientific research. Accordingly, one might expect university departments and public research institutes to shift their focus to these questions and provide the required knowledge. However, early on the environmental movement was not prepared to rely on this "standard path" of modern societies. There was widespread dissatisfaction with the speed of dealing with these questions and coming up with possible solutions. But, even more, there was widespread mistrust in the neutrality of provided findings. The conclusion drawn by activists in the environmental movement was not only becoming students and later professors in their respective fields and following up on the questions relevant for environmental protection; another path taken was the founding of independent research institutes.[38] These institutes did not claim scientific neutrality, but denied the possibility of scientific neutrality in general. Rather, they made their environmental bias explicit. Thus, they gained trust among the environmental movement and its sympathizers. Studies from these institutes were based mainly on meta-analysis assembling, in particular, voices critical of technological achievements and emphasizing risks for nature and humans. Furthermore, they were dedicated to applied research, trying to find practical and realistic solutions to environmental problems.

Studies from these institutes and from other concerned scientists were used as counter-expertise in public discussions, and also in approval procedures of various kinds. This scientific backing became an important asset of the movement, just as court procedures have been an important strategy to stop environmentally problematic projects.

These institutes and their expertise were first met with fierce resistance and accusations of general incompetence, irrelevance, and lack of seriousness by established researchers. As the staff of these institutes often lacked higher academic grades and the insignia of the classical academic system, the scientific quality of their work was refuted. This applies mostly to the highly controversial and polarized debate on nuclear energy. Again, the Chernobyl accident marks a turning point, as it fundamentally delegitimized the broad coalition of established scientists who claimed that large accidents in nuclear power stations were extremely unlikely and would not happen. Arguments about technological differences between Soviet and German nuclear power stations did not convince the German public, which was highly sensitized toward risk issues in nuclear technology. Rather, the independent environmental institutes were trusted as the only reliable source of information. Still it took another decade until this kind of counter-expertise was officially accepted, in the sense that critical voices were included in the most important advisory committee to the federal government on nuclear issues. In 1998, under the coalition government of the Social Democrats and the Green Party, representatives of the Öko-Institut, the oldest and largest of the independent environmental research institutes, were invited to this committee.

Today, environmental (counter)expertise has been institutionalized in the regulative, normative, and cultural-cognitive sense. Beyond environmental research in universities and universities for applied sciences, the independent environmental research institutes stemming from the environmental movement were able to establish themselves and many have grown considerably. They are able to finance their budgets largely by projects funded from public sources. They are accepted as contributing valuable work in the field of applied environmental research. Members of the independent environmental institutes are included in various influential consultative committees. In the wider public they are regarded as trustworthy sources of expertise.

Environmentalism in Everyday Behavior

The idea of restructuring private life according to political ideals is constitutive for many of the new social movements.[39] "The private is political" is among the catchphrases of the women's movement, the peace movement, and the environmental movement alike. From its inception, the environmental movement has addressed not only politicians or economic elites but also the wider public in hopes of changing everyday behavior.

In the early environmental movement some of these ideas were put to practice. Homegrown food, vegetarianism, environmentally friendly transport, and other measures were attempted. Though marginal at first, during the 1980s these practices gained ground. In its 1982 study on social milieus, the Sinus Institute identified a milieu of left-alternative younger people that they called the "alternative milieu," comprising 4 percent of the population.[40] This subculture established shops for natural food and consumption goods that fulfilled ecological standards.[41] A mostly small and local infrastructure developed in many cities.

In the following decades this infrastructure grew considerably. In particular, the availability of organic food expanded, receiving from time to time a push forward by food scandals, but also profiting from a supportive agricultural policy introduced by the national government and the European Union alike. Labeling for other types of ecologically friendly products was also introduced and received wide recognition. At the same time, the alternative milieu decreased and had disappeared by the study in 2000.[42] This is an effect of the normalization of the respective lifestyle. Buying environmentally friendly products and the respective values and attitudes have become widespread and are no longer limited to a specific milieu.

Today, environmentally friendly products of various kinds can be found in nearly all stores. Natural food, energy efficient appliances, and so forth, are among the products on offer. For a third of the German population, organic food plays a big role or a rather big role in their consumption habits. More than half of the population considers environmental matters when buying cleaning agents. 65 percent responded in the survey that they would buy energy-efficient electronic devices. At the beginning of the millennium, that share had increased to nearly 90 percent.[43] It is widely accepted that the use of environmentally friendly products might imply higher costs. In recent years, usually more than half of the population was somewhat prepared or very prepared to pay higher prices for environmentally friendly consumption. The preparedness to pay higher taxes has varied considerably over the last few years—from 60 percent (in 2000) to 33 percent (in 2008) with ups and downs in between. Overall, nearly 90 percent of the population thinks that ordinary people can help protect the climate (as the currently most pressing environmental problem) through their day-to-day behavior.

Looking at this data we should keep in mind that preparedness to act in certain ways and bear certain costs as stated in surveys does not directly translate into action when the situation actually occurs. Therefore, the survey results primarily document a normative institutionalization as the norm of environmentally friendly behavior is accepted in considerable parts of the population—though, of course, in many aspects a large minority or even a majority does not regard environmental protection as an important factor in their decision process. Beyond the reported actual behavior, this data shows that there is considerable potential for further environmentally friendly behavior. Diekmann and Preisendörfer have shown that environmental attitudes are influential in low-cost situations.[44] For example, as soon as several waste bins are available, people with environmental attitudes separate their waste. However, if an expensive product is purchased, environmental attitudes have no effect. These results underline once more the importance of facilitating environmentally friendly behavior, thereby transforming high-cost situations into low-cost situations. It is this interplay of policies that facilitates environmentally friendly opportunities and environmental attitudes in the population, which leads to widespread environmentally friendly behavior. Whether the rate of return is always appropriate in environmental terms may remain a matter of debate.

Due to the prominent role of nuclear energy in the German environmental movement, a final look at this issue is essential. In 2000, 76 percent of the population shared the opinion that Germany should phase out nuclear energy as soon as possible. It was in this year that the government decided to limit the maximum amount of energy produced by each nuclear power station, resulting in the end of nuclear energy in Germany by approximately the year 2022. This decision was widely accepted by the population. In the following years, support for an even faster end to nuclear energy was at around 30 percent. In the days after the accident in Fukushima, public opinion shifted dramatically.[45] Suddenly a majority opposed nuclear energy. While in autumn 2010 the federal government had decided to allow the nuclear power stations a longer period for energy production, this decision was taken back days after the Fukushima accident in light of this shift in public sentiment. Such an immediate political reaction to the nuclear accident has been unique globally and can only be explained by the constant resistance against nuclear energy by the environmental movement. Permanent criticism paved the way for this fundamental change in public opinion and policy alike that was even more pronounced than after the Chernobyl accident. The new legislation closed seven nuclear power stations immediately and puts a definite end to nuclear energy by 2022. This would fulfill a central demand of the environmental movement.

Environmentalism in German Culture

Environmentalism is widely institutionalized in Germany. Referring to the theory of institutionalism, three kinds of institutions can be distinguished: regulative, normative, and cultural-cognitive institutions. The first refers to explicit, primarily legal, norms and the setup of organizations; the second refers to norms and values; and the third denotes institutionalization in the sense of taken-for-granted categorizations, interpretations of reality and practices. In the domains of politics, knowledge production, and everyday life, these ways of institutionalization and stabilizing a social order have been found. Pushed forward by the environmental movement, environmental concerns have become the focus of founding specialized organizations in politics and science. Norms of considering the environmental impact are widely accepted in politics, science, and the population at large, with the respective implications for each domain. Practice of behavior that respects environmental concerns can also be found in all three domains, indicating a cognitive-cultural institutionalization. What Hoffman described for organizational rules in enterprises also applies to German society.[46]

Environmentalism is not only institutionalized in these different domains, it has also become part of the self-description of German society.[47] Germans regard themselves as an environmentally sensitive society that is among the forerunners of environmentalism in the world. Though this perception could be substantiated in the European comparison at least to some degree in the late 1980s, it is an invalid description of the present.[48] Still, the idea of Germans as environmentally friendly has become part of the German self-understanding.

The vitality of antinuclear protest as an integral part of the environmental movement is a German particularity. In most other European countries, antinuclear protest has at some point taken place, but probably nowhere is this issue similarly central to the whole movement and accounting for such a large share of protest.[49]

The institutionalization of environmentalism in a society should not be conflated with the successful social handling of environmental problems. Institutionalization refers to the firm grounding of an idea in society. Whether this leads to adequate behavior and social processes that successfully solve the problem of bringing social life in line with natural requirements is a completely different story. Skeptics are numerous. Still, the institutionalization of environmentalism is an important contribution to the goal of a society that values and protects its natural preconditions. And though firm scientific proof is impossible, it is fairly safe to regard this institutionalization as a success of the environmental movement.

Notes

1. Brundtland, ed., *Our Common Future.*
2. Jamison et al., *The Making of the New Environmental Consciousness.*
3. Clemens and Cook, "Politics and Institutionalism."
4. Scott, *Institutions and Organizations,* 48.
5. Scott, "Approaching Adulthood," 433.
6. Converse, "Nature of Belief Systems."
7. Questions that require a rating by the respondent might reveal interesting results as well, but are more in danger of producing pseudo attitudes than open-ended questions about relevant problems in contexts that do not insinuate a specific issue.
8. Reckwitz, "Toward a Theory of Social Practices."
9. Diekmann and Preisendörfer, "Green and Greenback."
10. cf. Baukloh and Roose, "The Environmental Movement."
11. Christmann, *Ökologische Moral.* For the history of the German environmental movement, see: Rucht, *Modernisierung und neue soziale Bewegungen,* 235; Rucht and Roose, "The German Environmental Movement at a Crossroads"; Rucht and Roose, "Neither Decline nor Sclerosis"; Rucht and Roose, "Institutionalisierung von Bewegungen"; Brand, "Umweltbewegung"; and Brand and Rink, "Institutionalisierung statt Mobilisierung?"
12. On the organizational structure of the German environmental movement, see also: Rucht and Roose, "Neither Decline Nor Sclerosis"; Rucht and Roose, "Institutionalisierung von Bewegungen"; and Blühdorn, "Campaigning for Nature."
13. Joppke, "Nuclear Power Struggles"; Koopmans, *Democracy from Below*; Roose, "Der endlose Streit um die Atomenergie"; Rucht, *Modernisierung und neue soziale Bewegungen,* 405.
14. Blühdorn, "Campaigning for Nature," 172.
15. Koopmans and Duyvendak, "Political Construction of the Nuclear Energy Issue."
16. On the lower Länder-level, environmental ministries were established much earlier.
17. Flam, ed., *States and Anti-Nuclear Movements.*

18. Rink and Gerber, "Institutionalization in Lieu of Mobilization."
19. Rucht, ed., *Protest in der Bundesrepublik.*
20. The local news was excluded in both papers.
21. This data has been analyzed on several occasions. See e.g.: Rucht, "Umweltproteste in der Bundesrepublik Deutschland"; Rucht and Roose, "The German Environmental Movement at a Crossroads"; Rucht and Roose, "Institutionalisierung von Bewegungen"; Rucht and Roose, "Von der Platzbesetzung zum Verhandlungstisch."
22. Schäfer and Metzger, "Was macht das Waldsterben."
23. It is a historical irony that the hypothesis of hyperacidity as the crucial influence on *Waldsterben* is incorrect. A connection between air pollution and illnesses among trees is still regarded as likely, but the actual mechanism of hyperacidity as the core of this problem seems to be rejected. Keil, "Chronik einer Panik."
24. Between 1980 and 1984, a fifth of all protests on ecological issues was violent with damaged property or injured people, while in all other time periods the share was below 10 percent. Rucht and Roose, "Institutionalisierung von Bewegungen," 286.
25. Kolb, "Castor-Konflikt."
26. Brand and Rink, "Institutionalisierung statt Mobilisierung?"; Roose, "Umweltlobby."
27. Rucht, "Umweltproteste in der Bundesrepublik Deutschland." The comparison of protest event time series based on different sources shows similar general trends, but singular peaks are time and again not found in exactly the same year. See also: Eilders, "Darstellung von Protesten."
28. Weidner, *25 Years of Modern Environmental Policy.*
29. Roose, "30 Jahre Umweltprotest."
30. This is documented inter alia in various Eurobarometer surveys.
31. Sources are results of the Politbarometer, a general survey on political issues in Germany.
32. Borgstedt, Christ, and Reusswig, *Umweltbewusstsein 2010*; Wippermann, Calmbach, and Kleinhückelkotten, *Umweltbewusstsein 2008*; Kuckartz, Rädiker, and Rheingans-Heintze, *Umweltbewusstsein 2006.*
33. Rucht and Roose, "Institutionalisierung von Bewegungen"; Roose, "Umweltlobby."
34. Kuckartz and Rheingans-Heintze, *Umweltbewusstsein 2004,* 71; Borgstedt, Christ, and Reusswig, *Umweltbewusstsein 2010,* 89.
35. Rucht and Roose, "Institutionalisierung von Bewegungen."
36. Brand and Rink, "Institutionalisierung statt Mobilisierung?"; Roose and Rucht, "Unterstützung der Umweltbewegung."
37. Jamison, *Green Knowledge.*
38. Roose, "Unterstützungslogik und Informationslogik"; Roose, *Made by Öko-Institut.*
39. Kriesi, "Neue Soziale Bewegungen."
40. Becker, Becker, and Ruhland, *Zwischen Angst und Aufbruch,* 86, 98.
41. Eder, *Die Vergesellschaftung der Natur,* 225.

42. Rainer Geißler, *Die Sozialstruktur Deutschlands,* 133.

43. All reported survey results refer to the study "Umweltbewusstsein in Deutschland," which is carried out biannually. Kuckartz, *Umweltbewusstsein in Deutschland 2000*; Kuckartz, *Umweltbewusstsein 2002*; Kuckartz and Rheingans-Heintze, *Umweltbewusstsein 2004*; Kuckartz, Rädiker, and Rheingans-Heintze, *Umweltbewusstsein 2006*; Wippermann, Calmbach, and Kleinhückelkotten, *Umweltbewusstsein 2008*; Borgstedt, Christ, and Reusswig, *Umweltbewusstsein 2010.*

44. Diekmann and Preisendörfer, "Green and Greenback. The Behavioral Effects of Environmental Attitudes in Low-Cost and High-Cost Situations."

45. Köcher, "Eine Atemraubende Wende."

46. Hoffmann, *From Heresy to Dogma.*

47. Goodbody, ed., *The Culture of German Environmentalism.*

48. Jänicke, Reiche, and Volkery, "Rückkehr zur Vorreiterrolle?"; Pehle, "Domestic Obstacles."

49. Rootes, ed., *Environmental Protest.*

Bibliography

Baukloh, Anja, and Jochen Roose. "The Environmental Movement and Environmental Concern in Contemporary Germany." In *The Culture of German Environmentalism. Anxieties, Visions, Realities,* edited by Axel Goodbody, 81–101. New York, Oxford: Berghahn Books, 2002.

Becker, Ulrich, Horst Becker, and Walter Ruhland. *Zwischen Angst und Aufbruch. Das Lebensgefühl der Deutschen in Ost Und West.* Düsseldorf: Econ, 1992.

Blühdorn, Ingolfur. "Campaigning for Nature: Environmental Pressure Groups in Germany and Generational Change in the Ecology Movement." In *The Green Agenda. Environmental Politics and Policy in Germany,* edited by Ingolfur Blühdorn, Frank Krause and Thomas Scharf, 167–220. Keele: Keele University Press, 1995.

Borgstedt, Silke, Tamina Christ, and Fritz Reusswig. *Umweltbewusstsein in Deutschland 2010. Ergebnisse einer Repräsentativen Umfrage.* Berlin: Umweltbundesministerium, 2010.

Brand, Karl-Werner. "Umweltbewegung." In *Die Sozialen Bewegungen in Deutschland seit 1945. Ein Handbuch,* edited by Roland Roth and Dieter Rucht, 219–44. Frankfurt/M., New York: Campus, 2008.

Brand, Karl-Werner, and Dieter Rink. "Institutionalisierung statt Mobilisierung? Dilemmata der Deutschen Umweltbewegung." In *Politik und Umwelt. Sonderheft 39 der Politischen Vierteljahresschrift,* edited by Klaus Jacob, Frank Biermann, Per-Olof Busch and Peter Henning Feindt, 499–517. Wiesbaden: VS Verlag für Sozialwissenschaften, 2007.

Brundtland, Gro Harlem, ed. *Our Common Future: World Commission on Environmental Development.* Oxford: Oxford University Press, 1987.

Christmann, Gabriele B. *Ökologische Moral. Zur kommunikativen Konstruktion und Rekonstruktion umweltschützerischer Moralvorstellungen.* Wiesbaden: Deutscher Universitätsverlag, 1997.

Clemens, E.S., and J.M. Cook. "Politics and Institutionalism. Explaining Durability and Change." *Annual Review of Sociology* 25 (1999): 441–66.

Converse, Philip E. "The Nature of Belief Systems in Mass Politics." In *Ideology and Discontent,* edited by David E. Apter, 206–61. London: Free Press of Glencoe, 1964.

Diekmann, Andreas, and Peter Preisendörfer. "Green and Greenback. The Behavioral Effects of Environmental Attitudes in Low-Cost and High-Cost Situations." *Rationality and Society* 15 (2003): 441–72.

Eder, Klaus. *Die Vergesellschaftung Der Natur. Studien zur sozialen Evolution der praktischen Vernunft.* Frankfurt/M.: Suhrkamp, 1988.

Eilders, Christiane. "Die Darstellung von Protesten in ausgewählten deutschen Tageszeitungen." In *Protest in der Bundesrepublik. Strukturen und Entwicklungen,* edited by Dieter Rucht, 275–311. Frankfurt/M., New York: Campus, 2001.

Flam, Helena, ed. *States and Anti-Nuclear Movements.* Edinburgh: Edinburgh University Press, 1994.

Geißler, Rainer. *Die Sozialstruktur Deutschlands. Die gesellschaftliche Entwicklung vor und nach der Vereinigung.* Bonn: Bundeszentrale für politische Bildung, 2002.

Goodbody, Axel, ed. *The Culture of German Environmentalism. Anxieties, Visions, Realities.* Ney York, Oxford: Berghahn Books, 2002.

Heinrichs, Harald. *Politikberatung in der Wissensgesellschaft. Eine Analyse umweltpolitischer Beratungssysteme.* Wiesbaden: Deutscher Universitätsverlag, 2002.

Hoffmann, Andrew W. *From Heresy to Dogma. An Institutional History of Corporate Environmentalism.* San Francisco: New Lexington Press, 1997.

Jamison, Andrew. *The Making of Green Knowledge. Environmental Politics and Cultural Transformation.* Cambridge: Cambridge University Press, 2001.

Jamison, Andrew, Ron Eyerman, Jacqueline Cramer, and Jeppe Laessoe. *The Making of the New Environmental Consciousness. A Comparative Study of the Environmental Movements in Sweden, Denmark and the Netherlands.* Edinburgh: Edinburgh University Press, 1990.

Jänicke, Martin, Danyel Reiche, and Axel Volkery. "Rückkehr zur Vorreiterrolle? Umweltpolitik unter Rot-Grün." *Vorgänge* 41 (2002): 50–61.

Joppke, Christian. "Nuclear Power Struggles after Chernobyl: The Case of West Germany." *West European Politics* 13 (1990): 178–91.

Keil, Günter. "Chronik Einer Panik." *Die ZEIT,* December 9, 2004, 39–40.

Köcher, Renate. "Eine atemraubende Wende." *Frankfurter Allgemeine Zeitung,* April 20, 2010, http://www.faz.net/aktuell/wirtschaft/wirtschaftspolitik/energiepolitik/umfrage-fuer-die-f-a-z-zur-atomkraft-eine-atemraubende-wende-1628015.html (Accessed July 1, 2012).

Kolb, Felix. "Der Castor-Konflikt. Das Comeback der Anti-Akw-Bewegung." *Forschungsjournal Neue Soziale Bewegungen* 10 (1997): 16–29.

Koopmans, Ruud. *Democracy from Below. New Social Movements and the Political System in West Germany.* Boulder, San Francisco, Oxford: Westview Press, 1995.

Koopmans, Ruud, and Jan Willem Duyvendak. "The Political Construction of the Nuclear Energy Issue and Its Impact on the Mobilization of Anti-Nuclear Movements in Western Europe." *Social Problems* 42 (1995): 235–51.

Kriesi, Hanspeter. "Neue Soziale Bewegungen. Auf der Suche nach ihrem gemeinsamen Nenner." *Politische Vierteljahresschrift* 28 (1987): 315–34.

Kuckartz, Udo. *Umweltbewusstsein in Deutschland 2000. Ergebnisse einer Repräsentativen Bevölkerungsumfrage.* Berlin: Umweltbundesministerium, http://www.umweltbewusstsein.de (Accessed July 1, 2012).

Kuckartz, Udo. *Umweltbewusstsein in Deutschland 2002. Ergebnisse einer Repräsentativen Bevölkerungsumfrage.* Berlin: Umweltbundesministerium, http://www.umweltbewusstsein.de (Accessed July 1, 2012).

Kuckartz, Udo, and Anke Rheingans-Heintze. *Umweltbewusstsein in Deutschland 2004. Ergebnisse einer Repräsentativen Bevölkerungsumfrage.* Berlin: Umweltbundesministerium, http://www.umweltbewusstsein.de (Accessed July 1, 2012).

Kuckartz, Udo, Stefan Rädiker, and Anke Rheingans-Heintze. *Umweltbewusstsein in Deutschland 2006. Ergebnisse einer Repräsentativen Bevölkerungsumfrage.* Berlin: Umweltbundesministerium, http://www.umweltbewusstsein.de (Accessed July 1, 2012).

Meadows, Donella H., Dennis Meadows, and Jorgen Randers. *Limits of Growth. A Report for the Club of Rome's Project on the Predicament of Mankind.* New York: Universe Books, 1972.

Pehle, Heinrich. "Germany: Domestic Obstacles to an International Forerunner." In *European Environmental Policy. The Pioneers,* edited by Mikael Skou Andersen and Duncan Liefferink, 161–209. Manchester, New York: Manchester University Press, 1997.

Pirani, Elena, and L. Secondi. "Eco-Friendly Attitudes: What European Citizens Say and What They Do." *International Journal of Environmental Research* 5 (2011): 67–84.

Reckwitz, Andreas. "Toward a Theory of Social Practices. A Development in Culturalist Theorizing." *European Journal of Social Theory* 5 (2002): 245–65.

Rink, Dieter, and Saskia Gerber. "Institutionalization in Lieu of Mobilization. The Environmental Movement in Eastern Germany." In *Pink, Purple, Green. Women's, Religious, Environmental and Gay/Lesbian Movements in Central Europe Today,* edited by Helena Flam, 120–31. Boulder, Columbia: Columbia University Press, 2001.

Roose, Jochen. "30 Jahre Umweltprotest: Wirkungsvoll verpufft?" *Forschungsjournal Neue Soziale Bewegungen* 19 (2006): 38–49.

Roose, Jochen. "Auf dem Weg zur Umweltlobby. Zur Vertretung der Umweltinteressen in Deutschland." In *Die fünfte Gewalt. Lobbyismus in Deutschland,* edited by Thomas Leif and Rudolf Speth, 270–87. Bonn: Bundeszentrale für politische Bildung, 2006.

Roose, Jochen. "Der endlose Streit um die Atomenergie. Konfliktsoziologische Untersuchung einer dauerhaften Auseinandersetzung." In *Umwelt- und Technikkonflikte,* edited by Peter Henning Feindt and Thomas Saretzki, 79–103. Wiesbaden: VS Verlag für Sozialwissenschaften, 2010.

Roose, Jochen. *Made by Öko-Institut. Wissenschaft in einer bewegten Umwelt.* Freiburg/Breisgau: Öko-Institut Verlag, 2002.

Roose, Jochen. "Unterstützungslogik und Informationslogik. Zu zwei Formen der Interessenvertretung im Umweltbereich." In *Interessenvermittlung in Politikfeldern. Vergleichende Befunde der Policy- und Verbändeforschung,* edited by Britta Rehder, Thomas von Winter and Ulrich Willems, 109–31. Wiesbaden: VS Verlag für Sozialwissenschaften, 2009.

Roose, Jochen, and Dieter Rucht. "Unterstützung der Umweltbewegung. Rückblick und Perspektiven." *Forschungsjournal Neue Soziale Bewegungen* 15 (2002): 29–39.

Rootes, Chris. "Environmental Movements." In *The Blackwell Companion to Social Movements,* edited by David Snow, Sarah Soule and Hanspeter Kriesi, 608–40. Malden, Oxford: Blackwell, 2007.

Rootes, Christopher, ed. *Environmental Protest in Western Europe.* Oxford: Oxford University Press, 2007.

Rucht, Dieter. *Modernisierung und neue soziale Bewegungen. Deutschland, Frankreich und USA im Vergleich.* Frankfurt/M., New York: Campus, 1994.

Rucht, Dieter, ed. *Protest in der Bundesrepublik. Strukturen und Entwicklungen.* Frankfurt/M., New York: Campus, 2001.

Rucht, Dieter. "Umweltproteste in der Bundesrepublik Deutschland: Eine vergleichende Perspektive." In *Politik und Umwelt. Sonderheft 39 der Politischen Vierteljahresschrift,* edited by Klaus Jacob, Frank Biermann, Per-Olof Busch and Peter Henning Feindt, 518–39. Wiesbaden: VS Verlag für Sozialwissenschaften, 2007.

Rucht, Dieter, and Jochen Roose. "The German Environmental Movement at a Crossroads." *Environmental Politics* 8 (1999): 59–80.

Rucht, Dieter, and Jochen Roose. "Neither Decline nor Sclerosis. The Organisational Structure of the German Environmental Movement." *West European Politics* 24 (2001): 55–81.

Rucht, Dieter, and Jochen Roose. "Von der Platzbesetzung zum Verhandlungstisch. Zum Wandel von Aktionen und Struktur der Ökologiebewegung." In *Protest in der Bundesrepublik Deutschland,* edited by Dieter Rucht, 173–210. Frankfurt/M.: Campus, 2001.

Rucht, Dieter, and Jochen Roose. "Zur Institutionalisierung von Bewegungen: Umweltverbände und Umweltproteste in der Bundesrepublik." In *Verbände und Demokratie in Deutschland,* edited by Bernhard Weßels and Annette Zimmer, 261–90. Opladen: Leske+Budrich, 2001.

Schäfer, Roland, and Birgit Metzger. "Was macht eigentlich das Waldsterben." In *Umweltgeschichte und Umweltzukunft,* edited by Patrick Masius, Ole Sparenberg and Jana Sprenger, 201–27. Göttingen: Universitätsverlag Göttingen, 2009.

Schatzki, Theodore R. "Introduction. Practice Theory." In *The Practice Turn in Contemporary Theory,* edited by Theodore R. Schatzki, Karin Knorr Cetina and Eike von Svigny, 1–14. Milton Park, New York: Routledge, 2006.

Scott, W. Richard. "Approaching Adulthood. The Maturing of Institutional Theory." *Theory and Society* 37 (2008): 427–42.

Scott, W. Richard. *Institutions and Organizations. Ideas and Interests.* Los Angeles, London, New Delhi, Singapore: Sage, 2008.

Weidner, Helmut. *25 Years of Modern Environmental Policy in Germany. Treading a Well-Worn Path to the Top of the International Field. Wzb-Paper Fs Ii 95–301.* Berlin: WZB, 1995.

Wippermann, Carsten, Marc Calmbach, and Silke Kleinhückelkotten. *Umweltbewusstsein in Deutschland 2008. Ergebnisse einer repräsentativen Bevölkerungsumfrage.* Berlin: Umweltbundesministerium, http://www.umweltbewusstsein.de (Accessed July 1, 2012).

Chapter 3

The Environmental Movement in Great Britain[1]

Christopher Rootes

Introduction

Great Britain comprises the greater part of the United Kingdom of Great Britain and Northern Ireland. Stretching almost 1,000 kilometers (600 miles) from south to north, and ranging in width from less than 100 kilometers (60 miles) to some 400 kilometers (250 miles) at its widest point, Great Britain lies just north of France and extends through almost 10 degrees of latitude north from 50 degrees North. Once a peninsula attached to the European mainland, the island was created by the rising sea levels that followed the end of the last Ice Age. Topographically, it ranges from mountains of volcanic origin in the west to submergent coastlines in the plains of the east and chalk downlands in the south. While Scotland in the north and Wales in the west are mountainous or hilly, most of England, to the south and east, consists of rolling hills and plains of low elevation. Despite its high latitude, Britain enjoys a cool temperate climate, with winter temperatures moderated and mostly held above freezing by the Gulf Stream, which carries warm water north from tropical latitudes. The result is that more than a quarter of the land area is arable.

With over 60 million people living in an area of 230,000 square kilometers (89,000 square miles), Britain is one of the most densely populated countries in Europe, with population densities much greater in the south-east than in the north and west. Britain is the most urbanized major country in Europe, with over 90 percent of its population living in towns and cities. Yet London, with over 8 million people (and nearly 14 million in the greater London conurbation), is by far the largest city, and less than 10 percent of Britain's land area is urbanized.

Constitutionally, the United Kingdom is a hybrid state, with devolved administrations in Scotland, Wales, and Northern Ireland enjoying variable but increasing degrees of legislative autonomy from the central state. There is, however, no devolved legislature for England, which has 84 percent of the U.K. population, compared with Scotland's 8 percent, Wales' 5 percent and Northern Ireland's 3 percent. A historically large measure of devolution to local government in England has during the past 60 years been steadily eroded to the point where local government, with severely limited powers to raise revenues, is mainly responsible for the local implementation of decisions made by central government.

Britain was the birthplace of the Industrial Revolution, and both its demographics and its population's relationship with the natural environment have been shaped by industrialism. The first and most general effect of industrialization was rapid urbanization. Most markedly in the early 19th century, small market centers such as Manchester and Sheffield grew rapidly into densely populated and seriously polluted industrial towns. London, until the 18th century a series of barely connected villages surrounding a small urban core, consolidated into a continuous urban settlement that pressed hard upon the still-rural common land within and around it.

Origins and Development of Environmentalism

It is not surprising, then, that Britain should have given rise to the first environmental conservation organizations or that they should have focused upon the conflicts between the pressures of urbanization, the interests of the increasingly urban population, and the customary rights of rural people.

From the final decade of the 18th century, Romantic poets, appalled by the visible impact of coalmines and factories upon the environment, celebrated natural landscapes. Soon concerns about pollution of air and water excited protests, reformers lobbied for protective legislation, civic initiatives created urban parks, and, in what was to become a hallmark of British attitudes to the environment, the countryside was idealized as a counterpoint to the squalor of the new industrial towns.

Also beginning in the late 18th century, great voyages of exploration embraced scientific investigation and enhanced interest in understanding the natural world. Even before the development of professionalized science, amateurs formed nature study groups and, later, influential conservation organizations. By the late 19th century what might, in retrospect, be called a conservation movement existed, but it was an elite rather than a mass movement that saw legislation as the means by which nature might be protected and whose successes were principally due to the influence of its socially privileged supporters.[2]

For centuries, local people had been struggling against landowners to keep open public footpaths and customary rights of way threatened by the increasing enclosure of what had been common land. Most of these campaigns remained local, but in the 19th century, urban elites began to be interested. Among the first environmental NGOs (ENGOs) to emerge in England was the Commons Preservation Society (1865), created to guarantee protection of public access to open land, and chiefly responsible for the preservation of the great London commons that survive to this day.

Campaigns to protect wildlife began early in the 19th century and extended before the end of the century to efforts to preserve beautiful landscapes. The three largest ENGOs today all date from the late-19th or early-20th centuries. The Royal Society for the Protection of Birds (RSPB) (1889) grew out of the campaign against the trade in feathers for ladies' fashion. The National Trust for Places of Historic Interest or Natural Beauty (NT) (1895) grew out of the Lake District Defence Society and the Commons Preservation Society. A 1907 act of parliament empowered it to declare its property inalienable and, by giving it protection from compulsory purchase, encouraged landowners to donate property to it. The forerunner of the present Royal Society for Wildlife Trusts, the Society for the Promotion of Nature Reserves (1912), listed areas deserving protection and raised money to purchase sites of special interest. Soon the RSPB and the newly formed regional wildlife trusts were acquiring land and managing nature reserves.[3]

The great social changes precipitated in Britain by the 1914–1918 war were reflected in the character and social bases of environmental concern. Prewar ENGOs were mainly started by resourceful, socially and politically well-connected individuals who sought royal or aristocratic patronage. After the war, new groups were formed upon different social bases. The Ramblers' Association demanded the restoration of the ancient rights of the public to roam the countryside, combined it with radical politics, and recruited many of its members from the working class.

While the Ramblers supported the demands of an increasingly urbanized population for access to the countryside, another new organization, the Council for the Preservation of Rural England (CPRE), aimed to protect the countryside from unplanned urbanization resulting from the massive increase in house building, extension of urban railways, and proliferation of automobiles. Perhaps the most influential of all English ENGOs during the 20th century, with sister organizations in Scotland and Wales, CPRE was established as an umbrella organization representing 40 NGOs including the National Trust, the Royal Institute of British Architects, the Royal Automobile Club, the County Council Association, the Society for the Preservation of Ancient Buildings, and the Central Landowners Association. Funded by architects and planners, it lobbied decision makers for the extension of land-use planning to the countryside, and the designation of protected areas, including national parks. Its leaders were middle-class professionals and "pillars of society,"[4] and CPRE's pressure for comprehensive land-use planning quickly resulted in the first of a series of Town and Country Planning Acts (1932). The 1939–1945 war marked a hiatus in the development of environmental protection, but CPRE's initial goals were fully achieved when the postwar Labour government passed the National Parks and Access to the Countryside Act (1949).

The 1950s in Britain were years of continuing economic austerity, but gradually interests in natural history were revived and popularized by BBC television nature documentaries and the attractive guidebooks made possible by the coming of cheap, high-quality color reproduction. New, specialized nature protection organizations were established, along with the Conservation Corps (now the British Trust for Conservation Volunteers—BTCV), which aimed to engage young people in practical conservation work.

The development that both marks the culmination of this period of development and foreshadows later innovations was the formation in 1961 of the World Wildlife Fund (WWF). Like the early nature conservation organizations, WWF was an elite initiative designed to raise funds for wildlife conservation, enjoyed royal patronage, and relied on wealthy sponsors for initial funding. But, like the new ENGOs that were to emerge a decade later, it appealed to the public through the mass media. WWF-U.K. was launched by an appeal for funds through the pages of a mass-market newspaper (the *Daily Mirror*) and it was, if only nominally, a mass membership organization from the beginning.

Another important development was the increasingly formalized networking between environmental NGOs. Cooperation and division of labor, rather than competition, had always characterized British environmental organizations,[5] but their increasing number and recognition of the need to integrate their various concerns stimulated the formation of a Committee (later Council) for Environmental Conservation (CoEnCo) in 1969.

By the end of the 1960s, Britain had a well-developed conservation movement that satisfied most of the criteria of a social movement:[6] there was a network of actors, including individuals and informal groups as well as organizations, and it was engaged in collective action in pursuit of shared concerns. However, unlike the archetypical social movement, it was rarely overtly in conflict with government. On the contrary, as the elite connections and patronage of many ENGOs suggest, its collective action almost always took the form not of public protest but of polite lobbying of generally sympathetic and accessible civil servants and government ministers.

When environmental concerns in Britain first began in the 19th century to take organizational form, concerns to defend human interests and promote human well-being were often combined with the protection of the natural environment. Perhaps because the 20th-century development of the welfare state had successfully addressed the former, by the 1960s the conservation movement had become more strictly focused upon the protection of nature. But change was in the air.

The Rise of the Modern Environmental Movement

Awareness of the environmental impacts of industrialism increased gradually from the 1950s onward. The great London smog of 1952, which killed thousands, led to the passage of legislation to protect urban air quality, and the wreck of the oil tanker, *Torrey Canyon*, in the English Channel in 1967 highlighted the risks to the marine and coastal environments. Other concerns were imported. Rachel Carson's book *Silent Spring*, which spotlighted the threat that indiscriminate use of insecticides posed to birdlife in the United States, was published in a U.K. edition and chimed with the increasing concerns of the RSPB about nonnatural bird deaths in Britain.

Britain was not immune from the stirrings of environmental concern elsewhere, nor from the social, intellectual, and political ferment of a period in which the constraints of postwar austerity were challenged and in which antiwar activism and student protests popularized a more participatory alternative to representative democratic politics.[7] At first, environmental politics was only lightly touched by the ferment, but it provided fertile ground for another transatlantic import—Friends of the Earth (FoE).

FoE was established in 1969 by Bill Brower, the former director of the leading U.S. nature preservation organization, the Sierra Club. Brower believed that to meet emerging environmental challenges, an environmental organization needed to embrace a broader spectrum of environmental issues than nature protection, and to be transnational in order to address environmental problems that overflowed national borders. After establishing FoE in the United States, Brower travelled to Europe to encourage the formation of affiliated organizations, and in London was introduced to student activists who had previously failed to interest the National Union of Students in environmental issues. Although Brower envisaged FoE as an international organization with subordinate national branches, he was persuaded to accept the formation in 1970 of an autonomous FoE organization in Britain.

FoE attracted little attention until 1971 when it organized a "bottle drop" on the pavement outside Schweppes's London headquarters to protest the drinks manufacturer's decision to switch to nonreturnable bottles.[8] This media stunt provided impressive photographs for the Sunday papers and FoE was deluged with calls from people wanting to become involved. To accommodate them, local FoE groups were set up, and by 1973 there were over 70. Meanwhile, the national office prepared briefings for the 1972 Stockholm UN Conference on the Human Environment. Even at this early stage, FoE was associated with concerns about global environmental justice and inequalities.

Determined to be free to take politically controversial positions, and to avoid the restrictions upon political action by registered charities, its founders incorporated FoE in England as a limited company. But in 1976, because charities enjoyed tax privileges, FoE set up a parallel fund-raising and research organization registered as a charity.

FoE was generally considered the vanguard of the new environmentalism, sharply distinct from the nature conservation organizations that preceded it, but there were important continuities. FoE insisted on the scientific basis of its claims, and several of its national activists were science graduates. Thematically, there were continuities: FoE was the first ENGO in the United Kingdom to campaign for whales, endangered species and tropical rainforests, and against acid rain, ozone depletion, and climate change.[9] In fact, FoE was more novel in the *style* of its actions than in the substance of its concerns. Much of FoE's activity was assembling, printing, and distributing information, but it also deliberately exploited the media attention attracted by its occasional public direct action. The conservation establishment disapproved of such "improper" publicity-seeking,[10] but FoE was committed to action that was not only nonviolent but also legal, and so campaigned within the system. Thus FoE expended much energy in making representations at the inquiry into the Windscale nuclear reprocessing plant. Critics considered this to be naïve, and when the 1978 inquiry report dismissed FoE's arguments, and FoE's long campaign against nuclear energy appeared to have failed, many supporters were disillusioned.[11]

By then, however, some frustrated FoE members and supporters, who wanted to be more directly active, had been inspired by the actions of Greenpeace in North America. In 1977 they set up a British branch of Greenpeace and immediately embraced the strategy of media-friendly calculated law-breaking that FoE had avoided. Thus as FoE patiently made an evidence-based case against nuclear reprocessing, Greenpeace activists blocked the pipes through which the nuclear power station discharged wastewater into the Irish Sea.

Thus it was Greenpeace, with its commitment to "bearing witness" and nonviolent direct action (NVDA) that, more than FoE, marked a step change in the development of environmental activism in Britain. Greenpeace's successes in exploiting media coverage of its actions to arouse public opinion and to put pressure on governments and, especially, corporations encouraged others to rethink their attitudes to mass media and inspired a subsequent generation of activists to adopt direct action. Its path in the United Kingdom was not all plain sailing, however. FoE was concerned with "getting the science right," but Greenpeace prioritized action, sometimes recklessly made claims it could not back up with scientific evidence, and so was several times forced to admit embarrassing mistakes. When Greenpeace U.K. encountered financial problems due to mismanagement, Greenpeace International, concerned about the brand image, intervened and restructured the organization. By 1984, the leadership, which was not accountable to Greenpeace's rapidly growing numbers of supporters, closely managed and coordinated all campaigns. As a result, Greenpeace became more cautious and attentive to detail in its use of evidence in its campaigns and public statements, and Greenpeace, which, like FoE, was incorporated as a limited company, established a parallel charitable trust to support scientific research and education.

In less than a decade, Greenpeace had developed from a small group of risk-taking direct activists into a formal, bureaucratically organized NGO. In fact, despite the rhetoric of political participation so much associated with new social movement politics from the late 1960s, the founders of neither FoE nor Greenpeace envisaged them as democratically accountable mass membership organizations. Instead, they saw them as vehicles for uninhibited campaigning by committed environmental activists, and they privileged campaigning effectiveness over democratic participation. But whereas Greenpeace was structured to ensure the autonomy of its leadership and was never a mass membership organization, FoE established a national membership system and became a relatively decentralized organization in which grassroots activists have a constitutionally recognized role.

FoE became a grassroots, membership organization by accident. In 1980 its 250 local groups demanded greater say in management and campaign strategy, and in 1981, against a background of financial difficulties, an alliance of national office staff and local groups successfully challenged the national leadership.[12] The result was that, even as it grew in size and was organized into specialized campaign departments, FoE became more decentralized and participatory.

While FoE and Greenpeace were becoming the public face of environmentalism, other new and quite different organizations were being formed. The Woodland Trust was a conventional organization both in its form of organization and in the way it raised funds. Neither the Woodland Trust nor another 1970s newcomer, Sustrans, best known for its promotion of cycle paths, was troubled by accountability to its supporters.

If the 1970s was the decade of new beginnings, for ENGOs the 1980s was a decade of rapid growth. Membership of the largest and longest established organizations, the National Trust and RSPB, grew fourfold between 1971 and 1981, and doubled again between 1981 and 1991. But the most spectacular growth was in the new campaigning organizations, FoE and Greenpeace; although they remained small by comparison with older conservation organizations, from the 1980s, when they began to use direct mailing techniques, the numbers of their supporters surged.

Both FoE and Greenpeace campaigned for protection of the natural environment, but conservation organizations were wary of alienating socially and politically conservative supporters. Nevertheless, impressed by the rise of FoE and Greenpeace, they also learned the value of high-profile public campaigns. So, despite conservation organizations' lingering reservations about their campaign methods, both FoE and Greenpeace were included in a new attempt to coordinate the activities of environmental NGOs. Thus in 1980 ENGOs as diverse as RSPB, RSNC, WWF, FoE, and Greenpeace joined to form Wildlife Link, which, with direct access to civil servants and regular meetings with ministers, greatly increased the political influence of ENGOs.

The 1990s: Consolidation and Challenge

The 1980s was a decade of remarkable organizational innovation: more than one-third of the ENGOs that existed in 2000 were established in the

1980s.[13] After 1989, the rates of growth of existing ENGOs slowed, and relatively fewer new ones were formed. FoE and Greenpeace grew little if at all in numbers of supporters or income, but several more strictly conservation-oriented ENGOs grew strongly. At first glance, it appeared that the public's growing concern with the environment was being channeled into politically uncontroversial conservation organizations rather than outspoken campaigners.

However, beginning in 1991, a new generation of radical environmental campaigners emerged, and the most remarkable decade of environmental protest in Britain had begun. But while Greenpeace and, especially, FoE never disappeared from dispatches, most of the reported protest was the work of others.[14] Many protests were mounted by local groups, and some by the likes of Earth First! and its urban offshoot, Reclaim the Streets. These were not formal organizations so much as banners under which younger activists, to whom FoE and Greenpeace appeared timid and bureaucratic, took direct action commensurate with what they believed to be the urgency of environmental issues. They were just as concerned as established ENGOs with protecting nature, but were more radically critical of capitalism and consumerism. Earth First!'s early actions targeted a nuclear power station and the importers of rainforest timber, but soon settled to a focus upon protests against the Thatcher government's massive road-building programme.[15]

Committed to grassroots participation in direct action, these "disorganizations" deliberately avoided establishing formal organizations that might sap the energies of activists and be exposed to legal sanctions. Thus they learned from the experiences of FoE and Greenpeace, which, as they became ENGOs rather than simply activist groups, became vulnerable to litigation. FoE played an important role in the campaign to obstruct the construction of the M3 motorway through the ancient and highly valued Twyford Down but, to the great irritation of direct activists, FoE was forced to withdraw when it was served with a legal injunction. Similarly, threats of legal action to seize its assets later forced Greenpeace to cease disruption of BP's oil exploration in the North Sea.

It is tempting to see the rise of the new radical environmentalism as a reaction against the institutionalization of the previous generation of new environmental activists. Certainly, to the new activists of the 1990s, FoE and Greenpeace appeared to be bureaucratic organizations governed and largely staffed by an older generation made cautious by experience and by having something to lose. Just as the first Greenpeace activists in Britain had been frustrated by FoE's preoccupation with representations within established institutionalized processes, Earth First! activists insisted on prioritizing the direct action for which their youth better suited them than the lobbying and science-based campaigning for which they were ill-qualified. But more important than any frustration they might have endured in FoE and Greenpeace, of which few can have had first-hand experience, was the fact that these younger activists were the products of their own more activist times.

The new informal networks formed at a time when perceptions of activism and themes of protest were strongly influenced by the campaign against another deeply unpopular policy of the Thatcher government—the poll tax. The poll tax (officially, "community charge") was a flat-rate per capita tax introduced in Scotland in 1989 and in England in 1990 to replace property-based local taxation. The introduction of the poll tax provoked the most widespread outbreak of civil disobedience in Britain during the 20th century, and before the end of 1990 the proponents of the tax, including Prime Minister Thatcher, had been forced from office, and the tax was abandoned. The apparent success of the campaign against the poll tax was widely interpreted as evidence of the efficacy of direct action, served as an example to activists, and influenced the movements and campaigns that came after it.[16]

Even anti-roads protesters, some of who were radically anticapitalist, did not act in complete isolation from other environmental groups. ALARM U.K. a coalition orchestrated by Transport 2000, a lobby group that originated in the railworkers' unions and had links with environmental NGOs, provided a national support network for anti-roads protesters. Moreover, even after it was forced to retreat from Twyford Down, FoE, along with other environmental NGOs such as WWF, gave advice and material support to anti-roads protesters.

The wave of environmental protest that began to rise in 1991 reached a crescendo in 1995 when protests against road building overlapped with others against the export of live animals, hunting with dogs, and Greenpeace-organized protests against French nuclear weapons tests in the South Pacific and the Shell oil company's operations in Nigeria and the North Sea. Protest subsided in 1996 when the Conservative government finally abandoned a road-building program made unpopular and more expensive by the direct action campaign against it, ebbed in 1997 when the election of a more environment-friendly Labour government was widely expected, and briefly rose again in 1998, when protesters against field trials of genetically modified crops and against building on greenfield sites tested the Labour government's commitment to its declared green principles.

Although the new environmental radicals have sometimes been represented as marking a shift from a merely reformist environmentalism to a more fundamentally principled ecologism, the differences between them were less those of principle, which Earth First! gatherings debated at length but which were perhaps inevitably compromised in practice, than of the forms of action that each adopted. Their actions were, in practice, less competitive than complementary. Just as the popularity and campaigning successes of FoE and Greenpeace had enhanced older ENGOs' ability to lobby successfully, so the rise of the new environmental radicals provided ENGOs with increased political leverage.[17] In a clear demonstration of the "radical flank" effect, the polite representations of "reasonable" ENGOs became more visible and audible in the corridors of power when there were radical activists outside, loudly demanding action.

By the mid-1990s, British ENGOs were responding to the new international agenda that emerged at the 1992 Rio Earth Summit (UNCED). To improve coordination in the UNCED process, ENGOs increased cooperation with aid, trade, and humanitarian organizations such as Oxfam. After Rio, a broadly inclusive Real World Coalition attempted to promote sustainable development with an agenda focused on social justice. This, however, was a step too far (or too soon) for ENGOs such as RSPB, CPRE, the Wildlife Trusts, and Greenpeace, and they withdrew before its formal launch in 1996.[18] Thus while WWF and FoE became increasingly concerned with social justice issues,[19] others such as RSPB and CPRE, fearing they were getting too far ahead of their supporters, reverted to narrower nature-protection agendas.

The Environmental Movement at the Turn of the Millennium

Already in 1991 it was claimed that Britain had "the oldest, strongest, best-organized and most widely supported environmental lobby in the world."[20] By 1999, almost 20 percent of Britons claimed to be members of one or more environmental organizations,[21] and in 2000 the combined membership of the 11 major ENGOs listed in the official statistical digest, *Social Trends*, was 5.5 million. Most—and all the largest of them—were conservation organizations; the campaigning organizations, FoE and Greenpeace, had fewer members or supporting donors than at the beginning of the decade.[22] Although it in no way diminishes their significance as campaigners, it is noteworthy that neither then nor later was FoE or Greenpeace ranked among the top 10 ENGOs in terms of income, staff numbers, or grant income from private foundations and trusts.[23]

Few countries boast a higher per capita density of membership of ENGOs, and the organizational complexity and diversity of British environmentalism is remarkable.[24] Considering only a dozen of the most prominent and/or best-funded ENGOs, some (notably the NT and RSPB) devote most of their resources to the acquisition and preservation of properties and reserves, and only rarely campaign. Some (notably WWF) raise funds and promote environmental education in Britain but conduct most of their practical work abroad. Others (e.g., FoE, Greenpeace, and CPRE) are principally campaigning organizations. Some are affiliates of transnational ENGOs, others are not. Some (including FoE and CPRE) have hundreds of largely autonomous local groups, others (notably Greenpeace) have local groups restricted to activities approved by the national organization; still others have no local groups at all. Some actively encourage local groups to campaign; others restrict them to providing volunteer labor for practical conservation work. Some are membership organizations in which members can hold officers to account, some have members but allow them no effective role in governance, and others have no formal mass membership at all. Some employ many staff and organize volunteers for practical conservation, others employ only core professional staff and intermittently mobilize supporters in public campaigns. Some (including WWF and BTCV) depend considerably upon government grants or corporate donations; others (notably Greenpeace) refuse them altogether. Some are registered charities; others have charitable status only for subsidiary activities.

If environmental movements are networks of actors, including organizations of varying degrees of formality,[25] they cannot simply be reduced to those organizations. Nevertheless, because organizations are generally the most visible and stable constituents of movements, the nature and frequency of the network links between them give us important insights into the character of the movement.

The only systematic survey of British ENGOs in the 1980s concluded that organizations tended to have network links either with a few "core" organizations or with others in their own thematic sector.[26] That, however, was before the new campaigning organizations became established and influential. When in 1999–2000 we conducted a survey of 144 national ENGOs identified from the database, *Who's Who in the Environment?* (*WWE?*),[27] it was FoE that appeared most central to the network, followed by WWF, Greenpeace, Wildlife and Countryside Link, CPRE, and RSPB, with secondary networks linking ENGOs specializing in "organic" and "transport" issues.[28] Of the six organizations that Lowe and Goyder identified as the core of the movement—CPRE, FoE, RSPB, the National Trust, CoEnCo, and the Civic Trust—only the first three appeared to be at or near the core of the network in 2000; the National Trust appeared marginal, and the latter two did not appear at all. Greenpeace, which appeared marginal in the early 1980s, and WWF, which was then identified as a noncore species protection organization, have become more central to the network than RSPB and CPRE.

Nature protection is still the predominant concern of most ENGOs, most of which remain niche players with specialized functions and narrow thematic concerns. While there are specialized networks of conservation NGOs, there are links between them and the larger conservation NGOs and, through them, with the campaigning ENGO, FoE. Large ENGOs such as the National Trust and the Wildlife Trusts appear marginal to the network, perhaps because so much of their work is focused on the management of their properties. The umbrella group, Wildlife and Countryside Link, has grown to embrace 38 ENGOs (including both FoE and Greenpeace as well as the National Trust and the Wildlife Trusts), and informal, ad hoc and bilateral cooperation has increased. Collaborative campaigns are common, and increasingly extend beyond nature protection to issues of human well-being. It is significant that FoE, despite being relatively small, should appear central to an environmental network in which most NGOs are primarily concerned with nature conservation, because FoE not only has an exceptionally broad agenda, a grassroots base, and strong international links, but has also gone furthest to incorporate social justice as a core concern.

The Decline of Protest?

In his campaign to rehabilitate the Labour Party after its shock defeat in 1992, Tony Blair deliberately reached out to environmental NGOs in order to appeal to a section of the electorate to which previous Labour leaders had been indifferent or even outright hostile. Thus in 1997 Labour assumed office promising to put the environment at the center of government. It created, in the Department of the Environment, Transport and the Regions (DETR) what the outgoing Conservative Environment minister conceded was the world's most powerful environment department, and it placed the Deputy Prime Minister in charge. Thereafter, although the Blair Labour government often disappointed environmentalists, it determinedly avoided antagonizing them, and this was a major factor leading to the lower profile of environmental protest after 1998. Another, however, was the exhaustion of the activists whose intense commitment had sustained the most radical forms of ecological direct action during the 1990s. This was especially so for the very small numbers who engaged in covert actions sabotaging property ("ecotage"); obliged to form secretive cells for their own protection, when burnout and economic necessity forced their retirement from action, they had not recruited or trained successors.[29]

A decline in highly visible environmental protest should not, however, be interpreted as evidence of any decline in the vigor of the British environmental movement. In the absence of any database that might facilitate systematic time-series analysis of their incidence, it is not possible to determine authoritatively whether the numbers of local environmental protests have declined or, indeed, increased. Nevertheless, local environmental protests against a wide variety of proposed changes in land use continued under the Labour government and into the period of protracted economic depression that has followed it. Insofar as many of these local protests are not well networked either with ENGOs or with environmental protests in other localities, they are only potentially part of the environmental movement, but some are clearly networked.[30] Local campaigns against airport expansion, which began in isolation, have since 2000 been networked by the umbrella organization, Airport Watch, whose members include several of the more important ENGOs. A national network of anti-incinerator campaign groups, UKWIN, was established in 2007 with assistance from FoE,[31] and claims more than 80 active campaign groups among its members.[32] Both of these networks, as well as the Campaign for Better Transport's assistance to local groups opposing new road schemes, have received funding from a private philanthropic foundation, the Manuka Club, which explicitly "supports networks seeking to build grass roots concern into a fully fledged social movement."[33] Thus local environmental protests not only persist but are increasingly effectively networked.

The networking of local environmental activism has not been at the expense of national ENGOs, which, during the early years of the 21st century, mostly experienced continued if generally modest growth. New ENGOs continued to be formed, but the rate of organizational innovation slowed. Indeed, since 2000, the more striking changes have been in the character of ENGOs than in their numbers or the numbers of their members and supporters. Most notably, the tactical repertoires and issue agendas of even some of the oldest and most institutionalized ENGOs have broadened. Thus CPRE, often considered the most "establishment" of the major ENGOs, in 2003 changed its name—from the "Council for the Protection of Rural England" to the "Campaign to Protect Rural England"—and thereby signaled a shift in its tactics from discreet lobbying to more active public campaigning.

Even RSPB, whose main activity remains practical measures to protect wild birds and their habitat, has broadened both its agenda—to campaign for reform of agriculture and even for solar power[34]—and its repertoire of tactics. From the late 1970s onward, RSPB, concerned about the increasing loss of wildlife habitat not only in Britain but along birds' migration routes through Europe and Africa, began to address governments more actively and to transform itself from a national bird protection organization into one concerned with global environmental change, albeit still with a sharp focus upon birds. It strengthened its links with other ENGOs, nationally and internationally, especially after 1992 when it played a key role in forming Birdlife International. Increasingly willing to mobilize its members, RSPB raised over 300,000 objections against a proposed airport adjoining wetlands at Cliffe in Kent in 2002, bussed 1,500 members to the 2006 Stop Climate Chaos rally in London, and was a prominent contributor to the 2009 climate march.

Even more remarkable changes occurred over the same period at WWF, which, from being a small, science-led fund-raising organization focused on endangered iconic species and habitat destruction, became a policy-oriented and politically engaged internationalist ENGO consciously attempting to balance the protection of threatened ecosystems with meeting the economic needs of the people who live in and around them. Overcoming its leaders' nervousness about alienating supporters whom it feared (wrongly) to be only narrowly interested in nature protection, after Rio, WWF worked to form a common sustainable development agenda with aid and development and environment NGOs. As WWF came to identify poverty, overconsumption, and climate change as the greatest threats to the environment, its spending grew on education, information, and regeneration projects to benefit people as well as the environment, while species protection shrank to account for less than one-sixth of its budget.[35] This brought WWF closer to FoE. Always international in its understanding of environmental issues, FoE, influenced by its partners in FoE International, became increasingly involved in campaigns to promote human rights and economic development in the global South, and grew its links with aid and development charities and organized labor.

These are simply the most striking examples of the continuing trend among Britain's ENGOs toward the adoption of a transnational agenda in which the environment cannot be isolated from a wider range of human concerns.[36] ENGOs, including WWF, RSPB, FoE, and Greenpeace, signed up to Make Poverty History and/or the Trade Justice Movement,[37] and when the Stop Climate Chaos (SCC) coalition was launched in 2005 it included the major aid and development charities as well as most larger ENGOs.

It has been suggested that conservation organizations are not part of the "green movement."[38] However, as we have seen, the distinction between them and other environmental groups and NGOs is often less one of principles or interests than of forms of action, and even these have been more complementary than in conflict with one another. Given the variety of ways in which many conservation NGOs and their more activist successors are linked by ties of information sharing, common membership of umbrella organizations, and, increasingly, participation in joint campaigns, it makes more sense to draw the boundaries of the environmental movement widely rather than narrowly.

Climate Change, Climate Justice

During the first decade of the new century, climate change emerged as the leading, unifying issue for the British environmental movement. Although both FoE and Greenpeace commissioned reports on climate change in the late 1980s, neither made it a campaign priority. Indeed, climate change was an issue raised by concerned scientists rather than ENGOs, and it was one to which U.K. government agencies were already responding. Thus, when ENGOs did address climate change, they rode a wave that was already swelling. For the most part, even FoE and Greenpeace left campaigning on climate change to the umbrella organization, Climate Action Network, and there were no reports in the national print media of the relatively few, generally small climate protests that were mounted in Britain before 2000.[39]

That changed after the 2000 UN climate conference (COP6), at which the United States, after failing to secure EU acceptance of the principle that forests could be treated as carbon sinks to offset greenhouse gas emissions, withdrew from the Kyoto process. FoE and Greenpeace, having protested at COP6, realized the need to campaign more directly on climate change, but new, more activist initiatives were the most immediately visible result.

Rising Tide, a grassroots network committed to building a movement against climate change, was formed out of the ad hoc coalition that protested at COP6. It drew on the activist milieu associated with Earth First! and was soon reproduced in the United States and Australia, but in England it never grew to more than 15 local groups. Much more visibly, the Campaign against Climate Change (CCC) was formed and from 2001 began organizing demonstrations. These were small at first, but in 2005 attracted 10,000 people in London and 400 in Edinburgh, and were repeated at least annually. Although CCC kept the flag of climate protest flying, its collaboration with the Stop Climate Chaos (SCC) coalition attracted much larger numbers.

SCC was formed in an attempt to raise the profile of climate change in the way that Make Poverty History had done for fair trade and poverty reduction in the global South. Its I-Count campaign persuaded 200,000 people to pledge to reduce their carbon emissions, and ended after a march of 30,000 people in a rally in London's Trafalgar Square in November 2006. Although a majority of participants in the march recognized climate change as a key issue of global justice, they were more often affiliated with ENGOs than with aid, trade, and development NGOs (Saunders 2008). This was the biggest environmental demonstration in Britain to that date, but it was small compared with the pro-hunting Countryside marches of 2002 and 2004, and the protest against the Iraq War in February 2003, or even the 2005 Make Poverty History march in Edinburgh.

SCC's I-Count campaign supported FoE's "Big Ask" for the adoption by parliament of a Climate Change Act legally obliging the government to secure an 80 percent reduction in the United Kingdom's greenhouse gas emissions by 2050. FoE's draft bill was introduced with cross-party support, quickly endorsed by a majority of MPs, adopted by the Labour government, and finally strengthened to include annual targets and aviation and shipping emissions. The campaign was a remarkable success, but it was achieved almost wholly by conventional means, some 200,000 people contacting MPs in person, by letter, or via an online petition. Thus the campaign was a striking example of orchestrated conventional lobbying that persuaded MPs to pressure government to take effective action in pursuit of its own declared policy ambitions. Neither in its tactics nor in its aims did it much resemble past social movement mobilizations.[40]

It did not, however, mark the end of protest on climate change. After 2006, CCC continued to organize marches and rallies but they attracted smaller numbers until in December 2009, in the build-up to the Copenhagen COP15 summit, SCC and CCC together marshaled more than 40,000 people through the streets of London. This was not only Britain's largest ever environmental demonstration but it was also inclusive, attracting members and supporters of the major aid charities and ENGOs as well as a wide variety of local groups and initiatives and even such strangers to protest as the Wildlife Trusts.[41]

In the meantime, a remarkable new mode of action emerged in the shape of the Camp for Climate Action. A mixture of information exchange, education, training, practical example, pre-figurative utopia, and protest, the Climate Camps were designed to facilitate direct action against climate change, "but also to be an exemplar of sustainable living and a site for alternative education."[42] They were deliberately temporary, did not attempt to establish a permanent organization, and were notable for their openness and tolerance of difference.[43] Visitors were welcomed, and attempts were made to involve neighboring communities; many who participated in discussions at the Camps did not participate in the associated protests, and only a minority was involved in direct action.

The first Climate Camp, in 2006 at Drax, the largest coal-burning power station in Europe, was followed in 2007 by a second at London's Heathrow airport, and in 2008 by a third at Kingsnorth in Kent, where it was proposed to build a new coal-fired power station on the site of an existing one. The Climate Camps gathered only a few hundred participants in 2006, and probably no more than 2,000 in 2007 and 2008, but they attracted substantial media attention to coal-burning and aviation, and highlighted contradictions between the government's rhetorical commitments and its failure to act on the largest and fastest rising contributors to carbon emissions. They were assisted in this by characteristically spectacular protests, at power stations, at airports, and even on the roof of the Houses of Parliament, by Greenpeace and the campaign group, Plane Stupid. The distinctively new forms of action were thus supported by the leading professionalized protest organization[44] of the previous two decades.

At local level, the rise of climate change as the most prominent environmental issue made many local campaigns against airport expansion, road building, and land-use changes easier to network, integrated them into the new dominant environmental discourse, and stimulated the development of new groups and networks, of which Transition Towns and the Greening Campaign are the most widespread.[45] The latter, however, are not closely linked to ENGOs or other environmental networks. Indeed, in their orientation toward practical action rather than contentious politics, they are not so much new forms of environmental activism as community-based alternatives to political action.

In some ways, climate change has presented the environmental movement with more challenges than opportunities. The public focus on climate change has overshadowed issues such as biodiversity or ground-level air pollution that are no less important to many ENGOs. Although climate change has concentrated attention on energy issues and put the burning of coal to generate electricity in question as never before, the search for alternatives has exposed divisions within the environmental movement. Greenpeace and FoE oppose new nuclear power stations but some prominent environmentalists argue that low-carbon nuclear electricity is preferable to burning fossil fuels. FoE advocates constructing wind turbines to exploit Britain's abundant wind resources but CPRE, wildlife groups and locals, concerned about their impact on the landscape, birds and people, have opposed many proposed wind farms. Similarly, ENGOs and local campaigners, fearing its impact on birds and the landscape, have opposed construction of a tidal barrage across the Bristol Channel.

More generally, because climate change is universal, its causes are deeply embedded in the taken-for-granted structures and practices of modern societies, and its particular local impacts cannot be unambiguously attributed to climate change, it is an issue that presents few promising targets for protest. As a result, local action on climate change often takes the form of support for the campaigns of national ENGOs, lobbying to get local councils and businesses to adopt climate-friendly policies, or practical energy-saving projects.

This leaves ENGOs in a difficult position. The rise to prominence of climate change and energy policy has drawn into the policy process pressure groups and vested interests that are more highly organized and vastly better resourced than ENGOs. Sympathetic governments and nervous corporations may invite ENGOs to advise or even collaborate, but are often frustrated when ENGOs are unable or unwilling to respond to their invitations. As a result, ENGOs often find themselves marginalized and may question whether continued dialogue makes best use of their limited resources, especially when governments and corporations profess good intentions but seem insufficiently willing to take the action necessary to reduce carbon emissions quickly, drastically, and permanently.

Conclusion

The long saga of the environmental movement in Britain has not been one of brave outsiders struggling to secure a place in policy-making circles. For most of its history, access to the powerful was relatively easily and quickly achieved, and even when fringe groups and newcomers staked new claims in less decorous ways, their input has generally been absorbed and has contributed to the renovation of the established environmental movement. Now, however, a mature environmental movement is confronted by the limits of its power. Governments, at best, expect the environmental movement to mobilize the public to support policies that ministers favor but are too nervous to introduce in the absence of evidence that the public is ready to accept them, but those same governments characteristically fail to appreciate the limitations that ENGOs' relatively slender resources impose.

It is possible that their frustrations may lead ENGOs that are unable or unwilling to act as handmaidens to governments and corporations to choose outsider strategies that leave them free to ask the radical questions that governments and corporations are unwilling to consider. That, of course, is the strategy that Greenpeace has, with brief exceptions, followed all along. FoE, focused upon climate change as an issue of social justice, is increasingly critical of the power of capitalist corporations. That bridges the gap between FoE and more radical environmental groupings and raises the possibility that grassroots direct action may be increasingly networked nationally and internationally. The conservation-oriented ENGOs are unlikely to follow, but the likes of Greenpeace, FoE, and even radical environmental campaign networks are no longer seen as cuckoos in the nest but as respected partners in the complex division of labor that now characterizes the British environmental movement.

Notes

1. Some parts of this chapter are adapted from Rootes, "Environmental NGOs."
2. Evans, *History of Nature Conservation in Britain,* 34.
3. Evans, *History of Nature Conservation in Britain.*
4. Lowe and Goyder, *Environmental Groups in British Politics,* 37.
5. Lowe and Goyder, *Environmental Groups in British Politics.*
6. Diani, "The Concept of Social Movement"; Rootes, "Environmental Movements."
7. Marwick, *The Sixties.*
8. Lamb, *Promising the Earth,* 38.
9. Lamb, *Promising the Earth,* ch. 7.
10. Evans, *History of Nature Conservation in Britain,* 104.
11. Lamb, *Promising the Earth,* 87.
12. Lamb, *Promising the Earth,* 97–9.
13. Rootes and Miller, "British Environmental Movement" Rootes, "Environmental NGOs."
14. Rootes, "Britain."
15. Doherty, "Paving the Way"; Wall, *Earth First!*
16. Rootes, "The Resurgence of Protest."
17. Rawcliffe, *Environmental Pressure Groups.*

18. Rawcliffe, *Environmental Pressure Groups,* 212–14.
19. Rootes, "Facing South."
20. McCormick, *British Politics and the Environment,* 34.
21. Johnston and Jowell, "Social Capital," 183.
22. Rootes, "Environmental NGOs."
23. Cracknell et al., *Where the Green Grants Went.*
24. Rootes, "Environmental NGOs."
25. Rootes, "Environmental Movements."
26. Lowe and Goyder, *Environmental Groups in British Politics.*
27. Environment Council, *Who's Who.*
28. Rootes and Miller, "British Environmental Movement."
29. Plows, Wall, and Doherty, "Covert Repertoires."
30. Rootes, "From Local Conflict to National Issue."
31. Rootes, "More Acted upon Than Acting."
32. http://ukwin.org.uk, accessed September 2, 2012.
33. http://www.manukaclub.org/strategicgrants.php, accessed September 2, 2012.
34. Rootes, "Nature Protection Organizations."
35. Rootes, "Nature Protection Organizations," 48–9.
36. Rootes, "Facing South."
37. Rootes and Saunders, "The Global Justice Movement in Britain."
38. Doherty, *Ideas and Actions in the Green Movement.*
39. Price, *How the Environmental Movement Influenced Climate Change.*
40. Rootes, "New Issues, New Forms of Action," 61.
41. Rootes, "New issues, New Forms of Action."
42. Saunders and Price, "One Person's Eu-topia," 117.
43. Saunders, "Reformism and Radicalism in Climate Camp"; Schlembach, Lear, and Bowman, "Science and Ethics."
44. Diani and Donati, "Organisational Change."
45. Rootes, "New Issues, New Forms of Action."

Bibliography

Cracknell, Jon, Heather Godwin, Nick Perks, and Harriet Williams. *Where the Green Grants Went 5: Patterns of UK Funding for Environmental and Conservation Work.* London: Environmental Funders Network, 2012.

Diani, Mario. "The Concept of Social Movement." *Sociological Review* 40 (1992): 1–25.

Diani, Mario, and Paulo Donati. "Organisational Change in Western European Environmental Groups: A Framework for Analysis." Environmental Politics 8, no. 1 (1999): 13–34.

Doherty, Brian. *Ideas and Actions in the Green Movement.* London: Routledge, 2002.

Doherty, Brian. "Paving the Way: The Rise of Direct Action against Road-Building and the Changing Character of British Environmentalism." *Political Studies* 47, no. 2 (1999): 275–91.

Environment Council. *Who's Who in the Environment?* London: The Environment Council, 1999.

Evans, David. *A History of Nature Conservation in Britain,* Second Edition. London: Routledge, 1997.

Johnston, Michael, and Roger Jowell. "Social Capital and the Social Fabric." In *British Social Attitudes: The 16th Report,* edited by Roger Jowell, John Curtice, Alison Park, and Katarina Thompson, 179–200. Aldershot: Ashgate, 1999.

Lamb, Robert. *Promising the Earth.* London: Routledge, 1996.

Lowe, Philip, and Jane Goyder. *Environmental Groups in British Politics.* London: Allen and Unwin, 1983.

Marwick, Arthur. *The Sixties.* Oxford: Oxford University Press, 1998.

McCormick, John. *British Politics and the Environment.* London: Earthscan, 1991.

Plows, Alexandra, Derek Wall, and Brian Doherty. "Covert Repertoires: Ecotage in the UK." *Social Movement Studies* 3, no. 2 (2004): 199–219.

Price, Stephan. *How the Environmental Movement Influenced Climate Change Debates and Action in Britain between 1987 and 2008.* PhD thesis, School of Social Policy, Sociology and Social Research, University of Kent, Canterbury, 2013.

Rawcliffe, Peter. *Environmental Pressure Groups in Transition.* Manchester: Manchester University Press, 1998.

Rootes, Christopher. "Britain." In *Environmental Protest in Western Europe,* edited by Christopher Rootes, 20–58. Oxford: Oxford University Press, 2003.

Rootes, Christopher. "Environmental Movements." In *The Blackwell Companion to Social Movements,* edited by David A. Snow, Sarah A. Soule and Hanspeter Kriesi, 608–40. Malden, MA: Blackwell, 2004.

Rootes, Christopher. "Environmental NGOs and the Environmental Movement in England." In *NGOs in Contemporary Britain: Non-state Actors in Society and Politics since 1945,* edited by Nick Crowson, Matthew Hilton and James McKay, 201–21. Basingstoke: Palgrave Macmillan, 2009.

Rootes, Christopher. "Facing South? British Environmental Movement Organisations and the Challenge of Globalisation." *Environmental Politics* 15, no. 5 (2006): 768–86.

Rootes, Christopher. "From Local Conflict to National Issue: When and How Environmental Campaigns Succeed in Transcending the Local." *Environmental Politics* 22, no. 1 (2013): 95–114.

Rootes, Christopher. "More Acted upon Than Acting? Campaigns against Waste Incinerators in England." *Environmental Politics* 18, no. 6 (2009): 869–95.

Rootes, Christopher. "Nature Protection Organizations in England." In *Protecting Nature: Organizations and Networks in Europe and the United States,* edited by C.S.A. (Kris) van Koppen and William Markham, 34–62. Cheltenham and Northampton, MA: Edward Elgar, 2007.

Rootes, Christopher. "New Issues, New Forms of Action? Climate Change and Environmental Activism in Britain." In *New "Participatory" Dimensions in Civil Society: Professionalization and Individualized Collective Action,* edited by Jan W. van Deth and William Maloney, 46–68. London: Routledge, 2012.

Rootes, Christopher. "The Resurgence of Protest and the Revitalisation of British Democracy." In *Social Movements and Democracy*, edited by Pedro Ibarra, 137–68. New York: Palgrave Macmillan, 2003b.

Rootes, Christopher, and Alexander Miller. "The British Environmental Movement: Organizational Field and Network of Organizations." Paper presented at ECPR Joint Sessions, Copenhagen, April 14–19, 2000.

Rootes, Christopher, and Clare Saunders. "The Global Justice Movement in Britain." In *The Global Justice Movement*, edited by Donatella della Porta, 128–56. Boulder, CO: Paradigm, 2007.

Saunders, Clare. "Reformism and Radicalism in the Climate Camp in Britain: Benign Coexistence, Tensions and Prospects for Bridging." *Environmental Politics* 21, no. 5 (2012): 829–46.

Saunders, Clare. "The Stop Climate Chaos Coalition: Climate Change as a Development Issue." *Third World Quarterly* 29, no. 8 (2008): 1509–26.

Saunders, Clare, and Stephan Price. "One Person's Eu-topia, Another's Hell: Climate Camp as a Heterotopia." *Environmental Politics* 18, no. 1 (2009): 117–22.

Schlembach, Raphael, Ben Lear and Andrew Bowman. "Science and Ethics in the Post-political Era: Strategies within the Camp for Climate Action." *Environmental Politics* 21, no. 5 (2012): 811–28.

Wall, Derek. *Earth First! and the Origins of the Anti-roads Movement.* London: Routledge, 1999.

Chapter 4

The Framing Processes and Outcomes of Environment and Conservation Campaigns in Ireland

Liam Leonard

Introduction: Framing Environmental Conservation in the Irish Case

This chapter will present an analysis of mobilization, framing processes, and outcomes of environmental conservation campaigns in Ireland. By locating contemporary disputes about the public space in Ireland within an understanding of the transition from a traditional rural and agrarian society into that of a modern industrialized and urbanized European state, a wider understanding of the circumstances surrounding such contests emerges.[1] This form of "issue history"[2] provides scope for a wider analysis that goes beyond a social narrative, and facilitates a fuller understanding of the consequences emanating from the framing and reframing of heritage within a context of a society where an obsession with history has been displaced by a zealous predilection for economic development. While this transition from traditionalist concern to modernist utility has placed great strain on the environment and heritage of the nation, it has also created opportunities for widespread political mobilization in successive decades during periods of economic growth or decline.[3] These outbreaks of collective action have provoked widespread debate and coverage in the media as several thousand people have taken part in heritage-based protests in Ireland over this period.

Nonetheless, a significant section of the Irish public have mobilized in response to threats to public spaces during a phase where people were characterized as being solely obsessed with financial gain and property. These sporadic outbreaks of collective action are a reminder that in Ireland as in most cultures, there are deeper concerns with space and place that go beyond material concerns or economics. Migrant and immigrant groups share this concern for spatially derived representation, rendering heritage as a most contested and culturally charged element of social construction. An instinctive or reflexive response to threats to public spaces has emerged during this series of heritage or community disputes. Such movements represent the articulation of indigenous monologues of communities who are embedded in the streets or landscapes of their towns and regions. Their rootedness may not always require a contemporary presence at the point of concern as the indigenous consciousness formed in the community inculcates an inherent compulsion to protect spaces rendered sacred through memory, loss, and shared expression. When such localities are threatened, a mutuality of concern for collective folk memories and meanings in the guise of a shared heritage emerges as a significant grievance in the mobilization of common cause. One consequence has been the emergence of an ongoing debate, alongside episodes of collective action around the ethical and moral considerations surrounding issues of heritage and conservation in a rapidly developing Ireland.

In attempting to understand heritage, many competing elements of nature and humanity may be evoked in an attempt to capture its underlying meanings. As with other cultures, Irish heritage contains a host of meanings and contexts carried through the portal of time as an inherent folk memory is rooted in the land or streetscape. While it is often stated that Ireland has a rich natural and built heritage, it can be difficult to find consensus on where this may be found, or what this heritage actually represents. What seems to be important is a sense of place and time embodying the natural beauty of God's creation or human endeavor, whether these are represented by buildings, battlefields, or bogs. Writing in 1996, David Brett presents an interrogation of the "concept of heritage" whereby locations and processes have come to be understood through history. Throughout the evolutionary patterns of societal engagement with our natural and built hinterlands, we have come to represent and define ourselves as an autonomous group by way of undertaking a reinvention of the norms and ways that have emerged throughout the course of Irish history. One result of this new economic direction was an increased antipathy toward our natural and built environment as polluting industries and multinational or state infrastructure began to impinge on communities, landscapes, and antiquities throughout the nation.[4] Furthermore, Hilary Tovey[5] has indicated that there has never been a period of significant prioritization of institutional conservation at any phase of Ireland's post-independence self-rule. It was as if the new state wished to erase the memory of the past by eradicating the cracked fanlights and façades of the nation's built environment while paving over its natural beauty and historical sites.

This dichotomy between the nationalist Irish state and those who were concerned with conservation can be traced back to the Unionist or pro-British leanings of early conservationists in Ireland. This group was concerned with maintaining the Unionist heritage of the Anglo-Irish ascendancy class, along with their large estates. Many in the local Irish population had been uprooted to make way for these large demesnes, and a sense of resentment against this form of heritage and conservation simmered long after Ireland gained nominal independence from Britain in the 1920s. This set the Irish conservation movement back, as anticonservation activities such as game poaching were seen as valid expressions of rebellion against the Unionist authorities. This residue of hostility remains apparent in the populist rural opposition to *An Taisce*, the Irish Heritage Trust, which was viewed as a largely urban and antirural organization by many local politicians and the popular press. While *An Taisce* has worked hard to address this residue, issues such as the location of houses in rural areas have tended to see battle lines redrawn along the traditional-urban/conservationists-versus-rural-populist divide.

As Tovey also notes, heritage groups such as the Irish Georgian Society took on the mantle of the "moral entrepreneurs"[6] of Dublin's built environment when the destruction of parts of the southern section of the city began in the 1970s. Faced with a hegemonic concern for development combined with attempts to wipe out unwanted aspects of the past, conservationists became the curators of a living history that was under threat from political elites and developers alike. Thomas Farel Heffernan sets the inception of this curator's vigil around the time of the excavation of the east end of the cross-block in Dublin Castle in 1961.[7] In the era of John F. Kennedy's presidency, Ireland's architectural heritage had been revealed in "an exciting glimpse."[8] Dublin's origins could be traced back to its ancient Viking and Norman past, with visible links to Celtic neighbors also evident. This discovery provided visible confirmation that the remains of Viking Dublin contained a site that could represent the past. For Duncan, "sites of representation"[9] present a dualistic signification of place. However in postcolonial Ireland, this representation was contentious, as the disputes about Woodquay and Georgian Dublin would later prove. In many ways the heritage of the past in Ireland became a version of Edward Saids's understanding of an oppositional form of *the Other*[10] in an Ireland gripped by a modernist and postcolonial zeal for state-led building. In such cases *the Other* can be reclaimed from past representations by providing a "familiar and useful"[11] utilization for a contested site.

These sites of contested heritage provide us with what is essentially a "cultural production"[12] that has come to be perceived in a variety of contexts throughout history. Moreover, this cultural production is at once linked to both structure and agency as society comes to comprehend culture and heritage through its potential uses rather than through any inherent value. Local authorities have been described as having certain methodologies that hamper any further valuation of heritage as all sites are invariably reduced to a set of uses such as "production or consumption related activity . . . private property or public service etc."[13] What follows is a utilitarian hegemonic order that reclassifies heritage within the cultural assumptions of temporary elites. This hegemony can be challenged only through the presentation and cultural project of conservationist movements.

In many ways Irish conservationists have been marginalized as part of a process David Lloyd describes as "the subalternity effect."[14] Following on from Gramsci, Spivak, and Homi Baba, Lloyd locates this effect within the contours of "historiographical discourses"[15] as part of the project of postcolonial cultural restructuring. From this perspective counter-hegemonic cultural production can be attributed to these curators of heritage based on "non-elite histories: histories of agrarian movements, local histories, social histories of the complex intersections of class and colonisation in rural Ireland, women's history . . . and social history."[16] It is within this array of conservationists, archivists, and antiquarians that we can find the curators of heritage, a disparate group who have rescued our heritage through vigilance and activism. For Tovey the contestation of heritage in Ireland is an "ideological conflict"[17] with problems on all sides of the divide. For instance, the project of modernization involves a debate about historically derived meanings surrounding built or natural environments. The demolition of Georgian Dublin was also part of a contested modernity.[18] Tovey further quotes MacDonald's account of a government minister on the destruction of Georgian buildings in Kildare Street as claiming "They stood for everything I hate."[19] The fact that much of Dublin's tenements were no more than slums in pre-independence Ireland added to the sense of ambivalence about the city's architectural heritage. Tourism has become marginalized in this development-led economy, in an era when tourism has become part of "the leisure industry" and part of "the rural experience."[20]

The outcome of this restructuring of the agricultural sector and increased reliance on multinational-led development has been further ecological degradation, with Ireland facing fines for noncompliance on a series of European Union (EU) environmental directives. In addition, the state has applied the industrialization approach to tourism. In the case of the Mullaghmore Interpretative Centre in County Clare on the west coast, the authority's plans to develop tourism through a series of visitors' centers across the island led to a significant campaign to protect the rare flora and fauna of the Burren in that County during the 1990s.[21] As the tourism industry developed, it came to be associated with certain environmental impacts as the infrastructure of tourism—hotels, camping sites, holiday homes, and marinas—began to dot the landscape. Further impacts were created from the increased road building, traffic, and waste or sewage plants associated with such developments.[22]

One such "gateway" chosen by the Office of Public Works (OPW) was Mullaghmore. It was the OPW's view that the interpretative center should be located within the heart of the Burren rather than in a nearby village such as Corofin, to facilitate greater visual access for tourists in order to provide a "first-hand experience of the park."[23] The result of this decision would lead to a sometimes rancorous dispute that affected both sets of local groupings that were for or against the project, and led to a series of legal challenges at national and European levels. The state's emphasis on an individualized consumer culture has also led to the commodification of transport, as public rail networks were allowed to diminish while private car ownership and motorways were prioritized. Over time, the focus on the Irish state's increased road-building capacity has led to three significant anti-roads protests on either side of the onset of the new millennium.[24] All three protests emerged at a time when the state was attempting to develop the commuter hub around Dublin city. The case of Tara/Hill of Skryne would lead to a heated debate about the manner in which heritage is dealt with in an age of rapid growth as the ancient site of Tara came under threat from the state's Critical Infrastructural Bill. This bill, which was introduced in the wake of a series of environmental campaigns, has been seen by many as an attempt to prevent social movement–based collective responses in the future.

Similarly, the Shell to Sea campaign has raised conservation issues as part of their campaign of resistance to the Corrib gas pipeline. Protesters have raised concerns about ecological damage to local cliffs, drumlins, and bog lands as part of their campaign. The area surrounding the Rossport site is in line for designation as a Special Area of Conservation and Scenic Importance.[25] The state's own conservation plans have also drawn criticism from heritage advocates in recent years.[26] The plantation of nonindigenous trees has become a contentious issue with serious consequences for landscapes and habitats. Local community infrastructure such as schools and health services has also had negative impacts on indigenous heritage.

Ecopopulism and Irish Conservation Campaigns

Ecopopulism—a term coined in 1994 by the American sociologist Andrew Szasz to describe the widespread, if not always closely coordinated, phenomenon of community groups coming together to protest against a variety of environmentally damaging threats—has long been a feature of Irish life. But in Ireland, more than in any other country, such conflicts take place in a context in which rural life and continuity with the past have historically been highly privileged. The exaltation of the peasant by the authors of the early 1900s literary revival, such as J.M. Synge and W.B. Yeats; the "rural, Gaelic, and Catholic" ideology of the new nation; the construction of an idyllic image of Ireland by tourism promoters have all worked to sustain a prominent role for the rural dimension of Irish life. For many, changes in Ireland over the last half century reflect an ongoing disengagement from the traditional patterns of life that had embedded a set of values and practices that allowed rural communities to coexist with their surrounding environment. The grievances of the wider population, which emerged from this perception of a loss of community, have contributed to the growth of populist environmentalism in Ireland, based on "rural sentiment"—a collective resistance born of the community's inherent understanding of its place in the hinterland, away from the urban center. During the pre-boom decades of the 1970s and 1980s, such ecopopulism was manifested within the rural-based discourse that characterized disputes over toxic substances and the activities of multinational disputes, among them lead and zinc mining at Tynagh, County Galway, and gold mining near Croagh Patrick, the County Mayo site linked with St. Patrick.

Some commentators have argued that the strong parochial rural sentiment contributes to the lack of acceptance of official environmental organizations in country-based disputes; as communities attempted to mobilize grievance based on local understandings and relationships, their concerns do not travel well. Such localism has also left many populist environmental groups open to the accusation of being merely cases of NIMBY—the familiar acronym for "not in my back yard"—when opposition to industrial or infrastructural projects in a community's "back yard," rather than in a larger systemic context, is identified as the primary rallying point for campaign mobilization. Populist campaigns often face the problem of translating rural discourses into normal legal hearings. In the "Rossport Five" campaign discussed later in this chapter, for instance, local farmers' attempts to explain that it was unsafe to construct on bog land were simply not understood by those from outside the region, while to the locals building anything on a bog was patently unsustainable. Their point, rooted in rural experience, was reinforced when part of the bog and local cemetery washed away after being disturbed by the development of the refinery site. While specific populist environmental campaigns come and go, the significance of each campaign's contribution to an articulation of community grievance has created a movement of sorts, where outcomes can be measured through an understanding of the extent to which populist fundamentalism has come to be seen as the very basis for traditional rural identities in the post-consumption, postmodern era.

The well of grievance that provides much of the underlying discontent for populist campaigns to exploit is the basis for an understanding of exactly how the various environmental campaigns that have occurred over recent decades can be characterized as components of an overall social movement. That movement constitutes part of the social capital of all rural and rural-urban communities. As such, it carries far greater significance than simply the impact of protest campaigns on policy implementation, such as the building of a pipeline, or the construction of a heritage center in the middle of a Special Area of Conservation such as the Burren in County Clare. The impact of rural populist discourse goes beyond moments of access to political opportunity structures, and contributes to the shape and nature of the populist Irish political system itself. Even before the recent economic downturn, communities in Ireland having enjoyed the benefits of extraordinary growth—the so-called Celtic Tiger—had also begun to witness the downside of the economic boom. Infrastructural projects had come to be perceived as threats to locals and their health and environment, or to the heritage of the nation itself. Such offshoots of rapid development as hyper-consumption, a buoyant property market, and increased car ownership led to further demand for roads, waste management sites, and water treatment facilities. The link between growth and infrastructure was something of a sacred cow: communities that voiced concerns about projects were widely deemed, in the press and in the statements of politicians, to be backward-thinking and against progress (see Table 4.1).

Table 4.1. Categories of Protest Event and Framing Analysis in Irish Environmental Disputes

Frame	Type of Campaign	Type of Protest/Tactics	Framing Outcome	Political Outcome
Political	Concerns for democratic deficit and party support/opposition; local political allies, national political allies exploiting political opportunities, electoral leverage	Locally alliances, national alliances, global alliances, embedded advocate framing issues, political opportunities	mobilization of grievance consensus at local and national level; heightened awareness of environmental issue and democratic deficit	Increased political access and leverage
Social	NIMBYism grassroots alternative energies new middle-class professionals, working class	Populism supported or opposed by local interests, health risks, marches, demonstrations	Evolving from NIMBYism into wider movement	Increased politicization of civil society
Cultural	Rural discourse/ecofeminist discourse, moral discourse/religions, use of interest-led expertise, nationalist discourse	External advocate, mobilized resources, cultural and heritage frames. Media and celebrity advocacy.	More expansive media coverage. Discursive interpretations of culture and heritage	Enhanced socio-political and cultural discourse
Legal	Oral hearing, High Court, European Court	Legal challenges to policy or projects	Establishing precedents or overturning laws or policy	Redefined policy agendas. Changes to legal framework

Institutional/ Agencies	EU – UN – Sustainable Development, Local Authority support or opposition, the State, An Taisce, An Bord Pleanála	Legal campaigns, expertise advocacy, institutional allies, political allies	Enhanced understanding of overall issue from grassroots grievance to institutional frameworks	Reinforced linkages between institutional sector and community
Economic	Pre-growth—pollution, resource protection; post-growth—infrastructural	Anti-multinational, anti-state, opposition to sitings	Establishing grievance over costs and location of projects	Projects delayed or shelved
Scientific	Global networks	Interest-led advocacy	Counter arguments to scientific viewpoints presented	State and industrial science challenged

Source: Leonard, 2006.

In the decades of rapid development in Ireland between 1996 and 2006, there was a re-emergence of rural populist discourse during the dispute that began in 2005, that of the "Rossport Five," local farmers who were charged with interfering with a Shell Oil pipeline. Their protests and convictions became something of a nation cause célèbre. Calls for local ownership of local resources were combined with concerns about the risk posed by onshore pipelines as part of the framing strategies of that campaign. One of the most important of the collective responses to development, the "Rossport Five" emerged in the wake of the attempt to build onshore pipelines on the west coast of County Mayo for the natural gas of the Atlantic Field. Shell Oil had proposed, and been authorized by the government, to build a £600 million high-pressure gas pipeline. Local farmers who lived near the proposed pipeline objected, initially on the grounds of safety, and resisted attempts by the multinational involved to gain access to their land. As a result, five local men were imprisoned for 94 days for refusing to agree to an injunction that would allow the agents of the multinational access to their property. What followed was a campaign that gripped the nation's imagination, as Shell to Sea protests sprung up across the island and beyond in response to the community's demand that the gas be processed offshore.

The imprisoned men's traditional lifestyle of farming in an Irish-speaking community represented to many observers an authentic and traditional way of life that had been abandoned in the rush to embrace the consumption-fuelled prosperity that had become prevalent in recent years. The fact that the Mayo farmers put family and community above profit or gain provided many observers with an indication of the wider context that had become the subtext of this dispute: the sense that Ireland had, in some ways, sold its soul in return for multinational investment and growth. The men were released in the summer of 2006, and their campaign continues. Despite persistent protests, the construction of the refinery is now nearly complete. The problems of waste disposal secondary to accelerated growth in Ireland are another flashpoint. The bourgeoning waste crisis, a consequence of increased consumption associated with accelerated growth, has led to community action.

Throughout the boom years of the 1990s, Ireland had fallen foul of European directives on waste management due to an overreliance on landfill, which received up to 93 percent of the country's solid waste. The Irish Environmental Protection Agency was set up in 1992 to deal with such issues, but many critics charged that it quickly became associated with business interests and government appointments, rather than with environmental protection. To deal with the solid waste crisis, the government introduced plans for regional incinerators across the country. Previous attempts to introduce an all-island incinerator in Derry, Northern Ireland, were opposed by both nationalists and Unionists. Following on the networks created from the Northern campaign, communities mobilized in such regions as Galway, Cork and Meath, and later, Dublin.[27] These regional campaigns benefited from links with the global anti-incinerator and "zero-waste" campaigns, and from the support of experts from the United States and Britain. Communities began to inform themselves of the dangers posed by the dioxins and furans that are released into the air and the food chain from incinerator emissions. The state's regional waste policy was seriously challenged by this campaign: one of the positive outcomes of community engagement with the waste issue was that regional recycling was introduced in many areas with great success.

The state has not abandoned its plans to introduce incinerators completely, as two major plants are still planned for Cork Harbour and Poolbeg in Dublin; these will deal with the bulk of municipal and industrial waste. Only two years ago, the construction of housing—which had come to provide the second cycle of the Celtic Tiger boom—caused another environmental crisis when the mass construction of new houses in rural areas led to the contamination of water supplies with effluent from archaic and suddenly overloaded treatment plants. In Galway, often referred to as "the fastest growing city in Europe," drinking water supplies were contaminated with the *Cryptosporidium* virus. The lax regulation of the construction industry by the state created a growing strain on the city's water and sewage systems. After heavy flooding in the winter of 2007, water supplies in Lough Corrib, the city's water supply, became contaminated.[28]

Hundreds of people fell ill, and a boil water notice was called by the local authority. Panic buying of bottled water ensued, and the boil water notice lasted from March until August 2007, costing retailers and hoteliers millions in extra expenses. The water crisis also became an issue during the general election called in May 2007, and was said by some commentators to have cost the city's Green mayor a seat in Dáil Éireann. Although the crisis predated the mayor's term, the minister for environment blamed him in many public and media appearances, and the high-profile criticism stuck. Rural communities have borne much of the cost of Ireland's transition from an agrarian society into an industrialized, globalized economy, particularly in terms of environmental degradation and an undermining of significant aspects of heritage. The state's policy of prioritizing rapid infrastructural development in order to underpin private consumption and multinational investment over local community services may prove to be ill-advised as fluctuations on the global markets continue (see Table 4.2).

Table 4.2. Campaign Outcomes

Campaign	Political Oucome	Social Outcome	Cultural Outcome	Legal Outcome	Institutional Outcome	Economic Outcome	Scientific Outcome
No Nukes Carnsore Point	Policy abandoned	Inter-group networking	Issue salience unions and environmentalists united	Moral and legal issues debated	Nuclear energy option abandoned	Energy policies altered	Nuclear arguments abandoned
Anti-Toxics Anti-Multinational Cork	State emphasis on TNC-led development maintained	Increased networking	Divergence between unions and environmentalists	Legal actions, legal threats against campaigners	IDA policy on TNC development maintained	State policies on TNC development maintained	Interest-led debates on toxics begin
Anti-Mining Tynagh, Donegal, Croagh Patrick	State resources sold off	DUC and Carnsore campaigners networking	Rural defence of space campaigns	Compulsory purchase laws enacted	State resource agencies established	Resources sold off	Debates about Uranium and Cyanide
Heritage Mullaghmore	State policy on interpretive centre defeated	Successful challenge to state policy	Raising of awareness about heritage	EU legal option utilized by campaigners	OPW challenged policy abandoned An Taisce's source	Tourism policy altered	Increased understanding of ecological issues

Anti-Incinerator GSE and NIA	Regional waste management policy delayed	Mobilization and networking of anti-incinerator groups	Increased participation by community groups	Legal challenges against incinerator licenses	EPA and An Bord Pleanála hearings	Waste plans shift towards recycling	Increased understanding of dioxins and waste flows
Anti-Incinerator CHASE	Contestation of regional waste policy	Re-emergence of Cork environmental campaigns	Increased community participation in Chase campaign	Legal challenges by CHASE	An Bord Pleanála hearings	Waste plans for cork pivotal to state	Increased understanding of wider issues involving incineration
Resources Shell to Sea	Developing campaigns of resistance	Emergence of rural political movement	Wider rural sentiment articulated	Imprisonment of Rossport Five	EPA ruling, human rights group interest	Gas resources held by multinational	Debates about resources
Glen of the Down, Carrickmines, Tara/Skryne Anti-Roads	Direct action campaign against infrastructural projects	Mobilization of new wave of environmental movement	Debates about development and heritage	Legal challenges against routing of motorways	Roads Authority developing nationwide roads network	Emphasis on private transport despite energy crisis	Ecological and energy issues raised roads & heritage issues linked with climate change

Source: Leonard, 2006.

Outcomes*: The Significance of Irish Environmental Activism

Throughout this examination of collective responses to environmental threats a number of framing processes, strategies, and types of campaign have been identified. But what are the consequences and outcomes of the various environmental campaigns witnessed in Ireland since the Carnsore Point protests? Without doubt, the Irish environmental movement has made an impact on Irish society. We can examine the impact of grassroots campaigners more effectively by applying the six most beneficial outcomes of environmental activism as identified by Freudenburg and Steinsapir in 1992.

In my book the *Environmental Movement in Ireland* (2008), I have outlined these beneficial outcomes in relation to the Irish case. They can be summarized as follows:

1) There is increased community control over public health.
2) There is the introduction of eco-efficient processes by the corporate sector.
3) Increased regulation or elimination of toxins from the production processes occurs.
4) The establishment of wider support networks for communities that were once isolated is achieved.
5) There is increased environmental and heritage awareness among the wider community.
6) The expansion of civic participation in environmental decision making occurs, underpinning civil society.

So, how best can we outline these six outcomes to the campaigns of the Irish environmental movement? In the first instance we can see that a number of potentially harmful projects have been delayed or abandoned due to environmental campaigns. We can include the nuclear power station planned for Carnsore as well as multinational industrial plants such as Raybestos Manhattan's asbestos plant and dump in the 1980s in addition to the more recent campaigns against incinerators or the Shell pipeline in Mayo in this category. And while these industries and their state sponsors have decried the abandonment or delay of such projects, all of the campaigns covered in this book have increased civic awareness of environmental, health, or heritage issues. The Mullaghmore dispute and the more recent anti-roads protests have led to substantial national and international debates about how we perceive and conserve heritage sites in an age of accelerated development. Furthermore, each of these campaigns

* With extracts from Leonard, *The Environmental Movement in Ireland.*

has contributed to the awakening of local knowledge and concern about hinterlands allowing for an augmentation of community derived "social capital" that politicians and commentators have come to value so highly.

Politically, successive environmental campaigns have strengthened the responses of civil society from the grassroots up, bolstering pluralistic discourses at a time when that aspect of democracy has come under threat from a variety of sources, including an increasingly technocratic state and a body politic damaged by a series of planning-related scandals. Environmental social movements have benefited mainstream politics in Ireland as witnessed by the rise of the Green party in local, national, and European elections over the last decade. One distinctive outcome of environmental conflict has been the establishment of the Environmental Protection Agency (EPA) in 1992. The EPA was originally criticized for its location of a regulatory framework within the context of the state's industrial development policies but the agency has come into its own over time and is now a leading component of institutional environmentalism in Ireland spanning a range of ecological, educational, and regulatory responses. However, some distance remains between the EPA and many grassroots campaigns and the bridging of this gap remains an outstanding issue for all levels of environmentalism in the Irish case.

Further understanding of "successful" outcomes for movements has been shaped by theorists such as Gamson or Burstein,[29] who have identified that "realization" and "influence on policy" (state action) are the key factors in campaign impact assessment. These concepts have been further developed by Amenta and Caren[30] who have perceived the distinctions inherent between actual achievement of stated goals and the achievement of certain "advantages for constituents." Inevitably, movement success has therefore become synonymous with state acknowledgment and response to movement grievance. However, the subsequent achievement of aspects of a campaign's aims has also been recognized as a long-term result of collective action, particularly when outcomes result from what Elias has termed "unintended consequences."[31] Changes to policy or adoption of movement aims as part of state policy can be part of this process. The formation of Irish environmental policy could be viewed as emerging in a dualistic response to bottom-up grassroots agitation on the one hand in addition to top-down EU legislation on the other. In the Irish case its populist political makeup lends itself to the attainment of some leverage for movements, particularly at the local level. However, the economic imperative of the state still dominates at policy level despite the occasional opening up of political opportunities at the time of elections.[32]

The Framing of Irish Environmental Campaigns

We can apply this understanding of a discursive opportunity structure to the Irish environmental campaigns covered in this study in order to examine the consequences of this form of collective action. We can apply understandings of the types of protest, tactics, framing processes, and outcomes to the different campaigns that have formed a part of the Irish environmental movements. A series of frames have been used by these movements, including political, cultural, social, legal, institutional, economic, scientific, and moral. Invariably the discursive framing patterns presented in cases of Irish environmental activism have emerged from a populist context.

Kitchings[33] examines populism from the perspective of developing societies and makes the distinction between populism and neo-populism. While both are presented as critiques of industrialization and mass production in favor of small and localized entities, "populism" is defined as being based primarily in a social and ethical critique, whereas neo-populism is seen as being "more ambitious" and "not primarily oppositional."[34] The process of migration from rural to urban centers during the industrial revolution is presented as a factor in the development of populism based on rural sentiment by Kitchings, creating an "anti-urban nostalgia" for the rural in elements of populism. In the Irish context a left-leaning anticapitalism and antimilitarism informed many of what have been termed "first-phase" environmental disputes in Ireland.[35] This has been combined with a strong sense of identity that is grounded in rural discourse creating a strong sense of opposition to first-phase, multinational projects in the 1970s and 1980s while contributing to the emergence of new middle-class opposition to second-phase state-sponsored infrastructural projects in recent decades.

The initial campaign that came to define first-phase disputes was the antinuclear protest at Carnsore Point. Campaign leaders were able to draw on existing moral and cultural frames in the process of creating a discursive response to plans for nuclear power in Ireland. The Carnsore Point protest benefited from wider symbiotic understandings of the perils of nuclear power created by U.S. and European antinuclear protestors in addition to the wider representation of the nuclear issue that was becoming part of popular culture. The linking of the Donegal uranium mining issue with the energy debates surrounding the state's plans for Carnsore brought together the antiwar and environmental camps at an early stage. However, while local populism played some role in the Tynagh mines dispute, the inability of the local population to create the momentum that would go beyond the economic rationale for mining at Tynagh contributed to the horrors that followed during the mining process and the scarred, poisoned site when it was eventually abandoned. The Tynagh mines case and the subsequent degradation that occurred at the site and across its surrounding waterways and hinterlands should serve as a reminder to those who argue that communities should comply and take whatever economic benefits accrue from such acquiescence.[36]

The one environmental campaign that can be seen as having a successful outcome for the movement involved was that of the Burren Action Group at Mullaghmore. Through a series of legal actions combined with a comprehensive understanding of the fragility of the Burren's ecosystems, the BAG campaign achieved its goal of preventing tourist infrastructure from impinging on the Burren. It also contributed to an ongoing debate about the nature of heritage, identity, and development both locally and internationally. The Mullaghmore campaign provides an illustration of the significance of expertise for any environmental group that finds itself in dispute with the interests of local authorities or the state.

The importance of new middle-class expertise has also been demonstrated as a key aspect of the anti-incinerator campaigns in Galway, Meath, and Cork. The ability of these groups to contest the scientific data of the host industry or the state may not prevent incinerators being built in Ireland, but this form of interest-led advocacy has informed concerned communities, industry, and state officials about the potential risks of any incineration process that does not comply with strict processes of waste separation. In addition, the cultural importance of the reuse and recycle component of waste management has been reinforced by the anti-incineration campaigns, all of which presented recycling-led alternatives that would challenge the "throw-away" syndrome so conspicuous in contemporary consumerism.

The debate about resources, which first emerged in the 1970s and 1980s, has resurfaced as part of the Shell to Sea campaign. This campaign has reinvigorated the sense of community in Mayo and across rural Ireland, an identity that can be traced back to the agrarian agitation of the Land League. The territorial response of the North Mayo community to the perceived threats of the gas pipeline set to run across their lands is an instinctive one. The injustice surrounding the jailing of the "Rossport Five" mobilized a wider response among the Irish populace that was growing angry at increased collusion between state and industrial interests at the expense of local communities.

The sense of grievance at state indifference to local or national heritage informs the anti-roads disputes. The current obsession with private car ownership over public transport has an ideological base. When combined with the destruction of heritage at Tara or Carrickmines or environmental degradation at the Glen of the Downs, a new ideological response has emerged that draws on wider discursive challenges to articulate a coherent position based on a sense of environmental justice and conservation of the past.[37]

Conclusion*

This chapter has examined the consequences of environmental movement collective actions in Ireland. We can locate the campaigns covered in this study within the context of what Tovey has termed "populist" environmentalism.[38] Emerging from a wider network of antiwar and moral campaigners, the antinuclear protestors who staged the festivals at Carnsore Point achieved enough leverage to have the state abandon its nuclear energy policy. Socially, a network of environmental actors was established that was all-Ireland and international in its makeup. Culturally, this network provided the fledgling Irish environmental movement with a unifying issue and the antinuclear issue maintained its salience for campaigners and the public over a number of years.

The campaign succeeded in opening a debate on a number of moral issues, many of which were covered in *A Nuclear Ireland?* With the abandonment of the nuclear energy option, alternative approaches were examined by the state and in time the Irish state became a long-time opponent of the British nuclear industry. Scientifically, in the Irish case, arguments in favor of the nuclear option were abandoned until the recent energy crisis engendered a resurfacing of this exchange with prominent environmentalists such as James Lovelock suggesting nuclear power as an alternative to fossil fuels.

The anti-toxics campaigns that occurred in Cork at locations such as Ovens, Ringaskiddy, or the Womanagh Valley and were written about by Adrian Peace emerged as part of the first phase of anti-multinational campaigning in pre-growth Ireland.[39] These successive events can be analyzed in the context of a number of anti-multinational campaigns in Ireland throughout the 1970s and 1980s involving multinationals such as Raybestos Manhatten, Merrill Dow, Merck Sharp and Dohme, DuPont, and Beechams.[40] These campaigns can be seen as part of the "cycles of protest"[41] that form part of the interconnected sense of grievance in many rural and

* Sections of this conclusion have been published previously in Leonard, "Rapid Development and Community Mobilization in the Republic of Ireland."

suburban areas. Campaigns established framing processes derived from this extensive sense of grievance at the location of multinational plants that were perceived as posing a pollution risk.

One unfortunate result of the emergence of community-based campaigns against multinationals was the opening up of a divergence between community-based environmentalists and the trade unions, a gulf that had previously been bridged during the Carnsore antinuclear campaign. Disputes about the siting of toxic waste plants opened up the now familiar legal action or oral hearing approach for environmental campaigners and led to the establishment of the EPA in 1992. While communities gained access to expertise about toxics as a result of their campaigns, the state never abandoned its policy of multinational-led development. However, the onset of the Celtic Tiger saw the emergence of IT-based technologies as a more popular form of multinational development with polluting industries facing constraints from the EPA on EU legislation and fines.

The issue of resources, either onshore or offshore, has never been satisfactorily resolved in the Irish case. The Shell to Sea campaign in some ways characterizes the ongoing dilemma of Irish resource management. While many European countries have developed energy policies that include resource partnerships with industry, the Irish state has sold off all of the resources in the hope of making short-term, small economic gains. As a result, approaches to the development of mines such as Tynagh, Donegal, and Croagh Patrick led to a great deal of community opposition and in the case of the Tynagh mines, extensive ecological degradation. And while the type of resource has varied from mineral to uranium to offshore gas, the heavy-handed approach of the state-backed industries involved, alongside the seemingly reckless sell-off of natural resources, led to a heightened sense of rural territorialism and a discursive process whereby communities reassessed their hinterlands when threatened by outside projects. The response of the state has been to introduce stringent laws to facilitate the compulsory acquisition of land in addition to the establishment of a series of agencies to deal with resource management, many of which seem to favor industry over local concerns.

The Mullaghmore heritage dispute provides us with the best example of a campaign that achieved a successful outcome. The campaign led to a reversal of the state's plans to build an interpretative center in the Burren. In addition, a wider debate about economic, social, and cultural aspects of tourism and heritage was entered into by the supporters and opponents of the center. When combined with the Burren Action Group's ecological expertise and ability to utilize national and European legal frameworks as sympathetic, discursive and strategic framing patterns that provided the campaigners with a successful outcome were established. Furthermore, national and international interest in the heritage and ecology of Ireland was rekindled. However, the dispute created a dichotomy between local groups opposed to or in favor of the interpretative center and exposed tensions between local interests that favor economic environment over such interests.

The anti-incinerator disputes in Galway, Meath, and Cork had mixed results for local campaigners. In Galway no incinerator has been built at this point. However, licenses for incinerators were granted for Meath and Cork in late 2005. While all three campaigns came to be involved in the 2002 general election campaign, GSE engaged with a wider group of candidates from both government and opposition parties. This, in addition to the concerns of local politicians who feared an electoral backlash on the issue, may have led to plans for a Connacht regional incinerator being shelved. However, the Meath and Cork plants were pivotal to the state's waste management policy that was based on a combination of incineration, landfilling, and recycling.

While recycling levels achieved in Galway would make incineration unviable, the multinational sector was demanding thermal treatment for industrial waste. The sites at Meath and Cork were adjacent to major urban centers thus creating the momentum for the granting of licenses in these areas. While GSE forged political links, NIA in Meath and CHASE in Cork took the legal route, a tactic that diminishes any sense of community-based sentiment in the context of any subsequent oral hearing. However, all three anti-incinerator campaigns as well as those in Clare and Dublin have raised awareness about the issue of dioxins.

The introduction of incinerators as part of the state's waste management plans is linked to the wider consumer society that produces waste flows that have led to the exportation of Irish waste to destinations in Asia. These moral issues must be addressed as part of any comprehensive solution to the issue of waste management in the future. Another framing issue for the new middle-class groupings was that of climate change. The framing process was extended in both the anti-incinerator and anti-roads movements to link the issues of toxic emissions and pollution from road usage as contributing aspects to the creation of greenhouse gases and climatic change. Although both campaigns only utilized the climate change frame in the wake of the popularization of the subject in the wake of media debates surrounding the release of *An Inconvenient Truth,* such framing processes became an effective part of those campaigns' tactical opposition and strategic approach, bringing a level of perceived global scientific authority to localized protests, an approach that was picked up on by the media at the time.

The Shell to Sea campaign has raised the complex issue of a community response to an infrastructural project based on both a sense of grievance about the intrusion of the project in the area and fears about the safety of the gas pipeline. The response of the North Mayo community has emerged from a historically derived sense of place that has been articulated as part of a wider discursive framing process that links cultural nationalism with community-based environmentalism. Politically the issue can be traced back to the long-standing state policy of selling off resources while facilitating land acquisitions by partners. Local political response has been mixed with some supporting the Shell to Sea campaign while mainstream councilors and national representatives were ambivalent or chose to favor onshore development. From a populist perspective, the rural or agrarian politics of the Land League has been rekindled, while the Shell to Sea campaign has developed an international support network from Ireland to Norway and Nigeria.

The initial phase of rural-based community politics that was synonymous with the campaign in its early years has been extended to wider links with a new wave of direct action environmentalists who emanated from the anti-roads or antiwar alliances. The jailing of the five men who challenged the onshore pipeline in their community became the defining event of the campaign as the "Rossport Five" and their families won a great deal of sympathy and support from the general public. Both the state and, subsequently, industry have come to recognize the extent of support Shell to Sea has achieved, and thus a mediation process was undertaken. This process, chaired by former trades unions' leader Peter Cassells, has been criticized by Shell to Sea spokespersons. Protestors have converged at the Rossport Solidarity Camp in the area and remain resolute in their stance. They are hopeful of a future where local voices will be heard on this issue. Their protest is part of a culture of mobilization that has become a feature of modern society, as seen in the "Occupy" movement or the antiglobalization protests that occur around major political summits.

Historically, the successive anti-roads campaigns that occurred at the Glen of the Downs, Carrickmines, and Tara/Skryne had their inception in the wider anti-roads protests undertaken by Earth First in the late 1980s. This form of direct action has become a feature of the new wave of environmental activism in Ireland; one that combines opposition to growth-based ecological degradation with wider issues such as opposition to the wars in Iraq and Afghanistan. However, the direct-action wing of the anti-roads campaign has also extended its wider networks with second-phase anti-infrastructural campaigns including anti-incinerator groups and the Shell to Sea protest. In addition, the anti-roads protestors have extended the debate about heritage in an era when even significant sites such as the Hill of Tara are coming under threat from development projects. A wider debate about the future direction of the nation in a post-growth, post-materialist phase has been embarked on in the wake of these second-phase campaigns.

The shift in policy direction undertaken by successive neoliberal Irish governments led to a move toward a primarily ecological modernization-orientated policy framework (as set out by the European Union). The emphasis on economic growth over environmental protection or concern for communities has subsequently become a priority. This has been exacerbated by the Eurozone crisis and the bailout of the Irish economy by the IMF since 2008.[42] Ultimately, the costs of economic growth have led to greater environmental degradation and an undermining of significant aspects of the nation's heritage. In particular, rural communities have borne the cost of Ireland's transition from a rural agrarian society into an industrialized and globalized economy. While communities have demonstrated a commitment to retaining some aspects of the traditional values and heritage of the "Emerald Isle," the state's policy of prioritizing rapid infrastructural development in order to underpin private consumption and multinational investment over local community services may prove to be ill-advised as fluctuations on the global markets continue. As the cycle of the Celtic Tiger gave way to the era of IMF-sponsored austerity in the wake of the post-2008 economic crisis, the values and concerns of these local communities may provide a new direction for the post-boom era.

Notes

1. These disputes can be seen as part of an ongoing contestation of lifestyles and hinterlands in a society that has seen the economy shift from a situation where agriculture provided 10 percent of GNP in 1983 (representing 1 in 6 of the population) move down to 2.6 percent of GNP in 2006 (representing 1 in 20 of the population), according to press reports (*Irish Times,* February 15, 2008).
2. Szasz, *Ecopopulism.*
3. Leonard, "Green Nation," and *The Environmental Movement in Ireland.*
4. Ibid.
5. Tovey, "Environmentalism in Ireland," p. 275.
6. Tovey, Share, and Corcoran. *A Sociology of Ireland* Second Edition.
7. Heffernan, *Woodquay,* pp. 5, 6. Discusses the events surrounding the Woodquay protests in detail.
8. Ibid.

9. See Duncan, *Place, Culture and Representation,* p. 39, for further discussion.

10. Said argues that concepts of "the Other" are used in the colonization process. See Said, *Orientalism.*

11. Duncan, *Place, Culture and Representation,* p. 39.

12. Duncan, "Elite Landscapes as Cultural (Re)Productions," p. 54.

13. Saunders, "Reflections on the Dual Politics Thesis"; Pickvance and Harloe, *Place, Policy and Politics.*

14. Lloyd, *Ireland after History,* pp. 77, 79.

15. Ibid.

16. Ibid.

17. Tovey, "Environmentalism in England," pp. 278, 280.

18. Ibid.

19. From Tovey, "Environmentalism in England," p. 279; from MacDonald, *The Destruction of Dublin,* p. 12.

20. See Marsden et al, *Constructing the Countryside.*

21. In an era characterized by high unemployment, tourism was prioritized by the state as providing 10 percent of GNP with revenues of over IR£1,000 million (McGrath, *The Mist-Covered Mountain,* p. 27).

22. Interpretative centers were seen as a necessary part of that infrastructural development in order that better understandings of Irish history and culture would be provided for visitors—"creating interpretive 'gateways' into our heritage" (O'Toole, from McGrath, *The Mist-Covered Mountain,* p. 29).

23. EIS Statement 1994 cited in McGrath, *The Mist-Covered Mountain,* p. 57.

24. These heritage campaigns occurred at the Glen of the Downs in County Wicklow, Carrickmines in County Dublin and at Tara/Hill of Skryne in County Meath (Leonard, *The Environmental Movement in Ireland*).

25. Connolly and Lynch, *The Great Corrib Gas Controversy*; and Leonard, *The Environmental Movement in Ireland.*

26. Writing in *Heritage Outlook,* Liam Lysaght called the Government Strategic Plan on forestry "one of the greatest threats to our countryside" (Lysaght, *Heritage Outlook* Vol. 1, 2000).

27. These issues are discussed at length in Leonard, *Community Campaigns for Sustainable Living.*

28. Ibid.

29. Gamson *The Strategy of Social Protest*; and Burstein et al., "The Success of Political Movements."

30. Amenta and Caren, "The Legislative, Organizational, and Beneficiary Consequences of State-Oriented Challengers."

31. Ibid.

32. Leonard, *Community Campaigns for Sustainable Living*; and L. Leonard, *Politics Inflamed.*

33. Kitching, *Development and Underdevelopment in Historical Perspective,* pp. 20, 21.

34. Ibid.

35. Leonard, *Community Campaigns for Sustainable Living*; and *Politics Inflamed.*

36. Tovey, "Environmentalism in Ireland."

37. Leonard, *The Environmental Movement in Ireland.*

38. Ibid.

39. Leonard, *Community Campaigns for Sustainable Living.*

40. R. Allen and Jones, *Guests of the Nation;* R. Allen, *No Global*; Leonard, *Politics Inflamed*; Adrian, Peace *A World of Fine Difference.*

41. Tarrow, *Power in Movement*; David A. Snow, and Benford, "Master Frames and Cycles of Protest," 133–55.

42. This is the subject of Leonard and Botetzagias, eds., *Sustainable Politics and the Crisis of the Peripheries.*

Bibliography

Bourdieu, P. "The Three Forms of Capital." In *Handbook of Theory and Research for the Sociology of Education,* edited by J. G. Richardson, 241–58. New York: Greenwood Press, 1986.

Brett, D. *The Construction of Heritage.* Cork: Cork University Press, 1996.

Commins, P. "Rural Social Change." In *Ireland: A Sociological Perspective,* edited by Prundy Clancy, K. Lynch and L. O'Dowd. Dublin: Sociological Association of Ireland, 1986.

Connolly, F., and R. Lynch. *The Great Corrib Gas Controversy.* Fiosrú Vol. 1, No. 2. Dublin: Centre for Public Inquiry, 2005.

Duncan, J. "Elite Landscapes as Cultural (Re)Productions: The Case of Shaugnessy Heights." In *Cultural Geographies,* edited by K. Anderson and F. Gale. London: Longman, 1999.

Duncan, J., and D. Ley, eds. *Place, Culture and Representation.* London: Routledge, 1993.

Flynn, B. *The Blame Game.* Dublin: Irish Academic Press, 2007.

Harloe, M., C. Pickvance, and J. Urry, eds. *Place, Policy and Politics: Do Localities Matter?* London: Unwin Hyman, 1990.

Harvey, D. *The Urbanization of Capital.* London: Basil Blackwell, 1985.

Heffernan, T. F. *Woodquay.* Austin: University of Texas Press, 1988.

Hourigan, N. *Escaping the Global Village: Media, Language and Protest.* New York: Lexington Books, 2004.

Irish Times, February 15, 2008.

Kousis, M. "Environmental Impact of Tourism." In *International Encyclopedia of Environmental Politics,* edited by J. Barryand E. G. Frankland. London: Routledge, 2002.

Kitching, G. *Development and Underdevelopment in Historical Perspective: Populism, Nationalism and Industrialisation.* London: Routledge, 1989.

Leonard, L. *Community Campaigns for Sustainable Living: Health, Waste and Protest in Civil Society.* Bingley: Emerald, 2011.

Leonard, L. *The Environmental Movement in Ireland.* Dordrecht: Springer, 2008.

Leonard, L. *Green Nation: The Irish Environmental Movement from Carnsore Point to the "Rossport Five."* Ecopolitics Series Vol. 2. Drogheda: Choice Publishing, 2006.

Leonard, L. "The Moral Framing of a Resource Dispute." *Studies: Irish Quarterly Review.* Dublin: Studies, Winter 2007.

Leonard, L. *Politics Inflamed: GSE and the Campaign against Incineration in Ireland.* Ecopolitics Series Volume one. Drogheda: Choice Publishing, 2005.

Leonard, L. "Rapid Development and Community Mobilization in the Republic of Ireland." *Community Development,* July 2008.

Leonard, L., and I. Botetzagias, eds. *Sustainable Politics and the Crisis of the Peripheries: Ireland and Greece.* Advances in Ecopolitics 8. Bingley: Emerald, 2011.

Lloyd, D. *Ireland after History. Field Day Essays Critical Conditions.* Cork: Cork University Press, 1999.

Lysaght, L. *Heritage Outlook* Vol. 1. Dublin: The Heritage Council, 2000.

MacDonald, F. *The Destruction of Dublin.* Dublin: Gill and MacMillan, 1985.

MacDonald, F. *Saving the City: How to Halt the Destruction of Dublin.* Dublin: Tomar, 1999.

McGrath, B. *The Mist-Covered Mountain: Change, Conflict and Identity in Ireland's Countryside—The Case of a Planned Development at Mullaghmore, Co. Clare.* Unpublished Masters Thesis. Galway: National University of Ireland Galway, 1995.

Marsden, T., J. Murdock, P. Lowe, R. Minton, and A. Flynn. *Constructing the Countryside.* London: University College Press, 1993.

O'Toole, F. *Blackhole, Greencard: The Disappearance of Ireland.* Dublin: New Ireland, 2004.

Peace, A. "Environmental Protest, Bureaucratic Disclosure: The Politics of Discourse in Rural Ireland." In *Environmentalism: The View from Anthropology,* edited by K. Milton. London: Routledge, 1993.

Peace, A. *A Time of Reckoning: The Politics of Discourse in Rural Ireland.* London: Routledge, 1997.

Pickvance, C.G., and M. Harloe. *Place, Policy and Politics: Do Localities Matter?* London: Routledge, 1990.

Putnam, D. *Bowling Alone: The Collapse and Revival of American Community.* New York: Simon and Schuster, 2000.

Said, E. *Orientalism: Western Conceptions of the Orient.* London: Penguin, 1995.

Saunders, P. "Reflections on the Dual Politics Thesis: The Arguments, Its Origins and Its Critics." In *Urban Political Theory,* edited by M. Goldsmith and S. Villadsen. Aldershot: Gower, 1986.

Smith, M. "Citizens, Denizens and the *Res Publica:* Environmental Ethics, Structures of Feeling and Political Expression." *Environmental Values* 14: 145–62.

Storey, D. *Territory: The Claiming of Space.* London: Prentice Hall, 2001.

Szasz, A. *Ecopopulism: Toxic Waste and the Movement for Environmental Justice.* University of Minnesota Press, 1994.

Tovey, H. "Environmentalism in Ireland." In *Ireland and Poland: Comparative Perspectives,* edited by P. Clancy, M. Kelly, J. Wiatr, and R. Zoltaniecki. Dublin: UCD Press, 1992.

Tovey, H. Review of Niamh Hourigan (2004), *Escaping the Global Village: Media, Language and Protest*; Lanham Maryland: Lexington Books. *Irish Journal of Sociology* 14, no. 1 (2004): 135–37.

Tovey, H., P. Share, and M. Corcoran. *A Sociology of Ireland* Second Edition. Dublin: Gill and MacMillan, 2007.

Varley, T., and C. Curtin. "Defending Rural Interests against Nationalists in 20th Century Ireland: A Tale of Three Movements." In *Rural Change in Ireland,* edited by J. Davis. Belfast: The Institute of Irish Studies, Queen's University Belfast, 1999.

Varley, T. "Power to the People? Community Groups and Rural Revival in 'Ireland.'" In *Rural Development Policy: What Future for Rural Ireland?,* edited by B. Reynolds and S. Healy. Dublin: Conference of Major Religious Superiors, 1991.

Chapter 5

The Green Movement in France: History, Ideology, and Action Repertoires

Sylvie Ollitrault

Introduction

Since the end of the 1980s, the political system in France has been challenged by a number of factors. Firstly, economic globalization and the European project have transformed national agendas, historically more state-centered, with even the Gaullists, traditional defenders of a strong national state, adapting their ideology in response to neoliberal influences to embrace the open market, reduced state intervention, and participation in European institutions. Secondly, under Chirac's presidency (1995–2007), French society was transformed as a consequence of economic and social globalization: public opinion is now increasingly informed by the Internet and by constant flows of images (such as those surrounding the events of September 11, 2001); protest takes new forms[1] and operates more at the level of the individual. People are increasingly mobile and well informed, yet new forms of poverty spread. The distance between the richest and the poorest grows wider as the capacity of the French state to protect the middle and lower classes has been attacked and reduced under Sarkozy's presidency. As a result, different types of protest have emerged in France, and the last decade saw huge demonstrations against liberal reforms.[2] Environmental movements and mobilizations must therefore be understood in this particular context. There is fierce debate in France around issues of nuclear energy, or genetically modified (GM) crops, but these "green" debates are partly inflected by criticism of free-market ideology. Whereas in the 1970s the French green movement fought the state and the decisions of the government, nowadays more and more green criticism is directed at economic liberalism.[3]

Naturally, the green movement is comprised of a variety of different groups. This diversity of political and social expression within the environmental movement is a reflection both of its history, of the various actors that take part in protest actions, and the range of interests it represents. The movement's key challenge since its inception is whether or not to participate in the political party system: should it function as a pressure group without allegiance to either left or right; or form a political party with a transformative agenda, supporting social movements such as feminism, minority groups, or antiracism. This dilemma becomes thornier as the movement comes also to be known for its criticisms of the effects of globalization.

Despite this radical position, the Green Party, especially in its most recent incarnation as Europe Écologie–Les Verts (EELV), is nonetheless part of the political party system, winning local and European elections, with particular support in urban areas. Recognizing that green issues are important to many of its voters, the Socialist Party (PS) is obliged to form alliances with the Greens in order to challenge Nicolas Sarkozy's right-wing Union pour la majorité populaire (UMP). A further challenge for the Greens is how to compete with the traditional parties, which try to capture new voters by greening their program.

The following analysis reviews the roots of the green movement in France, while maintaining a focus on the changing political landscape. Although the environmental movement first emerged in France at the end of the 1960s, we cannot hope to understand the new areas of protest or the nature of contemporary mobilizations without examining the ways in which it has been influenced by the international context. The environmental movement is a transnational movement;[4] thus, in the French case, my aim is to analyze this mix between global influences and local constraints. I also describe the difficulties encountered in the attempt to transform the green movement into a Green Party.[5]

But the green movement is more than just a political party; as a social movement it protests on a whole range of issues, such as climate change, genetic modification, nuclear power—as well as immigration, women's rights, and social welfare. I therefore outline French green ideology and values (de-growth, etc.) and describe how these ideas have increasingly inspired ordinary citizens to act at the local level in their role as inhabitant or consumer. This chapter concludes by analyzing this new form of everyday resistance, which involves ordinary French people, and illustrates the contemporary trend for very individual protest.

The Slow Growth of the Green Movement

From the 1960s to the Birth of the Green Party

In France, just as elsewhere in Western Europe, the Green Party and the environmental movement were created as a result of the protests of the 1960s. France in the 1960s was Gaullist and was in the process of constructing a modern urbanized state, with huge airports (such as at Roissy Charles de Gaulle), powered by nuclear energy. The contemporary drive to create a new modern France was powerful, and the capacity to protest against these programs was weak and not easily mobilized through the media. Resistance consisted only of nonviolent protest by Christian groups against torture during the Algerian War, and against nuclear energy. Protests were mostly generated by Marxist groups within the Communist Party, and subsequently by (extreme) left-wing groups such as the Maoists or the Trotskyists. The counterculture hippie movement reached France from the West Coast of the United States in the wake of May 1968. Books and articles in newspapers were published focusing on the philosophy of Henry David Thoreau and his ecological vision of the world, and on the counterculture movement that was leveling strong criticism against the overexploitation of the planet. In France, the green movement sprang from the antinuclear movement with the primary goals of pacifism and protection of the environment, and a utopian countercultural vision of how to reinvent society.[6]

At the same time, public opinion was becoming more and more concerned with environmental issues. In 1969 widespread support was mobilized to defend the Parc de la Vanoise,[7] a protected area in the Alps famed for the biodiversity of its fauna and flora, then threatened by the development of a winter sports facility. The campaign successfully mobilized scientists, voluntary associations, and the media in an attempt to inform public opinion, and became one of the most famous campaigns in French Green history.

In France, as in other Western countries, the green movement came about as a result of the 1960s protests and in the changing context of claims for a new type of society, one more respectful of people and the environment. The Stockholm Summit of 1972,[8] with its consensus regarding the limits of economic growth and natural resources, threw new light on global environmental problems. This new international agenda had consequences for the French administration. A new position of secretary of state was created, the environment secretary, and Georges Pompidou, then president of France, gave a well-known speech in Chicago in 1970 on the environment, and focused high-level attention on this new issue.

This combination of vigorous social movements and countercultural movements (such as neo-anarchist rural communities), a larger, well-educated middle class, and the demands of an international agenda thus provided the perfect conditions for the growth of a green movement capable of attracting media attention.

The Green Party: Contesting Elections

The Green Party was officially created on January 29, 1984, during the Congress of Clichy; however, political environmentalism has a longer history. The first "green" candidate appeared in 1973 representing the Ecologie et Survie (Ecology and Survival) Party; however, it was René Dumont's candidacy in the 1974 presidential elections, representing a new political movement with an original message, that definitively captured media attention.

René Dumont: The First Green Candidate for President[9]

René Dumont (1904–2001) was an agronomist who worked for international organizations such as the FAO (Food and Agriculture Organisation). As a professor in the prestigious Agronomist Institute, during the 1960s he wrote various books on the third world and the limits of natural resources. When he appeared on French television in spring 1974, Dumont surprised TV audiences and political leaders by wearing a red pullover and drinking a glass of water to underline that water is a very scarce resource. He was the first to criticize globalization and economic development that took no heed of long-term consequences.

Dumont's candidacy was supported by the local Parisian Friends of the Earth group and a network of environmental journalists. These two networks combined to create a positive image for him on television and in the press, surprising the media and public opinion with the originality of his message and personal style. But the media image did not deliver votes and René Dumont obtained only 1.32 percent of the vote. Nonetheless, the green movement had been born and it became increasingly political after 1974.

Before 1974, the green movement was shaped by various groups, in particular Les Amis de la Terre (Friends of the Earth), which had created a network of local groups in the main French cities. The year 1974 saw the first incarnation of the green movement as a political party—le Mouvement écologique. Then in November 1979, following the European elections, le Mouvement d'écologie politique (MEP) was created, constituting a precursor of Les Verts—Parti écologiste, which was created in November 1982 with few members. The Amis de la Terre network, with its more decentralized conception of the green movement, refused to develop formal structures until November 1982. At this date, this network chose the name of Les Verts—Confédération écologiste, and in January 1984, the two branches came together in one party, Les Verts–Confédération écologiste–Parti écologiste.

Les Verts: Organization and Activists

The organization of Les Verts[10] reflected its concern to preserve the autonomy of local groups. Members of the party (around 8,500) were invited to participate in vast regional general meetings (*Assemblées générales*). From these regional general meetings delegates were elected to become members of the *Assemblée fédérale* (federal assembly) of Les Verts. The interregional national council (CNIR) was composed of 120 members, with 25 percent of this council being elected by proportional representation by the federal assembly every two years, and 75 percent by the regional assemblies to represent (and reflect) the membership base. Two groups (colleges) are elected by the CNIR. The statutory college has decisional and legislative power, and the executive council names a national secretary and two spokespeople: one woman and one man—Les Verts is careful to observe parity between men and women at every level of the organization. This was a real difference between Les Verts and other French political parties, and it set a standard that is now recognized by the majority of political parties.

The executive council implements internal democracy, organizing referendums as a result of activist initiatives if 10 percent of activists from one-third of regions support the proposal, as was the case in the challenge to Antoine Waechter (leader of the Green Party) in 1990 or the project to support José Bové's candidacy during the presidential elections in 2007. This procedure aimed to exert control over the party leaders and allow grassroots members to participate in decision making, yet even with this very open democratic structure in which there are plenty of opportunities for members to speak out.

Green activists are very attached to a form of anarchism and delegating power is not their political culture. Most of them criticize the limits of direct democracy and believe that environmental battles[11] are local battles. They have a very negative view of centralized parties such as the Socialist Party or the previous Communist Party. Most would class themselves as on the progressive left, but are very careful to preserve their ideological differences, manifested in a belief in local democracy, direct action, and respect for individuals' beliefs.

French green activists are characterized by post-materialistic values. They are liberal because they defend social equality for women, homosexuals, and minorities and are concerned with the protection of individual rights. They tend to have the same profile: the majority have higher education qualifications—16.4 percent hold a PhD—and are members of the well-educated upper middle class, even if their incomes do not reflect their academic qualifications. All of which goes some way to explain their capacity for discussion and criticism, and their radically alternative vision for society, as well as their unwillingness to accept a party hierarchy and renounce their individual capacity to contribute to the debate. In 1999, 17.3 percent of green activists worked in the public sector (mainly teachers and university lecturers) and only 3.5 percent were manual workers. The majority of activists were male (70 percent), aged 40 or above, academically qualified, and working in an intellectual profession, whereas the majority of green voters were young women, academically qualified, and living in urban areas.[12]

In their support for regional claims for independence (such as those from Corsica and Brittany), and for the European project, green activists propose new ways of framing the world. Even when they opposed the new treaty on the European Union, their criticisms were aimed more at improving the treaty rather than rejecting the European Union per se. They are activists at local, European, and international levels, but highly critical of policy-making at the national level, especially on military or nuclear issues, or more recently on the protection of migrants (right to asylum, etc.).

Between Elections and Protest: Green Party or Green Movement?

Les Verts tends to win elections under proportional representation (local, regional, and European elections), and has made real progress even in presidential elections (scoring 5.25 percent higher than the Communist Party in 2002). It often chooses a tactical alliance with the Socialist Party, with Les Verts being the party of choice for voters in the first round of voting. In 2010—except in Brittany where the local context ruled out a strategic alliance between Europe Ecologie–Les Verts and the Socialist Party—all regional councils were won by the Socialist Party and Green Party alliance. Despite this electoral success, the Green Party is in competition with other environmental parties, such as Cap 21 on the center right, or other traditional parties that are greening their program to capture voters.

The Green Party is well known for its support of left-wing social movements, particularly the antiglobalization movement, and is a visible presence on their demonstrations.[13] Green Party members are also usually very active in voluntary associations in the fields of environmentalism, humanitarianism, and feminism, for example, or on behalf of poor people's movements, and are critical of the economic organization of society, which abuses both human beings and nature. Despite its strong protest identity, the Green Party has become increasingly pragmatic and focused on environmental policies, such as Agenda 21 or the pro–climate change movement.

Europe Ecologie–Les Verts: A New Party?

Cécile Duflot, the new leader of the Green Party, with the support of Dany Cohn-Bendit, a renowned leader in both French and German green parties, has ushered in a new era in the strategic organization of the Greens. Following the presidential elections of 2007, the Green Party's share of the vote fell, and its position as the second party on the left (after the Socialist Party) was under threat. To combat this, their strategy consisted of opening up the Green Party to people involved in social movements, or to very popular media personalities who espoused democratic or environmental causes. During its annual meeting (AG) in December 2008, the Green Party created Europe Ecologie to allow those who were not members of the Green Party to stand on the party list in the next (European) elections. The strategy was successful and Europe Ecologie obtained 16.3 percent of the vote and polled in third position, close behind the Socialist Party. This success reinforced the policy of a strategy of openness, and in 2010, the alliance obtained 12.5 percent of votes and was stable in third position behind the UMP and the PS.

In November 2010, Cécile Duflot succeeded in formalizing this new Green Party structure under the new name of Europe Ecologie–Les Verts (EELV). In July 2011, EELV comprised 14,869 members and 17,371 "co-operators." Alongside this new policy of openness within the Green Party, the internal structure has also been reformed and the CNIR has made way for a federal assembly. For the most part the organization of the party still reflects its commitment to direct democracy, and it enables a variety of channels for activists to express criticism. But, the split between the politicized left wing and the more neutral and environmental wing of the party is clearly visible and lies behind the party's difficulty in speaking with one voice, especially during the presidential election when a leader should speak for the whole party. Corinne Lepage, who has joined Europe Ecologie–Les Verts, prefers to continue a right-of-center strategy.

The new ideological focus of Europe Ecologie–Les Verts is to fight corruption in political and economic arenas. Their candidate for the presidential elections of 2012 is Eva Joly, a judge who made it her mission to root out economic corruption. Joly is well known to the French public and the party opted for her over Nicolas Hulot, who represented a more pro-environmentalism side of the party. The competition between Joly and Hulot for the Green Party nomination illustrates the two factions within the Greens: one very pro-environmentalism and concerned with climate change issues and the need to protect the earth's resources; the second focused more on finding solutions to social problems and negotiating a new political deal with new rules of engagement (e.g., less corruption, more women and ethnic minorities as political candidates), as well as envisaging a new economic system that protects both people and environmental resources. With the choice of Eva Joly, Europe Ecologie–Les Verts is signaling a left-wing direction to its election campaign. Eva Joly ran a very bad campaign with a weak media image. On April 21, 2012, she obtained 2.31 percent of the vote.

Nevertheless, the emergence of a new generation of leaders (Yannick Jadot and Cécile Duflot) has given fresh impetus to the organization, and equipped it to attract media attention and exploit new channels in order to create a new political socialization. Yannick Jadot worked as campaigner for Greenpeace producing facts and figures related to antiglobalization, economic crisis, and resistance to the financial power of the banks, and monitoring the international summit on climate change. With this influx of new blood, EELV is developing a program that speaks to the concerns of the urban, well-educated young people who vote for it. It is noticeable in France that, especially in local elections, those who vote green are predominantly urban dwellers, living in city centers that are more and more "gentrified." One challenge, therefore, is to capture more mainstream voters who are less seduced by the environmental issues. The second challenge is how the party will participate in the new alliance with the Socialist Party and take its place in the new government of the freshly elected president François Hollande. How will the EELV influence a very dominant Socialist Party in a context of economic crisis? How will it pressure to stop the nuclear program? How to be an EELV minister when the UN Earth Summit Rio+20 had so little audience?

Table 5.1. Green Performance in the European Elections

1984	3.4%	
1989	10.6%	8 elected
1994	2.75%	
1999	9.72%	9 elected
2004	7.2%	
2009	16.3%	6 elected

Source: L'Ecologie Les Verts: Archives des Verts.

Table 5.2. Green Performance in the Presidential Elections

1988	3.7%	1,150,000 votes for Antoine Waechter, the Green candidate
1995	3.32%	1,011,488 votes for Dominique Voynet, the Green candidate
2002	5.25%	1,495,901 votes for Noël Mamère, the Green candidate
2007	1.57%	576,666 votes for Dominique Voynet, the Green candidate
2012	2.31%	828,345 votes for Eva Joly, the EELV candidate

Source: L'Ecologie Les Verts: Archives des Verts.

Table 5.3. Green Performance in the Parliamentary Elections

1986	2.44%	
1993	7.9%	
1997	5.12%	6 elected
2002	4.46%	3 elected
2012		18 elected

Source: L'Ecologie Les Verts: Archives des Verts.

Table 5.4. Green Performance in the Regional Elections

1986	3.38%	0 elected
1992	6.8%	100 elected
2004	–	168 elected
2010		262 elected

Source: L'Ecologie Les Verts: Archives des Verts.

Table 5.5. Green Performance in the Departmental Elections

1985	4.08%	
1988	6.8%	
1992	11.06%	1 elected
1994	7.8%	5 elected
1998	7.66%	10 elected
2001	11.29%	13 elected
2004		11 elected
2010		14 elected

Source: L'Ecologie Les Verts: Archives des Verts.

Table 5.6. Green Performance in the Local Council Elections

1989	9%	600 elected, c. 15 mayors
1995	6.5%	650 elected, c. 20 mayors
2001	11.80%	33 mayors
2008		41 mayors

Source: L'Ecologie Les Verts: Archives des Verts.

Ecologist Parties: To the Right and Left of the Green Party

To the right of the Green Party are political groups that are more concerned with environmental issues and less critical of capitalism: Génération écologie, Ecologie bleue (the Gaullist environmental network), Cap 21, and the Mouvement écologiste indépendant are representative of this tendency. They focus their program on the preservation of the environment, and their objective is more to act as an interest group putting pressure on traditional political parties to pay attention to the environment than to propose a real alternative to liberal economy, the free market, and social inequality. These groups adopt a very open position on the green economy and the greening of business. Although they have very limited success, they nonetheless act to challenge Les Verts and can sometimes blur the message between environmental concerns and a more revolutionary vision of social change.

On the left, a more radical movement is developing around the idea of "degrowth." Its strategy is to put up candidates at local elections in order to find an audience for their ideas, central to which is the belief that economic growth cannot be unlimited because of the finite nature of natural resources. The policies of the *Parti de la décroissance* therefore seek to relocate the economy to the local level, through, for example, consuming products from the local economy (food, etc.), or supporting local craftsmen and boycotting industrialized products. The degrowth agenda is close to social ecology and, for the "degrowthers," the Green Party is too conventional; however, the degrowth program is more a criticism of capitalism and globalization than a real political party ready to propose policies in the public sphere.

Political ecology is now a key force in French politics. The Green Party has elected representatives at every level of government: local, national, European, as well as, more recently, in the Senate, a very conservative assembly. We may thus conclude that the Green Party has been accepted by voters and public opinion, and the Socialist Party can no longer afford to ignore it.

Green Mobilization: From Antinuclear Protest to the Antiglobalization Movement

The number of environmental organizations has exploded since the end of the 1960s. The events of May 1968 created a society in which urban well-educated people became increasingly active in protests against top-down decision making. The environment started to be recognized as a factor contributing to everyday happiness, and people reacted when contamination, motorways, or nuclear power stations disturbed their immediate surroundings. At the same time, scientists and global activists began to inform French citizens about the dangers that threatened the planet. The Stockholm Summit (1972) in particular represents a global event that contributed to the growth of an associative network. Illustrating this process, many small groups joined Federation Nature et Environnement (FNE), which represented a larger interest group, in order to ensure that the French voice was heard in the international arena.[14] Even before this, in France as in other countries, the first stirrings of environmental consciousness were visible in the networks of scientists centered on the Museum of Natural History. These networks actively promoted the notion of parkland to preserve wild flora and fauna (e.g., in the Camargue in the south of France), or to preserve birds (as in the Sept îles in Brittany). La Ligue de Protection des Oiseaux, similar to the Royal Society for the Protection of Birds in Great Britain, and part of the BirdLife International network, was established in France in 1911.

In France, then, the most powerful environmental network is France Nature Environment, which is an umbrella organization composed of 3,000 voluntary groups and recognized for its scientific expertise. It has helped to list endangered species and to provide scientific reports on the deterioration of nature. At the same time, the FNE has acted as a whistleblower for the French people, submitting appeals to the courts to ensure that legislation is respected. Recently, FNE has widened its sphere of action beyond the natural environment to include questions of environmental health or the impact of new scientific breakthroughs (such as the consequences of GMO on the environment). During winter 2011, FNE surprised public opinion and political actors with a new media strategy, closer to Greenpeace's modus operandi: the walls of the Paris metro were plastered with photos showing the contamination of French beaches by green algae as a result of intensive agriculture that destroys the quality of the water. This media campaign was launched on the first day of the Agriculture Salon in Paris, a huge event in France. Farmers' organizations reacted very strongly, as did the regional council in Brittany that defended both the intensive farming and the region's tourist image. This example demonstrates that French environmental organizations are increasingly provoking controversy in the name of protest. Although French environmentalists continue to participate in public decisions with the aim of defending nature and protecting the interests of the victims of contamination, even those organizations that are traditionally collaborative are transforming their protest repertoire and adopting more radical ways to bring home the responsibilities of political actors and economic interest groups, such as going to court to ensure that pro-environmental legislation is implemented.

Pacifism: From Larzac to Antiglobalization—The Rebellion of Local Farmers

Larzac was an emblematic space during the 1970s in France and a breeding ground for future activists. Jose Bové, the French farmer who attacked a McDonald's restaurant in the late 1990s,[15] was the young leader of this rebellion against the French state, and the neo-rural movement supported the emergence of an area of resistance. Because the Larzac Plateau, an agricultural area in the Dordogne region of south-western France, was considered a desert after the rural exodus, the French state had planned to extend the military camp there. Between 1970 and 1981, Larzac attracted huge gatherings of mostly pacifist activists from the antinuclear movement, third world movements, and the Occitan movement (an ancient language spoken in that part of France). Despite some violence at the beginning, the strategy of the local farmers and the incoming neo-rural farmers was oriented to nonviolence. The campaign was thus mainly pacifist, and the image of local farmers with sheep pitted against the French state and the army served to construct a collective identity within the movement and in media coverage.

In 1981, the president and socialist leader François Mitterrand decided to abandon the projected military camp, but this long campaign of resistance had served to form activists like the young José Bové—who continues to live in the area—and various leaders of the green movement, and Larzac now functions as a powerful symbol for large-scale, pacifist environmental mobilization in France. The Cun du Larzac is nowadays a place where activists are trained for nonviolent action, civil disobedience, and where antiglobalization discourse is preached. The struggle of Roquefort (the locally produced cheese) against international business symbolizes French resistance to forms of globalization that threaten local farmers. José Bové is not only a media-friendly activist but also enjoys public popularity in France as a veteran of the Larzac protest and a peasant David against the state Goliath.

Anti-GMO: Civil Disobedience against Globalization?

José Bové reappears in the second half of the 1990s in the anti-GMO protest coordinated by two major organizations: Greenpeace and the Conféderation Paysanne (Peasant Farmers' Union). GMOs are a very controversial issue in France, and one of the top priorities of the green agenda since 1990.[16] Bové is the Conféderation Paysanne leader and is charismatic and media-friendly; he also commits acts of civil disobedience and breaks the law in the name of universal value, aiming to defend public health and the future of farmers and appeal to democracy. According to Bové, GM crops are an industrial choice, supported by government and scientists: to resist the genetic contamination of other crops, the public must wake up to the facts, trash corn crops, and destroy fields. The *faucheurs volontaires* number perhaps 400 people, but a further 4,000–6,000 people have signed petitions indicating their support. A massive meeting (200,000–300,000 people) took place in summer 2003 in Larzac, recalling the demonstrations of the 1970s. At the heart of anti-GMO mobilization is criticism of globalization and the constraints imposed by the dominance of international business. WTO norms are contested because, according to anti-GMO activists, they protect firms and businesses against the choices and interests of citizens or small farmers. The Monsanto corporation sells seeds to peasant farmers in order to create a captive customer base; an example is India, where peasants were sold seeds by large companies without being able to make a choice of one seed over another. The same phenomenon occurs in Europe. Farmers from Confédération Paysanne defend their independence from the major firms and their right to choose how they cultivate their land. The GM issue also mobilizes political leaders, and the Socialists and the Greens have formed a political alliance to declare their regional constituencies GM-free and to implement the precautionary principle. Elected representatives have joined the movement, some of whom support the tactics of civil disobedience.

WWF, Friends of the Earth, and Greenpeace. From Global to Local: The Constraints of Being an International NGO in the Local French Network

The WWF was the first transnational NGO to appear with its panda symbol. It was a visible presence during the demonstrations against the Somport tunnel that threatened the biodiversity of the Pyrénées. It works in French territories such as Guyana and supports the European implementation of environmental rules (such as the Natura 2000 Directive, or inventories of endangered species). Because of its extensive networks within the international arena and its budget, the WWF has considerable influence on the international agenda; it takes the lead on the major issues and its environmental agenda in France is thus focused on sustainable development, the protection of biodiversity and, more recently, climate change. In France, the WWF is perceived as the NGO that supports an alliance with business, and for this reason the WWF is seen as an active actor in the greening of business, rather than an actor of antiglobalization.[17]

Friends of the Earth has a long history in France and can be considered as the first activist network to promote environmentalism. Established in 1970 in Paris as the result of the efforts of a small group of individuals who were included in the international network, Friends of the Earth/les Amis de la Terre then went on to found many local groups across France. Each group was very independent but with a shared objective to promote the environmental message and struggle. After the birth of the Green Party, local Amis de la Terre groups—who had mostly constituted the first members of the Green Party—disappeared or fell into relative decline until the antiglobalization protests during the 1990s kick-started a second wave of activism. Nowadays, Friends of the Earth are very active on climate change issues, and are at the center of big demonstrations. During the Copenhagen Summit, the French Amis de la Terre offered its counter-expertise on climate change and organized demonstrations. A member of both the Climate

Action Network and Climate Justice Now, the Amis de la Terre represents both sides of environmentalism: expertise devoted to putting pressure on states on the one hand, and support for new social models on the other.

Greenpeace is relatively new in the French green movement. Greenpeace was at the center of a major governmental crisis when the *Rainbow Warrior* was sunk in 1985 in New Zealand by the French foreign security services (Direction Générale de la Sécurité Extérieure). One activist was killed and France, having been recognized as guilty, gave $14 million to Greenpeace. Despite this fact, French Greenpeace was weak and found itself in financial crisis: in Europe, French Greenpeace represents one of the smaller and less active groups in the international network. However, the emergence of globalization as an issue for French Greens[18] has bolstered Greenpeace's message. Its main focus is the defense of the international environment. However, the antiglobalization issue represented a new opportunity for Greenpeace to increase its presence in France. New local groups were created in many French cities, attracting a new generation of activists (younger and more open to media strategies). The GM issue was the driving force behind the alliance brokered between Arnaud Apotheker (for Greenpeace) and José Bové (for the Conféderation Paysanne), the first time that the peasant farmers' network had worked with Greenpeace. Various factors explained this strategy. In France, there was a particularly controversial trial concerning blood contaminated by AIDS, and consequently people were increasingly worried about the impact of state decisions on their health. Arnaud Apotheker, director of Greenpeace France, and previously in charge of Vétérinaires sans Frontières (a veterinary NGO), recognized that GM food choice was the issue particularly troubling French citizens, and José Bové started a new campaign to defend traditional peasants and the quality of land. Greenpeace has begun to appear in local media, and activists attend markets to collect money and signatures on petitions; the organization has thus finally succeeded in becoming a part of the local green network.

Antinuclear Protest: Demonstrations or Expertise?

The French green movement, like the European green movement, is closely involved in antinuclear protest. The conflicts that appeared at the beginning of the 1970s brought together various groups: nonviolent groups, Christian groups, and more radical left-wing movements. Together they spread a new message about energy and society, and it is here that the green movement has its roots. It was a period of direct democracy. Christian groups preached solidarity with the third world and the need for a global vision of the world, and Marxist groups like Gauche ouvrière prolétarienne provided an insight into social inequality. With its emphasis on protection of land, green protest proposed a real new ideology. From Plogoff to Creys Malville, a message about preserving farmland and peaceful cultivation of the earth was developed to denounce nuclear energy, seen as the energy of death.

The first antinuclear demonstration took place in April 1971 at Fessenheim, with 1,500 people taking part.[19] But the demonstrations really took off in 1975, when the government unveiled a plan to build more nuclear power stations to ensure France's energy independence. Environmentalists, however, analyzed the situation somewhat differently: stating first, that France should explore other forms of energy; and second, that local people were excluded from the decisions as to where to site the power stations, even though a nuclear power station could increase the local risks of contamination, and even death.

The largest demonstrations took place in The Hague on July 3, 1976, with 8,000 participants, and 60,000 in July 1977 at Creys-Maville. German and Italian activists took part in the demonstration at Creys-Maville and battles with the police were particularly violent, leaving many activists injured and one participant, Vital Michalon, dead.

In Plogoff in Brittany, the situation was quite different. Local residents were at the heart of the mobilization and welcomed support from the environmental movement. Like Larzac, Plogoff symbolizes the small village resisting the power of the state, with even old ladies throwing stones at the police. Even now Plogoff is a symbol of this resistance and this alliance between local inhabitants and environmentalists. As he did with Larzac, François Mitterrand, incoming French president, decided to abandon the project.

After the Chernobyl disaster in April 1986, a new wave of protest began. The French government's reaction to the situation was different from that in other European countries, such as Germany or Italy, and it did not take measures to protect French citizens against nuclear contamination. CRII-Rad was created as an independent laboratory to identify the true level of contamination, and with its creation the green movement entered into a new era, the era of expertise.

Can we conclude that the era of massive demonstrations is now over? The European Pressurised Reactor (EPR) on the Flamanville site in Normandy is the major contemporary antinuclear issue. In spring 2006, 30,000 individuals protested against the EPR during a huge demonstration that brought together Agir pour l'environnement, Sortir du nucléaire, and Réseau Action Climat France to denounce the project of the industrial firm AREVA. The pro-nuclear lobby countered strongly by having the principal activist from Sortir du Nucléaire arrested for acquiring the confidential file on EPR. The Green Party reacted by publishing this confidential file to defend the activist, and denounced a lack of transparency. The French public is therefore aware of the risks of EPR, and, following the Fukushima catastrophe of March 2011, increasingly aware of the consequences of nuclear accidents, as the 20,000 demonstrators in October 2011 (at Rennes, Brittany) bear witness. In their pre-presidential election planning, Europe Ecologie–Les Verts put the nuclear issue at the top of the national agenda. Even the Socialist Party should have a political view on nuclear energy, especially after the German decision to close down their nuclear power program. During November 2011, the alliance between Europe Ecologie–Les Verts and the Socialist Party for the presidential elections was conditioning to the nuclear issue. The negotiations were harsh and finished with an agreement between the two parties that are under the hopes of the green movement. The Climate Action Network deplores the fact that France missed the opportunity to explore and adopt new types of energy that respect the environment and climate, and allow sustainable development of the land (radioactive waste contaminates the soil for centuries).

Contemporary Actions of "Ordinary Citizens": Civic Disobedience, Trials, and Expertise

Environmental action in France is not the sole preserve of activists: more and more actions are initiated by citizens, inhabitants, or consumers. From the beginning, green activism had its roots in local struggles, but the trend of individual action is becoming more and more pronounced.

People have more and more sources of information—the media, Internet, books—on potential health risks.[20] Whistleblowers too play an essential role in denouncing risk factors for the environment and the people. Green algae have been identified as a risk factor for health because they emit smells, and a gas that can be fatal. Scientists and the media combined to deplore this worrying situation and the environmental movement extended its focus on risk from the environment to human health. An increasingly common scenario is the mobilization of individuals against the siting of phone masts, motivated by the belief that microwaves have consequences for health, especially amongst children or vulnerable people (such as those with asthma). Numerous local residents' associations have sprung up as a result of the initiative of one or two individuals who gather information, organize neighborhood meetings, and create opposition to local government and the phone companies.

Involvement becomes more and more personally demanding because individuals join organizations that engage in civil disobedience, which has a long history in France: Lanza del Vasto[21] lectured on its use at the Cun on the Larzac Plateau in the 1970s. For the green movement, civic disobedience forms part of the action repertoire of the *Faucheurs Volontaires,* whose values are opposed to those of the state, and to the logic of business. In other words, because they believe that GMOs are a major danger for the environment and the individual, they also believe they have the right to attack property. Activists are frequently prosecuted; José Bové was jailed, which served to make him appear as a victim of repression. Young activists who block trains carrying radioactive waste, or who climb pylons, risk going to jail, even if their blockades are peaceful.

How can we locate the use of these new forms of direct action within an environmental movement that embraces Thoreau's philosophy on the one hand, and the media-friendly image of Greenpeace, an image particularly attractive to younger activists, on the other? Going to court is one way of denouncing the state's approach to green issues, "greening" some policies but not engaging in more radical reform in order to protect the environment, or taking emergency measures to prevent climate change. Civic disobedience attracts media attention and secures an audience for even the most niche groups. Green political parties support these citizen actions because they enact the famous phrase "act locally, think globally." Trials thus become platforms for publicizing a number of environmental conflicts.

French Green Ideology

The French green movement was very influenced by the translation of famous books by North American authors such as Rachel Carson, Barry Commoner, and Paul Ehrlich. The analyses of the Club of Rome on the limits of capitalism in 1971 constituted another inspiration for the environmental movement. From another perspective, Christian philosophy contributed some universal values regarding respect for the environment, and criticism of consumer society, such as in the writings of Bertrand de Jouvenel. In France, the green movement was also inspired by economic analyses: those by Serge Latouche and Paul Ariès being the most well known because they developed a theory of degrowth. For them, the economic growth myth sends our planet to its death.

Organic food consumption from local farmers is a good way to reconcile people to the idea that good food leads to good health, while allowing the consumer to avoid capitalist food channels. In 2001, a couple called Vuillon, farmers near Toulon in the south of France, discovered Community Supported Agriculture in the United States and imported this idea to France. The resultant AMAP (Associations pour le maintien d'une agriculture paysanne) is popular especially among the urban upper or middle classes.

The main difference between France and the Anglo-Saxon countries lies in the limited influence of animal rights movements and deep ecology. Some movements exist and lead hermetic lives in small groups, guided by anarchist philosophy. In this they follow the example of British groups:[22] renouncing society, sometimes being strictly vegetarian, supporting feminism and pro-animal action. But public sensitivity toward animal welfare is growing, and traditional entertainments such as bullfighting are increasingly discussed in the public arena; however, animal welfare is defended more by the Right wing, even the extreme Right wing (Front National) than by the green movement. In France, the green movement is more mobilized around environmental health, agricultural issues, the effect of globalization on the economy, and also around issues of democracy (e.g., corruption, oligarchy).

This overview of green activism in France must be seen within the context of two parallel trends. Firstly, the international arena is now so important that networks like the Climate Action Network and Climate Justice Now are present in France at both a national and local level. The financial crisis has sharpened criticism of the effects of capitalism on individuals' welfare and on the environment. Transnational NGOs have increased their presence in France and shaped the movement's action repertoire with more media-oriented protest; but key to the identity and history of the green movement is their espousal of nonviolent action. Civic disobedience, appeals to scientific counter-expertise, or appeals to the courts are their preferred strategies.[23] Activists' trials represent a real arena for imposing their view of society. At the same time, the French green movement is historically—and continues to be—closely associated with particular issues, reflecting the sociological profile of the activists. Agricultural issues are central to the green movement, and its urban middle-class supporters are prepared to pay more money for good "peasant" food, thus rejecting industrialized food and globalized products. Because it is hard for individual citizens to know how to protest effectively against globalization, individual French people have increasingly begun to resist through their consumption practices. The green movement is becoming stronger in France as a political movement, especially at the local and European level, with an increasing number of elected representatives. Thus, without concluding that the Greens have stolen the heart of the French electorate, it is clear that they do enjoy the support of many French citizens, especially those in urban areas and those who are part of farmers' networks (e.g., through the Confédération Paysanne, or organic farming associations).

Notes

1. Agrikoliansky, Fillieule, and Mayer. "Généalogie de l'altermondialisme français," 13–43.
2. Jossin, "How Do Activists Experience Transnational Protest Events," 42–63.
3. Ollitrault and Kernalegenn, "Le mouvement écologiste," 457–80.
4. Tarrow and della Porta, "Globalization, Complex Internationalism and Transnational Contention," 42–63.
5. Roche and Bennahmias, *Les verts de toutes les couleurs,* 1992.
6. Ollitrault, *Militer pour la planète,* 2008.
7. Charvolin, *L'invention de l'environnement en France,* 2003.
8. McCormick, *The Global Environmental Movement,* 1995.
9. Ollitrault, "L'invention de l'écologie politique: la candidature de René Dumont," 249–61.
10. Sainteny, L'introuvable écologisme français.
11. Villalba, "The French Greens," 43–60.
12. Roche and Bennahmias, *Les verts de toutes les couleurs.*
13. Agrikoliansky and Sommier, *Radiographie du mouvement altermondialiste.*
14. Fillieule, "France," 258–79.
15. Hayes, "Vulnerability and Disobedience," 831–38.
16. Hayes, "Collective Action and Civil Disobedience: The Anti-GMO Campaign of the Faucheurs Volontaires," 293–314.
17. Ollitrault, "Des plantes et des hommes," 443–65.
18. Derville, "Le combat singulier Greenpeace—SIRPA," 589–629.
19. Ollitrault and Kernalegenn, "Le mouvement écologiste," 457–80.
20. Lascoumes, Callon, and Barthe, *Acting in an Uncertain World: An Essay on Technical Democracy.*
21. Hayes and Ollitrault, *La désobéissance civile.*
22. Traïni, *La cause animale (1820–1980). Essai de sociologie historique.*
23. Ollitrault, *Militer pour la planète.*

Bibliography

Agrikoliansky, Eric, and Isabelle Sommier, eds. *Radiographie du mouvement altermondialiste.* Paris: La dispute, 2005.

Agrikoliansky, Eric, Olivier Fillieule, and Nonna Mayer. "Généalogie de l'altermondialisme français." In *L'altermondialisme en France. La longue histoire d'une nouvelle cause,* edited by Eric Agrikoliansky, Olivier Fillieule, Nonna Mayer, 13–43. Paris: Flammarion, 2005.

Charvolin, Florian. *L'invention de l'environnement.* Paris: La découverte, 2003.

Derville, Grégory. "Le combat singulier Greenpeace—SIRPA. La compétition pour l'accès aux médias lors de la reprise des essais nucléaires français." *Revue française de science politique* 47, no. 5 (1997): 589–629.

Fillieule, Olivier. "France." In *Environmental Mobilization in Comparative Perspectives,* edited by Christopher Rootes, 258–79. Oxford: Oxford University Press, 2003.

Fillieule, Olivier. *TEA Transformation of Environmental Activism, Final Report on France (1998–2001).* Canterbury, GB: University of Kent, 2001.

Hayes, Graeme. "Collective Action and Civil Disobedience: The Anti-GMO Campaign of the Faucheurs Volontaires." *French Politics* 5/4 (December 2007): 293–314.

Hayes, Graeme. "Vulnerability and Disobedience: New Action Repertoires and Environmental Protest in France." *Environmental Politics* 15/5 (November 2006): 821–38.

Hayes, Graeme, and Sylvie Ollitrault. *La désobéissance civile.* Paris: Presses de Sciences Po, 2012.

Jossin, Ariane. "How Do Activists Experience Transnational Protest Events? The Case of Young Global Justice Activists from Germany and France." In *The Transnational Condition. Protest Dynamics in an Entangled Europe, edited by* Simon Teune, 42–63. New York: Berghahn Books, 2010.

Lascoumes, Pierre, Michel Callon, and Yannick Barthe. *Acting in an Uncertain World: An Essay on Technical Democracy.* Cambridge: MIT Press, 2009.

McCormick, John. *The Global Environmental Movement.* Chichester, John Wiley & Sons, 1995

Ollitrault, Sylvie. "Des plantes et des hommes. De la défense de la biodiversité à l'altermondialisme." *Revue Française de Science Politique* 54, no. 3 (2004): 443–65.

Ollitrault, Sylvie. "L'invention de l'écologie politique : la candidature de René Dumont." In *La mutation du système partisan français après Mai 1968,* edited by Jacqueline Sainclivier and Richard Gilles, 249–326. Rennes: PUR, 2012.

Ollitrault, Sylvie. *Militer pour la planète.* Rennes: PUR, 2008.

Ollitrault, Sylvie, and Tudi Kernalegenn. "Le mouvement écologiste." In *France Rebelle,* edited by Xavier Crettiez and Isabelle Sommier, 457–80. Paris: Michalon, 2006.

Roche, Agnès, and Jean-Luc Bennahmias. *Des Verts de toutes les couleurs. Histoire et sociologie du mouvement écolo.* Paris: Albin Michel, 1992.

Sainteny, Guillaume. *L'introuvable écologisme français?* Paris: Presses universitaires de France, coll. Politique d'aujourd'hui, 2000.

Tarrow, Sidney, and Donatella della Porta. "Globalization, Complex Internationalism and Transnational Contention." In *Transnational Protest and Global*

Activism, edited by S. Tarrow and D. della Porta, 121–47. Oxford: Rowman & Littlefield Publishers, 2005.

Traïni, Christophe. *La cause animale (1820–1980). Essai de sociologie historique.* Paris: PUF, 2011.

Villalba, Bruno. "The French Greens: Changes in Activist Culture and Practices in a Constraining Environment." In *Green Parties in Transition: The End of Grass-Roots Democracy?*, edited by Benoit Rihoux, P. Lucardie, and Frankland E. Gene, 43–60. London: Asghate, 2008.

Chapter 6

The Russian Environmental Movement between 1985 and 2010: Changing Trends, Shifting Values

Oleg Yanitsky

Introduction

Russian environmentalism as a theory and a practice is deeply implanted in European science and culture. The majority of Russian biologists and natural scientists studied or did their postgraduate research at German and other European universities. The first book on social ecology, *Hunger as a Factor: An Impact of Hunger on Human Behaviour, Social Organization and Public Life,* was written by Pitirim Sorokin, the prominent U.S. sociologist of Russian origin, drawing on the evidence of revolution and civil war in Russia in 1917–1922.[1] Even in the totalitarian times of 1920–1950, the environmental movement had existed but it had been under the total control of the Communist Party of the USSR. After the fall of the Iron Curtain the contacts between scientists and environmentally concerned lay people of Europe and Russia extended even more.[2]

In its modern form, the environmental movement in the former Soviet Union and then, after its decay in Russia, has existed for about 50 years. It is one of the most stable, professionalized, and scientifically oriented social movements. Socially and culturally, the environmental movement is deeply rooted in the Russian intellectual milieu, that is, universities, research institutes, creative unions, and the numerous grassroots organizations across the country. Genetically and structurally, the modern environmental movement is connected with the student nature protection corps (or brigades) of 1960–1980. The environmental movement is currently and politically differentiated by the criterion of its relationships with the state and other civil society organizations. The environmental movement is horizontally network-structured, having a few umbrella NGOs that control resource distribution.

Why the Russian Environmental Movement?

To begin with, the environmental movement is the oldest new social movement in Soviet/Russian history. Secondly, this movement actually is a successor to pre-revolutionary ecological and local lore movements with their strong civic overtones. It is indicative that at the beginning of Russian transition to capitalism, the environmental movement was the first form of a political protest. Thirdly, the environmental movement has been deeply rooted in the Russian academic elite's culture from the very beginning of the 20th century. Fourthly, initially and up to now the environmental movement has been mainly rooted in universities and research institutes where students, researchers, and instructors closely collaborate with one another and with local initiative groups in remote provinces of Russia. These networks were and still are very important for maintaining the integrity of civil society given the growing gap between the state and civil society on the one hand and between rich and poor regions on the other. Finally, despite all the twists and turns of current Russian economic and political reforms, the environmental movement has not only survived but has also become an integral part of the international network community of environmentalists. Therefore, studying the environmental movement's transformations makes a deeper insight into the development of Russian civil society possible.

The Major Stages of Evolution

Stage I: The Late Bureaucratic Socialist Society (before 1985)

During the decade before 1985, Soviet society was relatively stable. However, this did not mean a stagnation of social life. From the beginning of the Nikita Khrushchev era in the mid-1950s onward, numerous creative civic initiatives emerged, with young as well as elderly people as their proponents and members. The six major types of milieus that generated these initiatives were universities, research institutions, artist unions (uniting writers, architects, filmmakers), municipal and community organizations, and newspapers.

Established in the late 1960s, the students' Nature Protection Movement (the *Druzhina* movement) was in the best position since the faculties of biology, soil studies, and geography at Soviet universities were very favorable milieus for the emergence of environmental initiatives. The concentration of hundreds of students sharing the same interests, teachers, libraries, and friendly youth organizations were the principal prerequisites for this.[3]

The *Druzhina* movement was mainly built *from without*, with the faculties acting as its driving force and resource suppliers. At the same time, self-identification and self-actualization were powerful motives enticing students to participate. The situation was different with other movements, but the role of intellectuals in shaping these cells of civil society in the pores of post-totalitarian society was always crucial. This type of social activity was not guided by the rational choice principle. On the contrary, the "at any cost" principle was the driving force, because the major motive behind this collective action was self-actualization. The prevailing types of motivation were mutual assistance and creativity with very strong civil overtones: "If not us, who then?" The striving for real deeds—organization, planning, protection, construction, public assistance to nature and people (that is, their social potential)—was extremely powerful. The sustainability of the Russian environmental movement is confirmed by the fact that most leaders of the *Druzhina* students' movement later became leaders of the Socio-Ecological Union (SoEU), the World Wildlife Fund for Nature–Russia (WWF-Russia), and many other environmental nongovernmental organizations (NGOs) across the country.

At this stage the mobilization of resources was not an issue since the state was the only resource provider. On the one hand, the universities played the role of resource transmitters, while on the other hand the initiatives' needs for resources in those times had been minimal. The major resource for a collective action had been the activists themselves. What they really needed were environmental knowledge and know-how, which were transmitted to them by their senior colleagues. The political opportunity structure (POS) defined by the state was very narrow, but the structure of social opportunities (SOS) widened day by day because masses of educated people were inclined to creative activity in their numerous research institutions, neighborhoods, and professional clubs. Formally, the Communist Party–state machine was the supervisor of all societal life. But in practice uncontrolled zones became ever more numerous and large. Initiatives and movements were slowly institutionalizing, again formally within the party-state system but actually more and more independently from it. Direct relations with sister movements in the country and abroad were still strictly forbidden.

But informal contacts by means of growing intellectual exchange with foreign scientific, cultural, and educational organizations were expanding day by day. The *Druzhina* movement was the first fully network-structured movement.

Stage II: The Beginning of Liberalization (1985–1991)

Although the economic context remained unchanged in this period, the political context began gradually to liberalize. Soviet society became more and more sensitive to party-state sanctions upon collective actions and to violations of human rights. The masses of lay people became seriously concerned about the environment. In this phase, political factors gradually acquired a decisive role in the establishment of civil initiatives and social movements. At the same time, the Soviet party-state apparatus was trying to preempt the initiative in order to advance its own green slogans.

After 25 years of existence (1960–1985), the *Druzhina* movement was a well-developed network of about 200 regional organizations. This network that still rested upon universities and research institutions covered almost the whole territory of the USSR. But at this phase the environmental movement's reproduction process became much more complicated since by this time the movement had shaped its own engendering milieu: the ex-members of the *Druzhina* who became teachers, researchers, journalists, or public figures formed the core of this milieu. The *Druzhina's* motto, "Nature has to have its people everywhere," shaped the large periphery of the movement's backers and sympathizers. Another important turning point was the environmental movement entering into the public arena. The countless informal organizations that emerged in the period became the environmental movement's secondary and quite different engendering milieu, because on entering this arena the movement's leaders were involved in the public debates. Discussion beyond the confines of the environmental movement became the core reproductive mechanism of green networks and of the environmental movement's backers and sympathizers.

Therefore, this type of the environmental movement's reproduction may be called *self-sustained.* All the efforts of its members—rallies, campaigns, manifestos, and so forth—were aimed at providing a niche for the environmental movement within the rather turbulent milieu of Soviet "informals" (i.e., other grassroots organizations). Key values that guided the reproduction of the environmental movement process were *self-organization* and *self-identification.* The latter became rather important when the environmental movement found itself among many other Greens, ranging from former Communist Party bureaucrats to radical anarchists. The prevailing type of social reflection of the environmental movement was self-centered, that is, dedicated to furnishing its own niche within the emerging civil society.

At this phase resource mobilization was not a problem yet. People were the main resource of the environmental movement. The initial engendering milieu continued to supply it with necessary human resources. It also gained other resources from the disintegrating party-state system, from friendly officials within it, and from other social movements. Also, over the previous 25 years the environmental movement had accumulated a great resource potential (knowledge, a set of tactics, action repertoire, networks). However, for the first time a new facet of resource production emerged—*the struggle for access* to resource pools.

The structure of POS rapidly expanded and its limits were increasingly eroded. More and more people and public organizations became environmentally concerned. The environmental movement gained support from many other protest groups and prominent public figures across the Soviet Union as well as unofficially (via personal contacts) from abroad. The environmental movement tried to keep a distance from the state, though their paths began to intersect. Contacts with foreign groups and global environmental communities remained rather sporadic.

Stage III: The Turning Point (1991–1993)

In this period the core values and structures of the former Soviet society were undermined. Instead of one unified society we see many societies, each with its own dominant aims and goals, collective actors, and pace of social change. In my view, the key words to define that phase must be these: conflict, diversification, decomposition, regionalization, mobilization. The major adversaries were the state versus civil society, the democrats vs the communists, and, the Westerners versus the patriots. A very high level of environmental mobilization was characteristic of almost all strata of society. The environmental movement's engendering milieu was again totally transformed. On the one hand, it shrank since many former eco-leaders became politicians and officials or turned back into teachers and researchers. On the other hand, it rapidly expanded at the expense of mass protest movements such as people's fronts, which began to advance their own ecological demands. In other words, these protest movements produced mass ecological protest that rapidly acquired a political overtone. But the principal shift was the actual *globalization of engendering milieu* in terms of contacts, mutual support, and—most importantly—finding new sources of financing.

The key goal of the environmental movement at this phase was to protect its identity. The goal has been achieved by two means. First, the environmental movement attempted to extend its constituency by advancing a democratic motto: "Those who are not against the Greens, join us!" The second means was this actor's option for financial aid from abroad. Self-preservation in a turbulent societal milieu made mobilization of financial resources necessary. Fund-raising came to the forefront. Therefore, some of the environmental movement's organizations were gradually transformed for the purpose of pumping financial resources from the West. Projects implemented by closed groups within the scope of grants became the main form of the environmental social movement organizations' (SMOs) reproduction. The type of social activity was also a dual one. On the surface it had a protest character, but in essence it was self-centered and self-limiting.

The structure of POS was losing its clear-cut outline at that time. Numerous eco-activists became politicians and officials. But after being elected or nominated they often forgot their ecological promises. The state–environmental movement relationship was ambiguous. Protest and collaboration were the two sides of the same coin. The environmental movement's attitude toward universities and research institutes revealed two trends: further distancing from the alma mater and efforts to shape a body of professional advisers and experts of its own. Finally, one could observe a substantial shift as to the environmental movement's relations with the outer world: *Westernization,* that is, more contacts with sister groups abroad, fewer with domestic ones, and an overall turn to the *Western mode* of maintaining an environmental movement.

Stage IV: Reproduction in a "Market Economy" Context (after 1993)

Liberalization and Westernization—as they were implemented in the Russian context—were again an ideological act, very much comparable to the imposition of the model of "scientific socialism" during the totalitarian period. Russia's "transition society" achieved and maintained its stability, but at the cost of keeping numerous utterly depleted and environmentally risky enterprises afloat. "First stabilization, then ecologization" was the motto of the new Russian government. In fact, the reforms were carried out by way of de-modernization and overexploitation of natural and human resources, and in some instances they were pushing the society back to the social order of feudalism. The democratic façade decayed in full public view; the declared democracy showed its authoritarian face. The widely publicized bottom-up system of self-government remained a mere formality. The ability of the public to participate in resolving urgent environmental problems was gradually eliminated by the state. Public involvement in the discussion of key environmental issues of Russia came to an end.

From 1991 onward, for millions of Russians, having some bread on their tables became more important than closing a nuclear power plant or a state farm. Alienation became the common denominator of the environmental movement activity's context in those times.[4] The two parts of a divided society, the wealthy minority and the poor majority, are both resistant to any "ecologization." Accordingly, the environmental movement split into two parts: those who had access to resources and those who did not. For the maintenance of provincial NGO units the existence of big umbrella NGOs was a necessary, but not sufficient, condition. After the regionalization of the country, local eco-NGOs remained face-to-face with the mighty regional authorities who became the major resource holders. Fund-raising became a matter of survival.

During this phase the environmental movement as a series of mass campaigns, blockades, and so forth, ceased to exist. It transformed into a loosely structured network of NGOs, each of which went its own way. The maintenance of the very network became a matter of prime importance. The environmental movement's existence in this phase can be called as *external* and *dependent*, since its networks and units began to function as subsystems of interregional or global green networks. The changes in the very structure of the environmental movement activity are also of crucial importance. In the previous phase a "final" success of a campaign was possible. Accordingly, the preparation of and participation in these events were the mechanism of the environmental movement's maintenance. In a hostile milieu a decisive success of the environmental movement's efforts became impossible. The environmental movement as a set of coordinated mass campaigns has been replaced by numerous long-term protest actions and litigations with the superior forces of state bureaucracy and corporations.

That is why the major goals of the above net-units system were *self-preservation* and *survival*. The prevailing type of social reflection was *rational choice*, cost-benefit analysis, and the like. The eco-NGOs were gradually getting bureaucratized and professionalized because this improved their chances in the competition for resources. All in all, the environmental net-units system became *marketized*. Hence, the end products of the environmental movement production process were information, managerial skill, and political know-how.

The hostile political milieu implied a reduction of the POS for environmental NGOs. The attitude taken by the federal and regional administration bodies toward the environmental movement was on the whole negative. The authorities viewed the NGOs as a source of labor to mobilize but not as partners in environmental policy-making. The state bodies strove to supervise foreign aid received by Russian eco-NGOs. Therefore the relationship between the state and the Greens became a combination of conflict and one-sided attempts by the Greens to establish dialogue and cooperation. As a result, the mainstream Russian environmental NGOs became deeply embedded in the global environmental network. To adapt to a hostile context, the environmental movement needed to be recognized by the state as a respectable and responsible force[5] able to meet its requirements. In the process of achieving these goals the environmental movement lost momentum: its core organizations were transformed into formal groups staffed by activists-turned-bureaucrats while other activists were co-opted by the federal, regional, and local authorities to staff new bureaucratic agencies.

There were some other causes for the environmental movement's bureaucratization. The dismissed or retired bureaucrats and politicians began to establish numerous quasi-NGOs to protect their own social status. The Soviet bureaucratic culture proved to be quite vital. Then, the bureaucratization of environmental NGOs was the irremovable by-product of their daily routine office work. Finally, the involvement of the NGOs' members in diverse recurrent processes such as the maintenance of global and domestic information networks, litigation, and decision making gradually transformed the environmental movement's amateurism into professional administering. In spite of all this, the eco-NGOs networks became an *archipelago of stability* for their members in the ocean of losses and stressful changes. These networks carry out important functions in aiding their members' adaptation to an uncertain and hostile context. The networks are also able to provide learning-by-doing, social protection, and emotional comfort.

Stage V: Building a Political Party?

In 2004, a politically active wing of the environmental movement based mainly on the round table of Russian environmental NGOs headed by prominent Russian environmentalist Alexei Yablokov began to build the

United Party of Russian Greens. On the one hand, it forced the environmental movement leaders to set up a political agenda for the party. On the other hand, this process brought about a great deal of uncertainty and tension because the majority of rank-and-file members of the environmental movement were not prepared to become political activists. The decision came from the top: in 2003–2006 the federal parliament adopted a set of laws that banned any social movements and public organizations from running in the elections. As to building a green party, after a number of unsuccessful attempts, an environmentalist faction was formed in the party named "Yabloko" (Apple), but currently the Yabloko has no representatives in the federal parliament.

For the Greens, this meant that the only way out was to act in the field of *sub-politics,*[6] the further the more—in the field of global sub-politics. And, as time has showed, this was the right decision. Up to 2008, the country was already dotted with local and regional protest actions of political nature. For instance, the protest action against the felling of an oak forest in a Moscow suburb has been multiplied throughout the country. "Stop felling in Moscow!" which actually meant "Stop destroying our living environment!" became the common motto of protesters.

The Value Basis of the Movement's Differentiation

The social bases of the Russian environmental movement's differentiation have been thoroughly studied.[7] However, insufficient attention has been paid to its ideological and value differentiation. After the end of the period of "democratic hopes," this differentiation has a still greater effect on the strategy and tactics of the environmental movement. In my prior work I have revealed seven main groups within the environmental movement that differ in terms of values, tactics, and action repertoire: conservationists, alternativists, traditionalists, civil initiatives, eco-politicians, eco-patriots, and eco-technocrats.[8]

Conservationists were the core of the Russian environmental movement. Their starting point was bioscientism ("Nature knows best"), which meant that they were globalists by definition. Their ultimate goal was the creation of a worldwide brotherhood of Greens and the construction of a society with modest material needs. In autumn 1988, conservationists founded the Socio-Ecological Union (SoEU), the Russian environmental movement's biggest umbrella organization. After the collapse of the USSR, the SoEU was transformed into the International Socio-Ecological Union (IsoEU), which embraced some post-Soviet and European countries.

The most ideologically oriented group in the environmental movement was the alternativists. Its leaders were professional ideologists of eco-anarchism. In the 1990s, the majority of alternativists were simultaneously members of green parties and of the anarcho-syndicalist movement, which was a unique group in the environmental movement that combined socio-political activities with a constant ideological work. The alternativists were principled adversaries of the state as a political institution. In their opinion, an ecological "turn" could be carried out only by means of implementing an alternative project for the entire society. The key ideas behind this alternative project were to be the decentralization of power and economic activity, self-organization, and self-provision.

Traditionalists, the successors of the local lore movement of the Russian past, represented the layer of the Russian intelligentsia oriented toward humanist values and eternal ideals of kindness, tolerance, and nonviolence. Traditionalists were past-oriented in the sense that they deeply appreciated the 19th-century culture, with its ideals of serving and enlightening the people and protecting nature. Educators, writers, journalists, and scholars constituted the main core of the group. The majority of traditionalists explicitly opposed the methods of Soviet-type industrialization. Traditionalists opposed both russification and Westernization of ethnic minorities in Russia. Despite their spiritual heterogeneity, traditionalists were united by their capacity to critically assess the environmental policy of the Soviet regime.

The various groups of civil initiatives (i.e., the grassroots) shared four critical types of values. The first one was their personal responsibility for the protection of their immediate milieu. The second one was independence in the sense of the capacity of self-organization of their creative activity. The third set of values comprised the need for self-actualization and for communication with like-minded people, the desire to be understood, and the wish to be useful and have support from one's peers. The fourth set consisted of values related to defending others in a safe and clean immediate environment. Taken together, these values helped to provide a sense of social protection, of emotional comfort and mutual support among these grassroots activists.

Eco-politicians were the most heterogeneous group in the environmental movement. This group included movement patriarchs who had a great influence on the group: theorists, who imparted the already well-developed ideological doctrines to different groups of the environmental movement; leaders of numerous small Russian green parties; former politicians ("people's deputies"), and practicing politicians, who originated from the milieu of civil initiatives and combined the role of professional politician with membership in an NGO. Eco-politicians were united by the idea that ecologically safe production should be at the top of the national political agenda. However, for some it was a deep conviction and their life-work, whereas for others it was merely a means for making a political career.

The ideology of eco-patriots was characterized by left radicalism, the idea of forceful ecologization of society, and implicit sympathy for state socialism. State ecological patriotism, a stiffly regulated market, restriction of private ownership of large plants and factories, an ecologically oriented socialist and private economy, and social justice were the benchmarks of Russian eco-patriots. Other eco-patriots stressed that forceful Soviet modernization in the past decades annihilated the natural and cultural landscape of Russia's core.

In the 1990s, eco-technocrats were the smallest group in the environmental movement. They believed that the solution to all environmental problems lay in the adoption of ecologically sound technologies. Strictly speaking, they did not promote a technocratic ideology. I call them technocrats because a naive technocratism was intrinsic to them—that is, they unconditionally believed in society's adoption of the technological innovations created by them. In those times, eco-technocrats were the least politicized segment of the environmental movement.

The Value Shifts

Today the conservationists are no more the core of the Russian environmental movement. When the period of informal bottom-up environmental politics ended, it became clear that the various environmental groups that had gathered under a big umbrella organization, such as the International SoEU or WWF-Russia, pursued different goals. But the crucial factor in the group's diversification was the changed mode of resource mobilization: each team of the group began to look for its own resources. Intergroup solidarity gave way to intergroup competition for resources. As a result, the environmental movement became a loosely structured network of NGOs. The idea of a brotherhood of Greens died away.

The idea of inevitable ecological catastrophe remained the prevailing mode of thinking of conservationists. As their leader Sviatoslav Zabelin put it, humankind continues to behave as a biological species. The leading forces behind this destructive behavior are the national resource extraction monopolies and transnational corporations. The conservationists' attitude toward the state was ambivalent. On the one hand, they continued to collaborate with government bodies on an individual basis (as advisers, experts, and even office workers). On the other hand, between 1999 and 2005, conservationists constantly accused the Russian federal government of "de-institutionalizing ecological policy." Inherently, conservationists were never inclined to radical action. The mode of this group's thinking and doing was becoming more and more reformist under the influence of a number of factors such as permanent scarcity of resources, the need to establish well-organized offices and to collaborate with experts and other professionals, to participate in parliamentary hearings and litigation, and to maintain contacts with donors and sister organizations abroad.

Conservationists still rarely appealed directly to the lay people, but the motivation behind this disengagement changed. The scientific ideology of the group and its purely conservationist orientation in no way presupposed such contacts ("we, the professionals, know better"). Conservationists were highly concerned about the decline of biodiversity in local ecosystems. As they were involved in the implementation of ambitious international nature protection programs, Russian conservationists presented themselves as producers of advanced ecological knowledge. To some extent this attitude was justified, given the unprecedented and potentially nonrenewable meltdown of Russian academic science and university scholarship.

The maintenance of the group's identity became a problem. In the end, the group has split into two main subgroups. The subgroup headed by or involved in the worldwide organizations such as the WWF-Russia and the Taiga Rescue Network usually identified itself as responsible professionals. The other subgroup represented by small units in remote provinces identified itself as advocates of particular local milieus. The mentality of the latter subgroup had much in common with that of civil initiatives.

Alternativists were still the most ideologically oriented group within the environmental movement. They were convinced that it is not nature but human beings with their unique local cultures who are at risk today. Alternativists never ceased to repeat that nature could be protected if society were based on a set of local self-sustained communities using environment-friendly technologies. Alternativists remained principled adversaries of the state as a political institution. They held that many of environmental problems can be solved most efficiently at the local level. They saw local democracy as a core value and rejected "managed democracy," which was the ideology proclaimed by Russian politically engaged theorists and some eminent politicians.

Though Russian alternativists joined the international antiglobalist movement, they did their utmost to protect their identity as an independent social force within the Russian environmental movement. At the same time, they rapidly developed contacts with the pacifist and communitarian movements around the world, especially in Central and South America. It is my belief that the Russian alternativists tried to survive by means of becoming part of the "localist" wing of the world antiglobalist movement.

Alternativists' attitudes toward the local population changed as well. Initially, the protest was a value in itself. Later on, their leader argued that "we must help any basic (i.e., local) movement anywhere to combine protest and creative activity." At the same time, he stressed that a "network of alternative settlements does not belong to any territory, because they are nodes of the international alternativist community." Therefore, these nodes can be inhabited by local people or anarchist activists as well as by migrants, homeless, and foreign citizens. Finally, alternativists describe themselves as "anti-scientists," since in the modern world science is an integral part of the existing social order. They recognize only applied science and those developments that are capable of producing environment-friendly know-how for alternative settlements across the world.[9]

Former traditionalists have dispersed among numerous local lore groups, and all-Russian cultural heritage protection associations do not exist any more as mobilizing social actors. Their social niche within the green community has been occupied by a rapidly growing and vigorous group of Russian "deep ecologists," whom I call "new pastoralists." The group represents a network of professionals, mainly biologists, that has the *Gumanitarnyi Ekologicheskii Zhurnal* (Humanitarian Environmental Magazine) as the heart of its ideological core and institutional structure. "Wild Nature First!" was the group's motto and the heart of its identity. Russian deep ecologists insist on the divine origin of wild nature. Given that nature is sacred, its protection is a top priority issue. Wild nature is considered a value in itself that needs no external justification. Wild nature, Ukrainian activist Vladimir Boreyko stated, "has the merit of being invaluable. Wild nature is the oppressed minority that has a moral right for liberty and existence."[10]

Politically the group's stand is intermediate. On the one hand, they were etatists because they needed a "strong hand" for their cultivation of the love of nature. On the other hand, they laid hopes upon local religious communities. In the activity of this group I see a manifestation of a persistent desire of the church (predominantly, but not only, the Russian Orthodox Church) to reassert control over as many spheres of public life as possible. The group's attitude toward the local population was openly paternalistic with a strong imperative and didactic overtone: "We must teach people to love nature."

Civil initiatives have also changed their social basis and value core. Having exhausted their potential of self-government and lost their faith in the "democrats," civil initiatives were no longer an integral part of the environmental movement. In that time, the group consisted mainly of poor and marginalized local people who were at risk across Russia. It also included those who did not have what they were supposed and entitled to have (deceived bank depositors and cheated home-buyers, victims of pyramid schemes, workers whose work had not been paid for months and the like). Furthermore, it included parents concerned about the life and health of their children who were at risk working or staying in crisis areas. The major value change was represented by the shift from the concern about the immediate environment to the defense of basic human rights and freedoms. According to the dominant worldview of the group, the world was a hostile and dangerous place making it necessary for them to defend themselves and their families here and now. This group's motto may be articulated as "Life first!" Or, putting it in another way, the mushrooming of such civil initiatives was a manifestation of the people's self-preservation instinct. The group can be called a risk-solidarity community because it struggled against all-encompassing risks such as corruption, deception, and violence.

The group's attitude toward the state was ambivalent. On the one hand, its members constantly appealed to executive power bodies including local administration in order to diminish risks and to meet their own urgent needs. More than that, the group's leaders did a lot to be accepted by the state bodies as partners. But, while preempting the initiatives' slogans, the state did not actually grant them access to policy-making. On the other hand, the group was among the strongest critics of the state's environmental policy. These civil initiatives were political opportunists because they appealed to any political force in power for aid and defense. The group was increasingly inclined to collaborate with the human rights movement.

Understandably, the group's members identified themselves with the impoverished and alienated majority of Russia's population: "They are the victims, just as we are." At the same time they saw themselves as a still-active and creative part of this majority, which was capable of resistance and protest. More than that, resistance and protest were often central to the group's way of life. As to science, the group had a strongly positive attitude toward independent experts and especially the scientist-advocates. The group did not trust the official science because they said it was "completely politically charged." Studying the group, I did not discern any definite attitude toward the church and various religious organizations. The key difference here, as my respondents reported, was between those "who help us to survive and those who do not."

Within the group of eco-politicians the situation has changed dramatically. Environmental initiatives and movements were the first form of political protest in the USSR in the late 1980s. Five years later all the Russian Greens except eco-anarchists were actively participating in shaping Russian environmental policy. Later, the Russian environmental movement, as already indicated, was politically marginalized. Hence, when I speak of "politicians," I am referring only to those activists who deal with environmental policy professionally and who present themselves as politicians in public. There are five subgroups of this kind.

Eco-politicians were the most influential subgroup with the others revolving around it. This subgroup's institutional forms included the Center of Russian Ecological Policy, the Ecological Commission of the Public Chamber of the Russian Federation (RF), and some other public organizations. The subgroup's core consisted of former biologists and other scientists who had acted as professional politicians and founders of numerous eco-NGOs and participants of international research programs during the 1990s.

The dominant worldview of this subgroup was political pragmatism with the motto "The eco-NGOs should share the ecological responsibility with business."[11] The ultimate goal of the group was to create an ecologically responsible market economy. But in practice the achievements were very modest. They came down to monitoring the drafting of environmental laws and their subsequent implementation. The attitude of the subgroup toward the local population was instrumental if not utilitarian: "We will mobilize them if it is needed." Nevertheless, attempts to build a bottom-up model of Russian environmental policy were obvious. At the same time, WWF-Russia launched a program to set up an "environment politics" in the oil and gas sector, which meant that WWF-Russia was ready to collaborate with this leading branch of Russian industry.

I would describe the second subgroup as political speculators or extra-utilitarians. During the 1990s when the West sponsored the Russian environmental movement, these quasi-eco-politicians mushroomed rapidly across the country. The majority of such groups have been established with just one goal, which is to secure resources from Western donors. Therefore, the subgroups took an utterly utilitarian attitude toward the state, business, church, and local population: it gained income simulating pro-ecological activity.

Members of the third subgroup were radical etatists. Their engendering milieu and nests were newspapers of patriotic orientation. In essence, this subgroup was authoritarian-oriented since its members highly valued the achievements of the Russian (tsarist) and Soviet empires. Their attitude toward current reforms aimed at transition to a market economy was negative, while toward the church and the local population was generally positive. Etatists presented themselves as defenders of the Russian nation-state and Russian culture.

The fourth subgroup (the communitarians) were moderate red-green eco-patriots. In their view, the Russian state should consist of a set of local communities based on the principles of justice and a socialist model of distribution. This subgroup emerged as one of the units of the *Druzhina* movement based at the Chemical-technological University in Kazan, the capital of the Tatarstan Republic. As late as 1980, Sergei Mukhachev, the subgroup's leader, organized a permanent seminar on how to resolve global environmental issues on the basis of communitarian principles.

Finally, the group of developers has emerged. Its political motto was and still is "Project first!" because members of this subgroup understood environmental policy as an endless chain of discrete projects and social technologies for their implementation. They were members of various NGOs that elaborated the methods of environmental impact assessment and developed technical devices for environmental monitoring. The subgroup shares many commonalities with the technocrats: both share the negative attitude toward the state and a positive attitude toward market economy and applied sciences. The emergence of this subgroup has marked an important shift in the mentality and motivation of a substantial part of Russian environmentalists characterized by a return to their former professional activity (e.g., as biologists, geographers, engineers).

There are two types of eco-technocrats. The first type can be termed quasi-technocrats who added the prefix "eco" to any type of business or public activity if it brings income or social capital. The second type refers to actual technocrats. This group included the semiprofessional Westernized Russian intelligentsia employed at numerous NGOs or in small intellectual businesses (e.g., consulting, auditing, and designing). According to the dominant worldview of the group, the environment could and must be governed. They believed that eco-technology and environmental management would save humanity. The group valued management technologies so highly that it practically identified environmental policy with environmental management. The group's motto was and still is "Management first!"

Eco-technocrats take a pragmatic if not utilitarian stance toward the state. During the 1990s, they presented themselves as liberals speaking in favor of a specific form of "greened" market economy. Eco-technocrats were proponents rather of the development of ecological services (e.g., eco-audit, eco-expertise, ecological standards of development for particular plants) than of pollution-free production. At the same time, they called for the strengthening of the state ecological institutions.

Eco-technocrats' attitude toward the local population was pragmatic as well. As a rule, technocrats kept a distance from the lay people whom they regarded as emotion-guided and unable to think and act rationally. In this respect, eco-technocrats contrasted with civil initiatives. Given their involvement in ambitious international projects, Russian eco-technocrats pretended to be the producers of advanced (and applied) ecological knowledge and neglected any form of "advocacy science." The blind trust in science and environmental management was one of the features of modern Russian technocratic thought.

Scientists as Advocates, Citizens as Experts

The science-society relationships are undergoing substantial changes. The transformation of many academics into advocates of the environmental movement and local communities is a distinguishing feature of recent Russian environmentalism. This kind of scholarship covers a wide spectrum of activities, ranging from participation in actions to protect civil rights and natural systems via independent scientific expertise, research and development on particular problems of human communities and ecosystems to the formation of alternative concepts of man-nature relations. In other words, advocacy science arises from mutual understanding and trust between academics and civil activists, that is, citizens-turned-experts. Advocacy science means research in vivo rather than in vitro. The fertile soil for advocacy science consists of groups and communities of academics with a sense of civic responsibility and moral debt.

There are five types of advocacy scientists. The first type is an "outsider" occasionally invited by an eco-NGO as an expert. These are usually scholars from research institutes. Working in accordance with existing standards and/or relying on their own professional experience, "outsiders" translate their conclusions into technical languages accepted in the given branch of academic knowledge.

The second type, the "aware," includes scientists whose regular contacts with NGOs help them acquire a better understanding of problems they are trying to resolve. As a result, they achieve mutual understanding and work out a common language as a medium of communication with politicians.

The third type, the "involved," comprises scientists who share the values of NGOs and systematically collaborate with them but remain staff members of state research institutes or universities. A distinctive feature of "involved" scientists is that they perform a certain professional function (expert or adviser), are interested in the success of a particular NGO's work, and sometimes take part in its activities without being its members.

The fourth type, the "partners," is represented by scientists who, while nominally working for academic institutions, actually are already activists of environmental NGOs. Thus, "partners" acquire a double institutional affiliation. This type of scientific advocates can also be described as multifunctional, because they alternatively participate in resolving scientific, social, and organizational problems.

Finally, the fifth type, the "fully integrated," is represented by researchers who have left their job in academic institutions and become members of an NGO. The "fully integrated" try to generate a kind of new knowledge based both on scientific research and on cultural rationality of local populations. A subtype of this type is a scientist who informs the public that he or she abandons his or her scientific career and becomes a leader of an environmental movement. In this case, a scientist (as a public figure) begins to formulate new rules of scientific research known in Western science as "participative research."[12]

Conclusion: The Inevitability of the Internationalization of the Russian Environmental Movement

The "Cities of Europe" project was just the beginning.[13] Simultaneously, the "Ecopolis" (i.e., eco-city) experimental project had been launched by a Russian-German team of academic scholars and postgraduates. An exchange of undergraduate and postgraduate students began on a regular basis. Russian universities, research institutes, and eco-NGOs were full of "Russists" (from the U.S. and European countries) as they called themselves. Following the collapse of the USSR in 1991, the SoEU was transformed into the International SoEU. The actual breakthrough in collaboration of Russian and foreign Greens happened in the early 1990s.

During the next decade, Russian Greens established contacts with their sister organizations all over the world. Conservationists participated in protest actions to protect the boreal (old-growth) forests in Fennoscandia. In South America, Russian eco-anarchists had assisted in a rehabilitation of rural communities destroyed by civil wars. Environmentalists from the United States, the Netherlands, Germany, Japan, and other countries participated not only in the development of strategy and tactics of defense for the lake Baikal but also in a similar work on a global scale. From the mid-1990s onward, Russian environmental NGOs together with their sister organizations in Finland and the Netherlands launched a set of model projects of environment-friendly agricultural production and timber industry, especially in regions populated by small ethnic groups. From 2000 onward, Russian Greens actively participated in the work of the Forest Stewardship Council (the FSC International) aimed at training national and international forestry businesses to work in accordance with the standards of free-will certification. In 2004, WWF-Russia officially became a Russian national nongovernmental organization. Up to this time, it already had run 150 field projects in 40 regions of the Russian Federation and in the newly independent states. In 2009, the international secretariat of the WWF commissioned WWF-Russia as supervisor of all WWF programs in the Russian Federation and Central Asian states.

In some cases Russian Greens mastered an action repertoire developed by their Western counterparts; in others, the Russians adapted the social know-how to Russian conditions; in still others (e.g., in some regions of China), Russian Greens have played the role of instructors. But in all cases it brought a mutual public benefit. Three other positive results should be mentioned. First, all the above projects and contacts have assisted greatly in the building of the international network of the Greens. Second, this network has become an alternative source of true and relevant information that exerted a growing pressure on the public opinion and government decision making both nationally and internationally. Finally, the phenomenon of localization is rather important. The WWF and Greenpeace, being initially international organizations, have step by step become involved in resolving local and regional environmental problems of Russia.

Structurally, the Russian environmental movement recently exists in two forms: as "transnationals" who advocate ecologization of extraction of gas, oil, timber, and other natural resources in keeping with international standards, and "localists" who defend their respective local ecosystems and local human communities from destruction by transnational megaprojects and by businesses seizing free public spaces (e.g., urban parks, suburban forests, various kinds of reserves).

Hence, two types of sub-politics develop. "Transnationals" supported by the WWF, Greenpeace, and other international organizations confront megaprojects with legal means: they propose draft laws to legislatures, offer advice and project-appraisal services, assist in voluntary certification of products for export, monitor the implementation of agreements with businesses, and so forth. Until 2004, "localists" could make use of such levers as public hearings, letters, and petitions to local executive power bodies. Since local authorities refused to listen to the arguments of the "localists," they resorted to mass protest actions to defend their basic human rights and freedoms all over Russia. The situation grew worse since businesses and government bodies began to generate and support countermovements and clone-movements whose actions were misleading the people. The prefix "eco" has lost its meaning, undermining citizens' trust in genuine environmental NGOs.

There are three more types of "ecological" actors that are new to Russia. The first type comprises environmental movements formed by the "educated poor" who became victims of unfair business practices. The second type is organizations that emerged as a result of conflicts between native inhabitants of cities and villages and immigrants. The third type is virtual communities whose members "emigrate" to the virtual space so as to avoid life in the real world or at least to minimize contacts with it. These communities represent a new form of international social-ecological organization with little connection to "traditional" eco-NGOs. The government's policy does not envisage grant-in-aid to support energy-efficiency projects for modernization of homes, for industrial area revitalization projects, forest fire prevention, habitat restoration, and other kinds of the Russian environmental movement's traditional activity.

The majority of Russians are preoccupied with basic survival issues. Their motto is "well-being first, ecology later." So far, the goals of most environmental protest actions of the local population are confined to preserving their habitat. But, as the events in the USSR 25 years ago showed, environmental protest can promptly transform into a mass political one.

Notes

1. Sorokin, *Golod kak factor.*
2. For a detailed history of ecological debates in Russia, see: Yanitsky, "The Shift of Environmental Debates."
3. Yanitsky, *Russian Environmentalism;* Yanitsky, "The Ecological Movement"; Weiner, *Models of Nature.*

4. See Yanitsky, *Russian Greens in a Risk Society.*
5. Shvarts, *We Share the Responsibility.*
6. Beck, *World Risk Society,* 91–108.
7. See: Weiner, *Models of Nature;* Weiner, *A Little Corner of Freedom;* Yanitsky, *Russian Environmentalism;* Yanitsky, "Industrialism and Environmentalism." Some of the material for this section has been extracted from a previously published article. See: Yanitsky, "The Value Shift."
8. Yanitsky, "The Ecological Movement in Post-Totalitarian Russia."
9. Fomichev, "Khraniteli Radugi," 48 and 50.
10. Boreyko, "Moral'ney i Religiozyne," 36.
11. Shvarts, *We Share the Responsibility.*
12. Yanitsky, "Dialogue between Science and Society."
13. Deelstra and Yanitsky, *Cities of Europe.*

Bibliography

Beck, Ulrich. *World Risk Society.* Cambridge, UK: Polity Press, 1999.

Boreyko, Vladimir. "Moral'nye i Religioznye Osnovaniya Okhrany Prirody [Moral and Religious Foundations of the Wild Nature Protection]." *Gumanitarnyi Ekologicheskii Zhurnal* 3, no. 2 (2001): 32–37

Deelstra, Tjeerd, and Oleg Yanitsky, eds. *Cities of Europe: The Public's Role in Shaping the Urban Environment.* Moscow: Mezhdunarudnye Otnosheniya, 1991.

Fomichev, Sergei. "Khraniteli Radugi i Grazhdanskoe Obszchestvo v Kasimove [Rainbow Keepers and Civil Society in the Town of Kasimov]." *Uchastie (Al'manakh)* 1(1999): 47–52.

Shvarts, Evgeni. *Prirodno-resursnye vedomosti/ Spetsial'nyi Vypusk.* [*We Share the Responsibility*]. State Committee for Ecology of the Russian Federation, 2004.

Sorokin, Pitirim. *Golod kak factor. Vliyanie goloda na povedeniye lyudei, sotsial'nyju i obszchestvennujyu zhizn.* [*Hunger as a Factor: An Impact of Hunger on Human Behaviour, Social Organization and Public Life*]. Moscow: Academia & LVS, 2003.

Weiner, Douglas. *A Little Corner of Freedom. Russian Nature Protection from Stalin to Gorbachev.* Berkeley, Los Angeles: University of California Press, 1999.

Weiner, Douglas. *Models of Nature: Ecology, Conservation, and Cultural Revolution in Soviet Russia.* Bloomington: Indiana University Press, 1988.

Yanitsky, Oleg. "Dialogue between Science and Society." *Social Sciences. A Quarterly Journal of the Russian Acad. of Sciences* 36, no. 2 (2005): 78–90.

Yanitsky, Oleg. "The Ecological Movement in Post-Totalitarian Russia: Some Conceptual Issues." *Society and Natural Resources* 9 (1996): 65–76.

Yanitsky, Oleg. "The Environmental Movement in a Hostile Context. The Case of Russia." *International Sociology* 14, no. 2 (1999): 157–72.

Yanitsky, Oleg. "Industrialism and Environmentalism. Russia at the Watershed between Two Cultures." *Sociological Research* 34, no. 1 (1995): 48–66.

Yanitsky, Oleg. *Russian Environmentalism: Leading Figures, Facts, Opinions.* Moscow: Mezhdunarodnye Otnosheniya Publishing House, 1993.

Yanitsky, Oleg. *Russian Greens in a Risk Society. A Structural Analysis.* Helsinki: Kikimora Publications, 2000.

Yanitsky, Oleg. "The Shift of Environmental Debates in Russia." *Current Sociology* 57, no. 6 (2009): 747–66.

Yanitsky, Oleg. "The Value Shift of the Russian Greens." *International Review of Sociology/Revue Internationale de Sociologie* 15, no. 2 (2005): 363–80.

Chapter 7

Pragmatic Ethics of Environmentalist Movements in Turkey

Gürcan Koçan and Ahmet Öncü

Introduction

"Environmentalism" covers a range of social and political movements and ethical perspectives that share a concern for the protection and improvement of the quality of the environment. Almost 20 years after the 1960s when sizeable environmentalist mobilizations had first come to light and begun to shore up critical public debates and projects for radical social change in Western Europe and North America, the environment was still not a very substantial worry across a wide spectrum of social movements in Turkey. In the 1970s small mobilizations for the environment had surfaced here and there in connection with the tide of student and working class radicalism; yet these were only sporadic groupings. None of them succeeded in reaching a scale capable of developing the capacity to influence the political context in their own right.

The situation changed drastically in Turkey with the military intervention of 1980. With the brutal suppression of democratic politics, all forms of mobilization for social change had either receded or gone underground. Paradoxically, after the return to the parliamentary system under the Constitution of 1982, environmentalism became popular and influential, not only among social activists, but also within official political circles as well as the least politicized sections of civil society.[1] These mobilizations made effective use of the opportunities provided by the new codes and regulations legislated and implemented by successive governments in response to an alarmingly deteriorating natural environment, in addition to the international obligations of the Turkish state as a result of signing bilateral or multilateral environmental accords.[2]

In what follows, we first briefly introduce a taxonomy that aims to distinguish particular environmental ethical positions. Here, by drawing from the literature of environmental ethics, we attempt to delineate some of the possible ethical stances that environmentalist mobilizations can take on while they are aiming for particular objectives as part of their agendas. Next we present a selective account of environmentalist movements with respect to four broad environmental issues that have been influential politically and have had lasting impacts on the nature of environmentalism constructed in Turkey. These are: the commons, pollution, energy, and animal rights. Finally we offer our interpretation of the ethical stance that is reflected from the actions and agendas of different environmentalist movements associated with those issues.

Taxonomy of Environmental Ethical Positions

Environmental ethical positions contain descriptions of states of the world, types of values, and normative claims for making or establishing evaluative judgments. They sort out things like obligations, permissibility of actions, duties, right/wrong, and good/bad in the human-environment relationship. They attempt to establish as well as challenge normative principles that facilitate people's judgments and decisions about competing choices of action, which may have direct or indirect impact on the environment. They enable agents to define normative environmental frameworks and value structures on what they are permitted to do in pursuing their aims. Normative environmental frameworks and value structures have intimate connections to people's preferences and evaluative attitudes. Here comes the impasse: is there a well-defined and consensually well-received normative environmental framework? If so, what is it? How is this framework defined, justified, and put in use in the theory and practice of environmentalist movements?

These questions, which are deeply tied to the issue of the nature of moral subjects, that is, subjects of moral rights and duties, raise issues regarding "what" and "who" matters in making judgments about the morality of environmentalism, and how the rules, once they are agreed upon, shall be enforced. Should we be concerned only about humans? If so, does the abstract notion of "humans" include everybody on the planet or only our fellow citizens or our local acquaintances? Should we be worried only about the interests of living humans or also of those who are yet to be born and those who have passed away? Should we expand the scope of our moral sensibilities to nonhuman animals as well? Or should our moral obligations lie in the interest of the living environment as a whole?

What is decisive in this ongoing ethical debate is its appeal to intrinsic values as opposed to extrinsic values.[3] Intrinsic value refers to a value as an end in itself, and thus something that exists for its own sake.[4] To say that something has an intrinsic value is to say that it doesn't get its value from another source.[5] Its value is nonrelational and noninstrumental. It is an objective entity that is unconditionally good.[6] Extrinsic or instrumental value refers to something valuable as a means to something else.[7] It is necessarily relational because it refers to another value.[8]

Environmentalist movements differ considerably in terms of their conceptualizations of moral subjects, resulting in noticeable divergences regarding their substantiation of intrinsic values in terms of morality of actions. Without making any claim to being exhaustive, we use the following two-by-two classification in Table 7.1 in order to be able to distinguish environmental ethical positions so that we can specify later the ideal type that environmentalist movements in Turkey seem to converge into. In the columns of Table 7.1 we classify ethical positions in terms of their judgments on the morality of actions (i.e., what matters). In the rows we categorize them according to their conceptualizations of moral subjects (i.e., who matters).

In terms of their judgments on the morality of actions we divide environmental mobilizations into two general categories: consequentialism and deontology. Consequentialists maintain that if actions lead to good or beneficial consequences they are morally right. Deontologists, in contrast, hold that it is not the consequences of actions but the motives behind them that are essential in making judgments about the morality of actions.[9] They claim that actions taken only in adherence to rules or duties are morally acceptable. Because of their deep-seated insistence on the rule-driven action as the basis of moral conduct, they argue for the formulation of a set of universal principles that must be defensible in all spaces and times. Consequentialists disagree with deontologists in that for them there are no or very few universal moral principles from which moral reasoning can depart. Adhering to moral particularism, they tend to evaluate actions in concrete situations by attending to the features of the context of an action.[10]

In terms of the nature of moral subjects, that is, subjects of moral rights and duties, we again argue that there are two broad categories: reductionism/individualism and holism. The former derives from an epistemological position that aims to gain the knowledge of a complex entity by selecting

Table 7.1. A Taxonomy of Environmental Ethical Positions

WHAT MATTERS
POSITIONS ON THE MORALITY OF ACTION

WHO MATTERS POSITIONS ON THE MORAL SUBJECTS	**Subjectivism:** There are no objective environmental moral facts **Consequentialism**	**Objectivism:** There are bjective environmental moral facts **Deontologism**
Reductionism/ Individualism	1. Anthropocentrism a. Ethical egoism i. A. Rand b. Utilitarianism i. J. Bentham and J.S. Mill c. Pragmatism i. J. Dewey and H. James 2. Non-anthropocentrism a. Sentiocentrism i. P. Singer b. Biocentrism (interest-based) i. K.A. Goodpaster	1. Anthropocentrism a. Duty ethics i. I. Kant 2. Non-anthropocentrism a. Animal rights i. T. Regan b. Biocentrism (rights-based) i. P. Taylor
Holism	1. Ecocentrism (interest-based) a. Tempered land ethics i. J.B. Callicott	2. Ecocentrism (rights-based) a. Deep ecology i. A. Naess

one of its individual constituents and analyzing it in relation to that individual constituent. Anthropocentrism, sentientism, and biocentrism are reductionist/individualist positions because they isolate human beings, sentient organisms, and all living organisms from the environment and theorize about moral obligations regarding the environment in connection with only those individual constituents. The normative implications of these three particular reductionisms are not difficult to see. While anthropocentrism reduces moral obligations to the interests and rights of past, present, and future humanity, understood as particular individuals or members of communities, sentientism and biocentrism do the same by expanding the scope of moral obligations to include all sentient organisms and all living organisms, respectively.[11] In this sense while the former sees only human beings as things that have intrinsic value, the latter two expand the set of intrinsically valuable things to all sentient organisms in the case of sentientism and all living organisms in the case of biocentrism.

In contrast to reductionism/individualism, holism rests on the premise that properties of wholes cannot be deduced from their constituent elements. In other words, the environment conceived as a whole is always bigger than the collection of its parts. Ecocentrism is the major strand of holistic approach to environmental ethics.[12] According to this stance, ecosystems and all living beings are intrinsically valuable.[13]

Of the four possible meta-ethical stances that can be derived from Table 7.1, ethical egoism is a perfect example of the anthropocentric consequentialist individualism, according to which every human individual as a rational agent should always act in his or her own self-interest and hence maximize his or her self-gain without being hindered by any authority or moral principle.[14] Resting on this maxim, ethical egoism encourages individual freedom and responsibility as the only way to deal effectively with issues pertaining to the environment. The difficulty with this stance is that the environment is such a complex matter, and as such conflicts of self-interest are inevitable; hence resolving these based on egoism is nearly impossible.[15]

Utilitarianism, archetypal anthropocentric consequentialist individualism, may offer a way out to the latter because it maintains that every human individual should act according to the rule that shall produce the greatest interest for all concerned.[16] Nevertheless, as utilitarian theorists themselves would accede, this is not an unproblematic objective to accomplish because what the good is and how it can be measured is not at all clear.[17] One side, known as hedonism, argues for the maximization of net pleasure over pain for all concerned, and the other side, known as preference utilitarianism, claims that the morally acceptable action can be specified with respect to the maximization of preference satisfaction for the current generation.[18] The implication of this for environmental ethics is, if not straightforward, understandable. Any act that causes a change in the environment may involve both costs such as pollution and benefits such as employment. The act will then be accepted as morally right if the preference satisfaction of all present is maximized.[19]

According to pragmatism, our third example for the anthropocentric consequentalist individualism, the utilitarian position cannot resolve the conflict of interest among concrete individuals. The primary reason for this is that pragmatism discards the utilitarian notion of "the greatest good of all concerned" because it conceives ethics as the practice of defining and achieving the common good.[20] It rejects the idea of the maximization of preference satisfaction for abstract individuals by summing up their utilities *ex nihilo*. For pragmatism human individuals gain their moral standing by being a member of a community.[21] Therefore through active participation in the practical construction of their collective good they can define what they accept as ethical, and hence good and right for them as a community. In other words, pragmatism leaves the decision regarding the morality of actions forever open by adhering not to a single, absolute criterion but to a multiplicity of positions from which communal construction of the meaning and value of the environment and its interests can be constructed. In other words, pragmatism privileges concrete ethical negotiations taking place among concrete individuals over any abstract ethical theory.[22]

One of the non-anthropocentric reductionist consequentialist positions is the kind of sentiocentrism that positions human beings as stewards of animals. As all sentientists, people adhering to this stance argue that sentient animals have intrinsic value. Yet, unlike some other sentientists, they also maintain that sentient animals are in many ways useful for humanity. Departing from this observation, they claim that human beings have a moral obligation to balance benefits from the usage of animals and harms done to them as a consequence. The leading theorist in this genre, Singer, sees sentience as a sufficient condition for having interest, and maintains that all creatures with interest must be equally treated.[23] This position approves only some uses of animals while being critical of unnecessary animal suffering.[24] For instance, Singer accepts medical research using animals as morally right in so far as it helps the well-being of all sentient beings.[25] The moral obligation of human beings from sentient beings to all living creatures has been enlarged by arguing that all living organisms ought to be morally "considerable" if not morally equivalent.[26]

Kant's ethics is the primary example of anthropocentric reductionist deontologism. He begins from the premise that human beings are the only rational beings who can have intrinsic value.[27] For this reason, he does not see any major problem regarding the exploitation of nature for the needs of humans.[28] Kant's problem appears to be one of finding a solution to the question of humane use of nature including all animals. Therefore, he argues for some moral limitations for humans who are the users of nature instead of the moral rights of nature. Regan in criticizing anthropocentric Kantianism constructs a non-anthropocentric reductionist deontological position, which is referred to as animal rights position.[29] He argues that animals just like human beings have moral rights, and hence cannot be treated as a means to human ends. In this sense, he strongly rejects any utilitarian claim to the usage of animals for the satisfaction of human needs.[30] He thinks that harming animals cannot be morally justified with reference to some notion of pleasure, gain, or well-being. He calls for the total termination of industrialized and commercial agriculture, medical research, and hunting. Taylor extends Regan's animal rights–focused non-anthropocentric position of reductionist deontologism to argue for an egalitarian environmental ethics for all living things.[31] He advocates a moral conscience of respect for life as a whole. His account involves the claim that life as a whole has a telos, that is, an end, whereby using living beings as a means is morally wrong.[32]

There are two major positions in the holist position. The first is the consequentialist holism that is best represented by the ecocentrism defined and defended by Callicott.[33] Callicott argues for the existence of a biotic community or an ecosystem that can be conceived as if it is a living organism in itself.[34] Borrowing from Leopold's land ethics,[35] which argues for the adoption of an ecological conscience, he conceives the value of individual beings as inseparably connected with their ecological function.[36] In his view the well-being or interest of an individual being cannot be detached from its function within the ecosystem and how and what it contributes to the ecosystem. His criticism of sentientism goes further than biocentrism in that under certain conditions a particular nonliving thing's well-being may be more decisive for the well-being of the ecosystem than a sentient being's pain or the extinction of a single plant.[37] In other words, Callicott goes beyond the criticism of all forms of anthropocentrism as well as individualist biocentrisms.[38] Having said this, however, he seems to be a holist utilitarian, which is sometimes referred to as "practical holism" or "tempered ecocentrism," because he calls for the maximum good of the ecosystem, according to which we are asked to define the morality of our actions. Naess who coined the term *deep ecology* disagrees with this practical holist-type ecocentrism because he wants to change the terms of debates in environmental ethics radically by going beyond the "shallow" or "preservationist" interests of the individuals of consumerist capitalist societies.[39] He wants us to focus "deeply" into our relationship with nature to understand and feel our oneness with nature. Naess's deontologist holist ecocentrism wants us to live as simply as all other things of the global ecosystem.[40]

Environmental Mobilizations and Their Ethics in Turkey

A Brief Historical Background

Modern Turkish society constitutes a unique case for late and dependent industrialization.[41] The founders of the Republic of Turkey, established in 1923, proclaimed that modernization tied to the dual developmental objectives of political and economic independence was the raison d'être of their revolution.[42] Given the absence of internationally competitive large-scale private industrial enterprises, the result of the attempted industrialization endeavor on the basis of private sector initiatives turned out to be an epic failure.[43] With the onset of the Great Depression both foreign capital inflows and foreign trade plummeted considerably, leaving the leaders of the state with no other option than to pursue a radical change in the direction of industrialization. From the failure of private sector–led industrialization, there emerged a successful model of state-led industrialization in the 1930s, later to be called *etatism* in the Turkish political economy literature.[44]

The etatist model that lasted until the beginning of the 1950s helped to build the foundations of large-scale mechanized industry owned and managed by the state. From this time on the land known as Turkey has come to be seen widely as a means for the creation of wealth or an asset to be used to its fullest potential for the development of the nation and the affluence of its people, not only by the state elites and the capitalist class but also by a large number of intellectuals. This has been very much in line with the pragmatic nature of modernization in Turkey noted by one of the leading theorists of the Turkish social formation.[45] In short, an ethos of industrialization a la Dr. Faust began to invade the horizons of the leaders of the society, an ethos that treats nature as a means to an end or, rather, as a factor of production. Not surprisingly, the 1950s wit\ nessed another major "move forward" in the industrialization "battle." This time, alongside the public enterprises, large-scale private industrial concerns gradually came to occupy and control the manufacturing sectors of the economy. Thus, the etatist model of industrialization came to a sudden end, and another cycle of industrialization started, this time on a higher and larger scale, one that deepened in the 1960s.[46] This was the milieu of the mixed economy period. The Faustian discourse of the etatist era was taken on board from tip to toe by the owners and managers of private companies.[47] In the 1970s the country seemed to be divided along two separate yet intertwined lines.[48] In addition to class divisions, the country as a whole was divided regionally as the developed and industrialized west and the underdeveloped and rural east.[49] These divisions were the underlying reason of the political and social cleavages that led to the legitimacy crisis of the state coupled with a deep economic crisis in the late 1970s.[50] With the military intervention in 1980 the Turkish society was legally, politically, and economically reconstituted. The result was full-fledged private sector–led industrialization without any genuine concern about the conservation of the environment in almost all parts of the country, which has continued even up to the present day.[51]

The Commons

All natural and manufactured things shared together are called the commons. Some examples are forests, rivers, air, pasture lands, beaches, Internet, literature, music, arts, design, film, video, television, radio, information, streets, and parks and public services such as health, education, and transportation systems. Using the term coined by Linebaugh, movements for the preservation and conservation of the commons can be called *commoning*.[52] Of all the commoning initiatives, the one that has been focusing on forests is the most established and proliferated in Turkey. The first serious attempts in modern times to raise the awareness of the public to the worsening conditions of forests go back to the 1940s. İnal's manuscript is one of the early publications that directly addresses the conservation of forests as a responsibility of both the state and the citizens.[53] İnal, ahead of his time, introduces the idea of a national park as a means of conserving and improving the forests and their ecosystems.

A comprehensive history of forestry in modern Turkey has been recently published by Yücel Çağlar, a forest engineer and an environmental activist.[54] In his account, the Chamber of Forest Engineers (CFE), founded in 1954 as a professional union, advocates the idea of conservation of ecosystems as a socially very sensitive issue, and has become an active organization politically critical of the exploitation and occupation of forests as ecosystems made up of plants and animal species. The CFE is still an active participant in the conservationist movement in general and calls for governments and nongovernmental organizations to take steps to prevent the overuse of forest resources, to watch closely the harmful effects of privatization of forest lands, and to develop efficient and economic methods of using natural resources according to sustainability criteria and community development—especially the peasant communities whose life rests solely on forests. Besides the CFE there are a vast number of nongovernmental organizations, labor unions, and associations formed by businesses in the forest sector, which are actively involved in the ongoing debates on deforestation and soil erosion.

Among the nongovernmental organizations TEMA (The Turkish Foundation for Combating Soil Erosion, for Reforestation and the Protection of Natural Habitats) is one of the most influential organizations with the most public visibility and followership. TEMA's activities include projects, legal battles, education, and building international partnerships. Most notably, by mobilizing thousands of individuals and other small and big NGOs, TEMA has organized over 100 demonstrations across the country on issues of sustainable rural development, reforestation, biodiversity, conservation, and sustainable land management.

Sustainable rural development projects pioneered by TEMA have set a new trend for ecological agricultural practices and helped change the attitudes toward the ecological hazards and public health risks of fast-growing industrial farming. In recent years farmers' markets where organic products are supplied have come into view in big cities, and become a part of the material life and popular culture. TEMA's rural development projects rest on the understanding that poverty alleviation in rural areas is indispensable for the protection of the soil. In connection with these projects the president of TEMA explains in a 2012 speech how TEMA conceives of the value of soil: "Soil is so valuable because a soil surface of one square centimeter comes into being in 1 million years." He adds that "the most precious treasure in the twenty first century is the productive and high quality soil and water." He then goes on to compare the Netherlands with Turkey to argue that a country as small as the Netherlands compared to one of the big cities in Turkey can produce more and better quality agricultural products because of the advanced land and water management systems it has. Then, he calls on the public authorities to build water and land management systems to prevent soil erosion.

When one looks at the field of commoning as a whole, there seem to be some exemplary mobilizations in the last 30 years or so, which have helped structure certain forms of action and interaction characteristic of environmentalism in Turkey. In 1986, while the watchful eyes of the junta behind the civil government were still being felt, the people of Ankara were shocked and saddened by the metropolitan municipality's cutting of age-old plane trees in Zafer Park in order to construct a parking lot. They at once formed a spontaneous group of concerned citizens to stop the cutting of trees. When the group filed a case at the Ankara administrative court against the municipality, Zafer Park, a symbol of Ankara as well as a site with historical, political, and cultural significance, became a subject matter of law. In the court's decision the municipality was defined as the authority that must be responsible for the conservation and expansion of green areas within the city. This was the first time in the history of administrative law in Turkey that citizens won a legal case against the state.

Thus the Zafer Park mobilization was a milestone that set a precedent for future mobilizations.

Increasingly, more of such similar campaigns using the same strategies and tactics have sought to protect Turkey's cultural, natural, and historical heritage since the 1990s. Some of the most influential and internationally backed examples of note were the mobilizations for preventing the conversion of the historical building of Istanbul Technical University at Taşkışla to a five-star hotel; stopping the demolishing of historical buildings and structures in the Tarlabaşı quarter in Istanbul as part of a highway project; protecting the original architectural form of the historic Galata bridge in Istanbul (when it was under reconstruction planning); the Arnavutköy district initiative against the construction of a bridge connecting the European and Asian regions of Istanbul due to the destructive effects that the bridge would have on the area's natural and cultural endowments as well as the community life in the district;[55] and perhaps finally, saving the loggerhead sea turtles (Caretta Caretta) in Köyceğiz, Dalyan, from tourism developments. These mobilizations resulted not only in a significant change in the attitude of society at large regarding the value of the commons but also in the actions of public authorities. Today no public authority involved in any public project goes ahead without taking into account the concerns of citizens about the perceived environmental consequences of those projects.

Taken as a whole, the majority of commoning movements perceive the environment in terms of a set of human interests. Yet there are also signs emitted in and by their actions that the environment exists beyond any notion of human interest in that it has an existence in itself. In line with their dominant pragmatic consequentialist ethical standing, however, the *weltanschauung* of commoning movements posits ideas as tools whose meanings and values should be worked out according to whether they are viable or not. If ideas are not practicable, they are treated suspiciously. Just like the manner by which dominant agendas have been imposed on or introduced to the larger society by the ruling elites of Turkish politics, economy, and society, commoning activism has been promoted and engaged in with reference to some "progressive" discourses that are not derived from systematic theoretical reflections. As their approach is pragmatic, their campaigns are relativistic. The language of conservationism in Turkey seems to consist of a collection of opinions that are not well-articulated except for defining some strategic goals to achieve. In other words, reminiscent of pragmatism in the larger social context, the issue-driven framework of commoning activism is not principled but inherently anthropocentric, situationalist and hence subjectively oriented.[56]

Pollution

Since the 1980s environmental pollution has constituted another important concern of the mobilizations across the country. Here the main focus has been on the undesirable change in the chemical and biological quality of air, land, and water due to both natural pollutants such as forest fires as well as manmade discharges such as carbon monoxide, sulfur dioxide, oxides of nitrogen, lead, arsenic, asbestos, and the like. The first outstanding mobilization of this kind was the grassroots campaign organized against the construction of a lignite-run thermal power plant station at the rim of the famous tourist destination of Gökova Bay in 1984. The owner of the project was a public authority, Turkish Electricity Authority, which faced the annoyed citizens in close proximity, this time the villagers of Gökova, who were concerned about the pollution of their fertile farmlands and the negative consequences of the potential harms to the coastline posing a direct peril to the tourism sector of the region, one of the mainstays of the small family businesses.

The other two early influential antipollution mobilizations were against the constructions of the Aliağa and Yatağan thermal power plants. In the late 1990s, the case of the mobilization of a group of people from 17 villages in the hills of Bergama in the northern Aegean region of Turkey in opposition to a multinational consortium operating a gold mine in the region was, to be sure, the longest and most high-profile antipollution movement in the short history of environmental activism in Turkey.[57] The Bergama movement struggled effectively against the gold mine that used cyanide for washing the gold; the cyanide had been leaking into the most fertile lands of the whole Aegean region. The Bergama movement has helped change some of the laws and regulations concerning mining, and turned the Bergama versus the state legal battle into an international case when the local villagers appealed to the European Court of Human Rights against the unlawful operation of the mine backed by the opportunistic governments whose main interests lay in the inflow of foreign capital and investment.

In all these and similar mobilizations the interactional structure was almost the same. Typically a group of locals led by the local elected administrators such as mayors or the heads of villages joined by the chambers of leading professional unions such as engineers, lawyers, and physicians and sometimes individuals from the Green Party of Turkey and Greenpeace of Turkey as well as locally founded small environmental NGOs with such names as Friends of Bodrum formed a coalition, organized demonstrations, petitions, panels, and seminars to make their cause public. Time and again they came up with novel and creative ideas and forms of action to gain media attention, such as forming a human chain or bicycle tours starting from the region of contention to reach the capital city or the next major city where the legal case was under review in the local administrative court. Hunger strikes appeared to be another common form of action for publicizing the agenda of the movements.

As a result of successful mobilizations awareness and "stock of knowledge" about air, soil, and water pollution in Turkey have become widespread across the country as a whole. Today if the public at large is at all familiar with the causes and sources of pollution, this is perhaps not because of the formal education at different levels of the Turkish school system but because of these movements, their agendas, and the coverage they have succeeded in receiving in the media. Even the least environmentally concerned individuals may know today that water pollution is mainly caused by the more or less standard practice of unchecked wastewater discharges of municipalities of coastal towns, which do not have any efficient wastewater management systems. The majority of people in industrial regions may have heard of acid rain as one of the main sources of soil pollution, which is produced by the waste gas emissions of industrial facilities using lignite burning power plants. The hazards of using underground water resources for agricultural purposes, the use of polluted water from the rivers for irrigation, leakages of cyanide, arsenic, and the like, from mines may watchfully be observed by the public at large concerned with the quality of soil thanks to the hundreds of convivial demonstrations of thousands of individuals who joined the cause of mobilizations against pollution.

Assessed overall, the ethical position of the environmental mobilizations formed in connection with the issue of pollution has also taken on a pragmatic character. The underpinnings of the ethics they emit in action imply that circumstances modify issues, that is, actual problems. Hence they may even assert that the situation changes rules and principles. This is mainly because they tend to focus on ongoing economic, social, and political negotiations, and thus are relatively less attentive to the problematic nature of interests and rights of species and ecosystems. They seem to have become practically aware of deep ecological concerns going beyond human interests without being seriously involved in elaborate ethical theorizing. This might be the outcome of their participation in concrete rather than ethics-based agendas that not only lead to the recognition of environmental interest as a common interest of a plurality of stakeholders but also unify particular interests surrounding the issues for the cause of a collectivity. Participants of environmental movements tend to come in actions in every respect equipped with ethical truisms of their communities and their customs and traditions. Because of this environmental interest as depicted in the practice of environmentalist movements appears to be the telos of an emergent community. Not surprisingly, the ethical message received by the wider public from these mobilizations is reflective of the pragmatic consequentialist framework. A survey research conducted in 2010 reports that 63.7 percent of the respondents believed the government should pass laws so that people would protect the environment.[58]

Energy

There is no question that the majority of environmental problems are associated with the search for finding cheap and abundant energy sources. Turkey is not an exception. Although Turkey's energy resources are quite diversified, oil and natural gas reserves are limited. Moreover, because of the country's fractured substrata, deposits are often contained in small pockets, which make exploration and extraction difficult. The rapid industrialization and urbanization in the past two decades have unexpectedly increased energy consumption, which has exacerbated the already big gap between supply and demand.[59] It is in this context that nuclear energy plants and hydroelectric power plants as alternatives in the deepening "energy bottleneck" came to the forefront.

The first recorded antinuclear action was a protest movement by a group of fishermen from Mersin, Silifke, in 1977. These people paraded by their boats in the harbor of Mersin-Akkuyu upon the rumors about a governmental decision regarding the construction of a nuclear power plant in the region. In 1993 the first Anti-Nuclear Congress was convened in Ankara by nearly 50 NGOs, followed by an eventful Anti-Nuclear Week". The Chamber of Mechanical Engineers (CME) was the chief organization within the emergent antinuclear coalition. In the same year the CME organized the Nuclear Technology Congress with the stated aim of stopping the construction of the nuclear power plant in Akkuyu. One of the consequences of these congresses and meetings was the media interest that they created. Given the clout of the professional organizations the antinuclear movement eventually reached the lives of those in Akkuyu. Finally a local antinuclear coalition was formed in the region of the construction, which is still active today. Over the years since then local and national movements have gained the support and backing of a non-negligible section of people across the land and established themselves as politically powerful stakeholders that could influence public deliberations at local and national levels. In tandem with the pragmatic ethical discourse of the other environmentalist mobilizations, the antinuclear coalitions have depicted the nuclear power plants as serious threats to the economic and social well-being of the citizens. The Chernobyl disaster in the neighboring Ukraine has always been used as a case in point for alerting the people to the dangers of nuclear energy production and consumption.

From 2003 until 2010 the government introduced a series of regulations and decrees that would allow the private sector to construct hydroelectric power plants (HEPP) and dams for energy production. The response of businesses was overwhelming. As of today the government has approved the construction of 1,738 dams and HEPPs to be built by 2023. The total length of river systems that will be enveloped by the HEPPs is approximately 10,000 kilometers. There has been a growing public concern throughout the country about the possible hazardous implications of these projects on the biodiversity and the well-being of the people living along the rivers where the HEPPs will be built. Not surprisingly, in the 2000s, in the midst of growing environmental activism and with the presence of a pragmatic kind of non-negligible public awareness about the environment, various anti-HEPP coalitions have swiftly been formed by the participation of NGOs from all over the country. Despite the financial difficulties, these coalitions filed 83 legal cases against the HEPPs. Although only half of these cases were finalized, in the majority of them the courts ordered the cancellation of HEPPs on the grounds that the Environmental Impact Assessment (EIA) reports were inadequate or misleading regarding the implications of HEPPs for biodiversity and the well-being of people. Meanwhile, various anti-HEPP coalitions got organized under the umbrella of Turkey Water Assembly and in the beginning of 2011 launched the campaign called "We Won't Give up Anatolia," and called for the "Great March for Anatolia."

Animal Rights

Honoring the rights of animals has historically been viewed as important in Turkey. The history of the animal rights movement in modern times can be traced back to the early 1910s. According to a recent study,[60] the first animal protection organization in Turkey was founded in 1912; and since then the number of nongovernmental organizations aiming to protect and augment the well-being and rights of animals has increased to 155. Today all these NGOs, spreading over 53 provinces, have a membership base of more than 5,500. Gürler, Melikoglu, and Osmanagaoglu[61] conclude that the activities of these organizations have mainly focused on the protection of animals from various public or private abuses, which have been carried out as "the devoted work of 'benevolent' people." One of the most common demands of these people has been the release of animals from animal shelters referred to popularly as "animal prisons" to the streets after being neutered and vaccinated.

If animal shelters have been an important spotlight of the animal rights organizations for decades, the other major meeting point of them has been the cruelty against stray dogs and cats living in the streets. In recent years more and more campaigns calling for an end to the killing of street animals have come into view in public life. In these campaigns animal rights have been equated with human rights on the basis of the claim that animals have no nationality, and hence they have every right to be protected by international laws and regulations for which government is asked to be respectful.

It was within this context that *Hayvan Haklari Federasyonu,* also known as the HAYTAP Animal Rights Federation, was established in 2008 as the first federation in Turkey to unify all animal rights associations and organizations. HAYTAP declared its mission as raising awareness on the violation of animal rights. Some of HAYTAP's activities include reaching out to parliament through court trials, providing training activities such as education programs at schools, and using public relations to spread their message. HAYTAP's main concern is to change the law in Turkey regarding crimes against animals. As it sees the current circumstance, the present law does not contain a strong enough punishment for animal abusers. One of the primary objectives of HAYTAP is to instill in people the intrinsic value of animals. According to HAYTAP, only those who do not have compassion toward animals can abuse them. Thus, all animal lovers and animal rights activists must do everything to sensitize people to care for animals. Nevertheless HAYTAP must move to the center of politics with larger objectives than being an influential public relations campaigner concerned about the well-being of animals. For the activists of HAYTAP, all ecosystems are connected to one another to make up a whole, and hence every living "tissue" has the right to life in balance with the rest.

As a result of the increased visibility of animal rights activism in Turkey in recent years, in 2012 a party for animal rights was established with the name Animal Party (HP). HP declared its mission as the protection of animal rights and welfare by participating in politics. Yet they are not aiming to be represented in parliament; however, they want to work together with the parties in the parliament and have a say in the new constitution. HP, by calling for certain attitudinal and lifestyle changes, seeks to establish animal rights issues as part of regular politics and policy-making. Thus it aims to extend ethical frameworks centered on human beings to include nonhuman individuals and make them an integral part of democracy. The party is against hunting, pet shops, abusive animal shows such as dolphinariums, and the use of fireworks.

Conclusion

An overview of the short history of environmentalist movements in Turkey shows how a pragmatic environmental ethics has emerged in response to the environmental dynamics in a particular sociopolitical context. The tendency of environmentalist activism in Turkey has been to identify issues that have just cropped up in the daily lives of people and to deal practically with them by attempting to actively participate in mostly noninstitutional politics at the local, national, and international levels. The only visible setting in which they have established direct links with institutional politics has been the justice system, that is, the legal front. In spite of this they have not only succeeded in having an impact on governance and legislation process but also discerned new issues demanding new responses. Furthermore they have creatively responded to those new situations. What they have learned in practice have become later tools for handling new issues that arise. They have moved away from their once limited focus on narrower local issues such as the preservation of trees, forests, and coastlines or the contamination of farmlands toward more diversified and broader agendas such as biodiversity, animal rights, and the privatization craze of neoliberal capitalism. Because of being issue-based, however, the emergent pragmatic environmental ethics in Turkey failed to constitute a self-contained, fixed, or solid moral standing. It rather produced a flexible ethical point of view without a substantive conviction.

Although environmental activism has had an ever-expanding horizon that has undeniably empowered the forces of change against the forces of the status quo, except perhaps for the animal rights movement, it has remained chiefly driven by the "logic of practice" that judges the rightness or correctness of a thought or action on the basis of viability rather than principled moral obligations. As a result the questions regarding the motivational bases of environmental activism such as ethical ends, metaphysical aspirations, and moral ideals are largely left unanswered. Like most other political and social mobilizations in Turkey, environmentalism is by and large unequipped with the theoretical depth necessary to satisfactorily answer deep moral questions posed by holistic deontologist positions. To be sure, over time environmentalism has enriched its vocabulary, discourse, and scientific knowledge and insights into ecological problems that the earth has been facing. This has come about via the links that it has established with global movements, international and local scholars studying different dimensions of the environment, international and national media and "green" politicians mostly from advanced capitalist countries where environmentalism has produced rich theoretical traditions. Yet even these linkages did not change the issue-oriented form of environmental ethics in Turkey. Today the majority of the leading figures of various movements look more like political activists responsive to the demands of the case in hand than principled ethical philosophers or theoreticians in political action. Their moral standing has meaning that habitually makes sense in the context of their own community where their particular discourse has been constructed and put in use as critically shrewd tactical formulae for the environmental strategy of a particular movement. Beyond the given social and political context the emergent moral standing may be found lacking. Thus, although environmental mobilizations usually aim at making universalistic objective claims, the dominant form of environmentalist ethic in Turkey still remains subjective, and hence particularistic.

There appear to be two damaging implications of their subjectivist moral standing that is also mostly individualistic. The first derives from the utilitarian perspective that dominates their pragmatic ethics. Utilitarianism as a form of consequentialism privileges expediency over moral motives; and maintains that the end justifies the means. The difficulty with this position is that it is very hard to judge actions on the basis of their consequences simply because actual consequences are for the most part unknown, and hence disputable. For example, it is always controversial to be against the investments in productive activities that may provide employment and income for a less-developed region on the basis of environmental consequences. In countries such as Turkey where millions are in need of jobs, those who are controlling the commanding heights of the economy can easily sweep aside consequentialist environmentalist arguments by emphasizing the economic benefits of their projects against some "unknown" hazards. The second, which is very much related to the first, stems from the anthropocentric values emphasized by pragmatic ethics.

As noted in environmental ethics literature, anthropocentric notions of value are evidently not capable of justifying environmental opposition to the interests of capital. In the current neoliberal context, by drawing from the utilitarian discourse of economic growth and development, governments can vigorously dispel anthropocentric environmentalist criticisms by incorporating the environment into their market as the only solution to all problems.

Having said this, however, we must emphasize once again the important contributions of environmentalist movements to the ever-growing democratic pressure from below on the government that runs the country as if it manages a business corporation. Environmentalists as vibrant forces of democratization ask and call for the greater participation of citizens in policy-making processes. The environmental knowledge and insights enacted in their practice enable various powerless sections of civil society to stand up for their rights and fight for them.

Notes

1. Aydın, *The State, Civil Society, and Environmentalism,* 59–66; Özdemir, *Environmental Consciousness.*
2. Kaygusuz and Arsel, *Energy,* 162–63.
3. Callicott, *Nonhuman Species.*
4. Carlsson, *States of Affairs.*
5. Rabinowicz and Rønnow-Rasmussen, *A Distinction in Value.*
6. Moore, *Principa Ethica.*
7. Garrett, *Extrinsic/Intrinsic.*
8. Bradley, *Extrinsic Value.*
9. Johnson, *Morally Deep World.*
10. Cooper, *Teleology.*
11. Brennan, *Thinking about Nature.*
12. Calow, *Ecosystem Health.*
13. Callicott, *Intrinsic Value.*
14. Rand and Branden, *The Virtue of Selfishness.*
15. Goodpaster, *Egoism.*
16. Bentham, *Morals and Legislation.*
17. Mill, Bentham, and Ryan, *Utilitarianism.*
18. Johnson, *Morally Deep World.*
19. Singer, *Practical Ethics*; Johnson, *Morally Deep World.*
20. Dewey, *The Public,* 327–28.
21. Dewey, *Human Nature,* 314–32.
22. Ibid.
23. Singer, *Animal Liberation.*
24. Ibid.

25. Singer, *Practical Ethics.*
26. Goodpaster, *Morally Considerable*; Goodpaster, *Egoism.*
27. Kant, *Groundwork.*
28. Ibid.
29. Regan, *The Case*; Regan, *Possibility.*
30. Regan, *The Case.*
31. Taylor, *Superior.*
32. Taylor, *Respect for Nature.*
33. Callicott, *Elements.*
34. Callicott, *The Conceptual Foundations of the Land Ethic.*
35. Leopold, *Sand County Almanac.*
36. Callicott, *In Defense of the Land Ethic*; Leopold, *Sand County Almanac.*
37. Callicott, "Review of Tom Regan, the Case for Animal Rights."
38. Callicott, *Non-Anthropocentric.*
39. Naess, *Long-range Ecology.*
40. Naess and Sessions, *Platform Principles.*
41. Boratav, *Türkiye İktisat Tarihi.*
42. Adaman and Arsel, *Environmentalism in Turkey.*
43. Birtek, *The Rise and Fall of Etatism in Turkey,* 407–38.
44. Ibid.
45. Mardin, *Religion,* 44.
46. Boratav, *Türkiye İktisat Tarihi.*
47. Buğra, *State and Business.*
48. Schick and Tonak, *Class Struggle.*
49. Sönmez, *Bölgesel Eşitsizlik.*
50. Öncü, *Dictatorship Plus Hegemony*; Savran, *Türkiye'de Sınıf Mücadelesi.*
51. Kalaycıoğlu and Gönel, *Role of Business.*
52. Linebaugh, *Magna Carta Manifesto.*
53. İnal, *Doğa Korumacılığı.*
54. Çağlar, *Ormancılık Tarihi.*
55. Voulvoli, *The Identities.*
56. Fletcher, *Situation Ethics.*
57. Öncü and Koçan, *Democratic Citizenship.*
58. Çarkoğlu and Kalaycıoğlu, *Türkiye'de Çevre.*
59. Kaygusuz and Arsel, *Energy.*
60. Gürler, Melikoğlu, and Osmanağaoğlu, *Animal Protection Effects.*
61. Ibid.

Bibliography

Adaman, Fikret, and Murat Arsel. *Environmentalism in Turkey: Between Democracy and Development.* Burlington: Ashgate, 2005.

Aydın, Zülküf. "The State, Civil Society, and Environmentalism." In *Environmentalism in Turkey: Between Democracy and Development,* edited by Fikret Adaman and Murat Arsel. Burlington: Ashgate, 2005.

Bentham, Jeremy. *Introduction to the Principles of Morals and Legislation*. New York: Oxford University Press, 1996.Birtek, Faruk. "The Rise and Fall of Etatism in Turkey, 1932–1950: The Uncertain Road in the Restructuring of a Semiperipheral Economy." *Review (Fernand Braudel Center)* 8, no. 3 (1985): 407–38.

Boratav, Korkut. *Türkiye Iktisat Tarihi 1908–2009* [*Economic History of Turkey 1908–2009*]. Ankara, Turkey: Imge, 2003.

Bradley, Ben. "Extrinsic Value." *Philosophical Studies* 91, no. 2 (1998): 109–26.

Brennan, Andrew. *Thinking about Nature: An Investigation of Nature, Value, and Ecology*. Athens: The University of Georgia Press, 1998.

Buğra, Ayşe. *State and Business in Modern Turkey: A Comparative Study*. Albany: State University of New York Press, 1994.

Çağlar, Yücel. *Türkiye Ormancılık Tarihi*. [*History of Forestry in Turkey*]. Ankara: ODTÜ Gelişme Vakfı Yayıncılık, 2012.

Callicott, J. Baird. "The Conceptual Foundations of the Land Ethic." In *In Defense of the Land Ethic*, edited by J. B. Callicott. Albany: State University of New York Press, 1989a.

Callicott, J. Baird. *In Defense of the Land Ethic: Essays in Environmental Philosophy*. New York: State University of New York Press, 1989b.

Callicott, J. Baird. "Elements of an Environmental Ethic: Moral Considerability and the Biotic Community." *Environmental Ethics* 1, no. 1 (1979): 71–81.

Callicott, J. Baird. "Intrinsic Value, Quantum Theory, and Environmental Ethics." *Environmental Ethics* 7 (1985): 257–75.

Callicott, J. Baird. "Non-Anthropocentric Value Theory and Environmental Ethics." *American Philosophical Quarterly* 21, no. 4 (1984): 299–309.

Callicott, J. Baird. "On the Intrinsic Value of Nonhuman Species." In *The Preservation of Species: The Value of Biological Diversity*, edited by B. G. Norton, 138–72. Princeton: Princeton University Press, 1986.

Callicott, J. Baird. "Review of Tom Regan, the Case for Animal Rights." In *In Defense of the Land Ethic*, edited by J. B. Callicott. Albany: State University of New York Press, 1989c.

Calow, Peter. "Ecosystem Health A Critical Analysis of Concepts." In *Evaluating and Monitoring the Health of Large-Scale Ecosystems*, edited by D. J. Rapport, C. L. Gaudet, and P. Calow. New York: Springer-Verlag, 1995.

Çarkoğlu, Ali, and Ersin Kalaycıoğlu. *Türkiye'de Çevre 2010*. [*The Environment in Turkey 2010*]. Istanbul: Istanbul Policy Center, Sabancı University, 2011.

Carlson, E. "The Intrinsic Value of Non-Basic States of Affairs." *Philosophical Studies* 85, no. 1 (1997): 95–107.

Cooper, G. "Teleology and Environmental Ethics." *American Philosophical Quarterly* 35, no. 2 (1998): 195–207.

Dewey, John. *Human Nature and Conduct: An Introduction to Social Psychology*. New York: The Modern Library, 1930.

Dewey, John. *The Public and Its Problems: An Essay in Political Inquiry*. University Park, PA: Pennsylvania State University Press, 2012.

Fletcher, Joseph F. *Situation Ethics: The New Morality*. Philadelphia: Westminster Press, 1966.

Garrett, Brian. "Extrinsic/Intrinsic." In *A Companion to Metaphysics*, edited by J. Kim and E. Sosa, 162–63. Oxford: Blackwell Publishers, 1995.

Goodpaster, Kenneth E. "From Egoism to Environmentalism." In *Ethics and Problems of the 21st Century*, edited by K. E. Goodpaster and K. M. Sayre, 21–35. Notre Dame, IN: Notre Dame University Press, 1979.

Goodpaster, Kenneth E. "On Being Morally Considerable." *The Journal of Philosophy* 75, no. 6 (1978): 308–25.

Gürler, Ayşe Menteş, Berfin Melikoglu, and Şule Osmanağaoğlu. "Evaluation of Animal Protection Efforts of Non-Governmental Organizations in Turkey." *Kafkas Üniversitesi Veteriner Fakütesi Dergisi* 17, no. 6 (2011): 901–8.

İnal, Selahattin. *Tabiatı Koruma Karşısında Biz ve Ormancılığımız.* [*Us and Our Forestry before the Conservation of Nature*]. Istanbul: Osmanbey Matbaası, 1949.

Johnson, Lawrence E. *A Morally Deep World: An Essay on Moral Significance and Environmental Ethics.* Cambridge: Cambridge University Press, 1991.

Kalaycıoğlu, Selma, and Görel Feride. "The Role of Business in Environmental Policy-Making." In *Environmentalism in Turkey: Between Democracy and Development*, edited by F. Adaman and M. Arsel. Burlington: Ashgate, 2005.

Kant, Immanuel. *Kant Groundwork of the Metaphysics of Morals*, edited by C. M. Korsgaard, M. Gregor, and J. Timmermann. Cambridge: Cambridge University Press, 2008.

Kaygusuz, Kemal, and Murat Arsel. "Energy, Politics and Policy." In *Environmentalism in Turkey: Between Democracy and Development*, edited by Fikret Adaman and Murat Arsel. Burlington: Ashgate, 2005.

Leopold, Aldo. *A Sand County Almanac: With Essays on Conservation from Round River.* New York: Ballantine Books, 1970.

Linebaugh, Peter. *The Magna Carta Manifesto: Liberties and Commons for All.* Berkeley: University of California Press, 2008.

Mardin, Şerif. "Religion, Society, and Modernity in Turkey." In *Modern Intellectual and Political History of the Middle East*, First Edition. Syracuse, NY: Syracuse University Press, 2006.

Mill, John Stuart, Jeremy Bentham, and Alan Ryan. *Utilitarianism and Other Essays.* Harmondsworth, Middlesex, England; New York, NY, U.S.A: Penguin Books, 1987.

Moore, George E. *Principia Ethica.* Second Edition. Revised Edition. Cambridge: Cambridge University Press, 1993.

Naess, Arne. "The Shallow and the Deep, Long-Range Ecology Movement. A Summary." *Inquiry* 16, no. 1–4 (1973): 95–100.

Naess, Arne, and Georgel Sessions. "Platform Principles of the Deep Ecology Movement." In *The Deep Ecology Movement: An Introductory Anthology*, edited by A. Drengson and Y. Inoue, 49–53. Berkeley: North Atlantic Books, 1995.

Öncü, Ahmet. "Dictatorship Plus Hegemony: A Gramscian Analysis of the Turkish State." *Science & Society* (2003): 303–28.

Öncü, Ahmet, and Gürcan Koçan. "Democratic Citizenship Movements in the Context of Multilayered Governance: The Case of the Bergama Movement." *New Perspectives of Turkey* 26 (2001): 29–57.

Özdemir, İbrahim. "The Development of Environmental Consciousness in Modern Turkey." In *Environmentalism in the Muslim World*, edited by R. Foltz. Boston: MIT Press, 2003.

Rabinowicz, Wlodek, and Toni Rønnow-Rasmussen. "A Distinction in Value: Intrinsic and for Its Own Sake." *Proceeding of the Aristotelian Society* 100, no. 1 (2000): 33–51.

Rand, Ayn, and Nathaniel Branden. *The Virtue of Selfishness: A New Concept of Egoism*. New York: New American Library, 1965.

Regan, Tom. *The Case for Animal Rights*. Berkeley: University of California Press, 1983.

Regan, Tom. "Nature and Possibility of an Environmental Ethic." *Environmental Ethics* 3, no. 1 (1981): 19–34.

Savran, Sungur. "The Legacy of the 20th Century." In *The Politics of Permanent Crisis: Class Ideology and State in Turkey*, edited by N. Balkan and S. Savran. New York: Nova, 2002.

Savran, Sungur. *Türkiye'de Sınıf Mücadeleleri: 1919–1980*. [*Class Struggles in Turkey: 1919–1980*]. Turkey: Kardelen Yayınları, 1992.

Schick, Irvin C., and Ertugrul A. Tonak. "Turkish Politics and Class Struggle: 1950–1975." *Middle East Report* 84 (1980): 14–18.

Singer, Peter. *Animal Liberation*. New York: Avon, 1975.

Singer, Peter. *Animal Liberation*. New Revised Edition. New York: Avon Books, 1990.

Singer, Peter. *Practical Ethics*. Cambridge, New York: Cambridge University Press, 1979.

Singer, Peter. *Practical Ethics*. Third Edition. New York: Cambridge University Press, 2011.

Sönmez, Mustafa. *Bölgesel Eşitsizlik: Doğu-Batı Uçurumu*. [*Regional Inequality: East-West Divergence*]. Istanbul: Alan Yayıncılık, 1998.

Taylor, Paul. "Are Humans Superior to Animals and Plants?" *Environmental Ethics* 6 (1984): 149–60.

Taylor, Paul. *Respect for Nature: A Theory of Environmental Ethics*. Princeton: Princeton University Press, 1980.

Turner, Jonathan H. *Human Emotions: A Sociological Theory*. New York: Routledge, 2007.

Turner, Jonathan H. *A Theory of Social Interaction*. Stanford: Stanford University Press, 1988.

Voulvouli, Emilia. "Leftism, Secularism, Transnationalism and Localism: The Identities of an Urban Protest in Contemporary Istanbul." *Ethnologia* 1 (2010): 1–30.

PART II

Asia

Chapter 8

Movement against Protected Areas in India: Social Networks and the Role of External Actors

Ashok Swain

Forestlands are increasingly being declared as protected areas in the developing countries. Claims are that these protected areas not only preserve forests and biodiversity but also trigger conflict and promote cooperation between countries and communities. However, there is concern regarding the transfer of this Western conservation approach to the South. Protected areas have become the sources of local conflicts in many countries, including India. In many places, affected people are protesting against this policy, and some of them have been successful in their mobilization and some have failed. This chapter aims at finding out how social networks influence the mobilization of the anti-protected area protest movements in the southeastern Indian state of Odisha (Orissa). Furthermore, we argue that the network structure of a society is an evolving process and the impact of external actors and ideology can change the existing interpersonal relationship and with it, the mobilization dynamic of protest movements. In the inland part of this Indian state, for many years people were unable to organize effectively against the antipeople protected area policies of the state. However, the protest capability and method of these people has considerably improved recently mainly due to inroads of external actors, such as left-wing radicals as well as Hindu and Christian missionaries, into the region.

Protected Forest Areas and Affected Forest Community

The world has recently witnessed several initiatives to address unsustainable deforestation. Since the United Nations Conference on the Environment and Development (UNCED), popularly known as Earth Summit, in 1992, the forest conservation issue has featured prominently in the global agenda. At the national level, several European countries have put some supply side restrictions on tropical timber import with high tariff and labeling by country of origin and species. There are also global initiatives to promote social forestry and community-based forest management. Community-based management of forest area is being encouraged, whose aim is inclusion of the local community to participate in limited activities that lead to incentive generation, thus providing secured livelihoods and the need for sustainable forest utilization.

However, special attention has been paid by the international community to the establishment of forest reserves, national parks, and other protected areas in the South. This measure is being increasingly adopted and encouraged for the protection and restoration of the forest and wildlife. The Earth Summit has endorsed the goal that countries protect 12 percent of their area to conserve the natural flora and fauna. Pressures from international agencies such as financial institutions and aid organizations tend to support this measure.

International environmental groups and conservation agencies are also promoting the establishment of protected areas. Sizeable amounts of foreign aid to the developing countries are also provided for this effort. Besides this incentive, the North is also putting pressures of various types on the South to place larger areas under this scheme. Protected areas include sizeable tracts of land and forest resources. Thus, the creation of newly protected forest areas in many developing countries has become one of the most controversial and challenging issues. Conflict between developing and industrialized countries is not limited to the idea of creation of protected areas only. Moreover, when developing countries fall in line, the establishment and management of protected areas also become a source of conflict between affected forest people and the state.

The income derived from the protected area, and the attachments people form with the area, often becomes an important component of the local community.[1] When the establishment of the protected area adversely affects the local population or is perceived to be guided by external interest, it leads to resentment and conflict. Usually it imposes serious restrictions on the local people's customary livelihood activities, and they tend to perceive that society at large reaps the benefits at their cost.[2] Among the affected people, the group that gets the hardest hit is indigenous communities. They get prohibited from grazing, hunting, fishing, and gathering food, wood, and fodder. In several cases, conflict erupts between hunters, gatherers, loggers, miners, fishermen, and tourism operators or protected area staff or environmental advocates. Indigenous inhabitants in the protected forest area are dependent on shifting cultivation and use the forest as a source of firewood, construction material, and medicinal plants. Their exclusion from the decision making process and the restriction imposed on their livelihood result in discontent between the forest authority and local people.

Protected areas increase the state control over the forest while simultaneously decreasing the community control.[3] Though protected area initiatives make regular references to the involvement of local communities, generally they are only used as a labor pool. External authorities, which are entrusted to protect the forest, usually lack the local knowledge of environmental interactions.[4] In some cases, people are relocated. According to Geisler, "global land conservation efforts are creating a growing class of invisible refugees in the third world."[5] In a densely populated country like India, protected areas have displaced 600,000 indigenous populations, who are commonly referred to as tribals. Mostly, these people are left to themselves to find alternative sources of survival. Even when the state does undertake the responsibility of resettlement, it does not get properly carried out by its inefficient and corrupt institutions.[6] The transfer of the people to a different socioeconomic environment and the provision of land unsuitable for agricultural activities bring frustration and anger. The loss of various services and their traditional source of livelihood upsets local communities and brings economic instability to the local population. Thus, in many cases, forest people mobilize against the state to bring organized opposition to the protected area policy. India has been experiencing many of these movements in recent years.

Social Movement Theory and Protest against Protected Areas

The last two decades have witnessed a remarkable transformation in the political and social fabric of the South. Developing countries are increasingly witnessing free elections, parliamentary politics, independent judiciary, and social autonomy. Coinciding with the wave of democratization, protest movements have also gripped a large number of countries in the South.[7] These movements are concerned with forcing greater responsiveness, representation, and participation from state institutions and elites.[8]

They not only express the voice of dissent, as Ponna Wignaraja argues, "they are also providing some basis for developmental and democratic alternative to the system as it now works."[9]

Some movements are more successful in mobilizing and sustaining support than others. As a form of struggle, movements sometimes succeed and sometimes fail. At times they are able to get concessions from the state; at other times they are ignored or suppressed. For broader and more successful social action, there is a need for greater and well-sustained popular mobilization. The diffusion of the movement is necessary in order to keep up the pressure on authorities. Alongside the political process approach in which the state plays the most influential role, a number of studies have brought out the importance of social networks through which people are recruited to the movements. Nearly three decades ago, Snow et al pointed out in their seminal article the importance of social networks in explaining the variation in the mobilization of movements.[10] The concept of social networks has recently gained popularity among the researchers of social movements. In a study of environmental activism in the Italian city of Milan, Diani found that 72 percent of people had joined a movement via social networks.[11] The social network is the objective existence of social capital while ties of trust and the norm of reciprocity represent its subjective part.[12] Ostrom has argued that the network fosters norms of reciprocity and trust.[13] The social networks, both formal and informal ones, enhance a group's capacity to join together in collective action, to address common problems or to pressurize the authority to address these issues. For successful mobilization, a movement depends primarily upon the social network, both formal and nonformal ones.[14]

Different types of social networks can be used for recruitment to be part of a movement.[15] In some societies, people belonging to one social group have close connections with people of other groups and the social ties cut across two or more groups. To McAdam and Paulsen, the ideal network structure of a social movement is the "one in which dense networks of weak bridging ties linked numerous local groups together by means of strong interpersonal bonds."[16] In his study of social capital in the United States, Putnam writes that social networks are not always divided between "bonding" and "bridging" ties, rather for a successful mobilization, ideal group structures should "bond along some social dimensions and bridge across others."[17] However, the social ties and network structure of any society are not static.[18] External issues and actors may help to provide new information and resources to individuals to develop connections outside their bonding circles. A change initiated by outside agents may help to cover the structural holes between diverse groups and build connections among them in a society where none previously existed. The newly created ties can possibly extend the sphere of social interaction and networking.

The majority of literature on social movements focuses on the success or failure of the movement on the basis of their stated goals.[19] However, the outcomes of the protests are not only shaped by the strength of the mobilized support, but it can also be affected by third-party intervention, environmental changes, and nonmovement politics.[20] In this context, this chapter aims to analyze the evolving mobilization process and the influence of a changing social network structure on the diffusion and sustenance of protest movements in India.

Study of Protest Movements in India

India has been successful in maintaining its democratic structure since independence in spite of its multiethnic, multireligious, and multilinguistic society. India has always been a surprise for theorists of democracy. It has consistently defied those who prophesied its imminent demise. In spite of a fissiparous society and considerable socioeconomic challenges including abject poverty, widespread illiteracy, and a deeply hierarchical social structure, all of which, in theory, are uncongenial for a flourishing democracy, India's democracy shows no signs of withering away. Most analysts agree that crucial to the survival of democracy in India has been the nature of the Indian state, and its willingness to bargain and accommodate varying interests. This has been seen as key in maintaining the democratic system despite the deep divisions in society.

This "Indian wonder" has been for some time passing through a phase of "democratic churning."[21] A substantial number of people in the Indian population are increasingly resisting accepting an exploitative state structure or a development model that they presume excludes them. The democratic awakening seems to have led to the increased self-assertion and political participation of these silent groups.[22] As a part of these new organized protests in India, forest movements have increased in frequency and intensity.

These forest movements are not being bound or guided by any particular philosophy. One movement differs from another in its way of expressing demands, organizing the struggle, and also in its perceptions about development. These movements represent a wide spectrum of groups and activities, which cut across specific regions and issues. They mobilize large masses of people, organize popular resistances that cut across political and social barriers, and create new sociopolitical actors in the country. The movement represents a wide range of collectives that fall outside the existing institutionalized groups of society.[23] This ideological plurality in Indian environmentalism has helped to widen the sphere of the forest movements and sharpen the terms of the debate. Moreover, it has also facilitated the formation and collaboration of a varied number of protest organizations in Indian society.

The 67.83 million hectares of forest, which constitutes 20.64 percent of India's geographical area, is the second largest form of land use in India.[24] After establishing the forest department in 1864, the colonial regime took control of India's large forest areas. After destroying the forests in their own country, the colonial rulers embarked upon exploiting Indian forest resources to meet the need of timber for shipbuilding and railway tracks.[25] Almost everywhere and throughout the colonial period, the takeover of the forest was bitterly resisted by the local population.[26] However, all the protest movements meant to preserve the rights for forest use were usually regarded as part of a grand national movement or an anticolonial movement.

With independence from colonial rule, the conflict over forest resources was expected to diminish. However, the forest use policies in postcolonial India (the National Forest Policy of 1952) continued along with the old colonial pattern of the 1894 Forest Policy. This reinforced the right of the state to exclusive control over forest protection, production, and management.[27] The concept of "reserved forests" was justified in the name of "national needs."[28] The needs of the commercial-industrial sector now replaced the strategic needs of the colonial power to guide the forest policy in independent India. Powerful corrupt alliances between the bureaucrats, politicians, and businessmen emerged in all states and they involved themselves in the large-scale destruction of tree cover.[29] As Madsen argues, the sins of commission (exploiting the forest resources) as well as the sins of omission (inability to protect the forest) by the forest department resulted in massive deforestation in India.[30] Protest movements in different parts of the country came as a response to the threat of the survival base and as a demand for the conservation of natural renewable resources.

The Chipko (Hug the Trees) movement is undoubtedly the most well known of environmental movements in modern India. It emerged in 1973 as a movement of the hill people in the northern state of Uttar Pradesh to save the forest resources from exploitation by outside contractors. They effectively stopped commercial felling in the nearby forest by threatening to "hug the trees." The success of this movement led to a wave of popular protests against commercial logging in the Himalayan region. Due to its novel technique and Gandhian philosophy, the movement acquired quick fame and also inspired the Appiko movement in the southern part of India. These movements actively resisted illegal felling of Indian forests and at the same time worked to replant forestlands with multipurpose broad-leafed tree species. Movements focused on the protection of forests to ensure their survival have also emerged in the tribal belts of the country, particularly in central India.

Each of these protest movements had different demands in different phases but the most common theme was that control of forest ownership and management must revert from the state to communal hands and the local communities should be actively involved in the reforestation programs. Opposition to the Draft Forest Bill of 1982 brought these protests together to form networks at various levels. The proposed Bill aimed to further strengthen the power of the forest department, but was successfully resisted by the different forest-based environmental movements all over the country. The government was therefore forced to reconsider its decision and did not bring the bill to parliament for ratification. The movements were also successful enough to force the government to cancel its commercial reforestation policy in the existing forest areas.

An upsurge in popular protests has brought significant changes in government policy. When India's Prime Minister Rajiv Gandhi launched the National Wastelands Development Board in 1985, he not only favored the people's participation but also emphasized that only through a people's movement could the catastrophic destruction of forests be arrested and reversed.[31] The movements were ably assisted by a large number of NGOs in their efforts. The success of this collaborative struggle was clearly reflected in the Government of India's National Forest Policy of 1988 and the Circular on Joint Forest Management of 1990. While revising its National Forest Policy in 1988, India for the first time declared that forests were not to be commercially exploited for industries but must contribute to soil conservation and environmental protection and meet the survival needs of the local population. The revised policy also emphasized the role of the local people's participation in the protection and development of the forests.[32] In June 1990 the central government issued guidelines to state governments on involving village communities and voluntary agencies to aid in the regeneration of degraded forestlands on a usufruct sharing basis. In recent years, the forest-based popular protests have successfully asserted the people's right to manage forests and brought a significant change in the existing forest policy to facilitate that. The Scheduled Tribes (Recognition of Forest Rights) Bill of 2005 has become a law, which provides mining rights in addition to food-fuel and fodder rights to forest communities. This new development has received strong protests from the wildlife conservationists and environmental activists. They fear that the regaining of these powers by the forest communities will expedite the deforestation process in the country.

In India, at present nearly 29 percent of the country's forest areas are declared as protected forest areas.[33] These 668 protected areas, extending over 161,000 square kilometers, cover 4.9 percent of its total land area, and have directly displaced a large number of people; they have taken away the control of the forest resources from the forest-dependent communities and given it to the forest department. Since the 1980s, several movements in different parts of the country have come up against protected area policy as a response to the threat to the survival base.[34] The participants of these movements are demanding that they should get back ownership, control, and access to *jal, jungle aur jamin* (water, forest, and land) in the protected areas.[35]

However, the emergence of this type of protest against protected areas in India's southeastern state of Odisha is a new development.[36] Odisha lies on the eastern side of the country, facing the Bay of Bengal. It comprises 4.7 percent of the country's landmass and 3.7 percent of the total population. Odisha is the poorest state in the country, with more than 47 percent of its population living below the poverty line. While 43 percent of India's tribal families are under the poverty line, for the tribal population in Odisha this figure reaches 73 percent.[37] Deprivation and denial of access to rights over land seem to be the major reasons for poverty and disempowerment among the tribal people who own only around 13 percent of the total land area. The state owns approximately 74 percent of land in these regions where the majority of the tribal people either remain landless or are marginal landowners.

Odisha is the third largest state in India in the context of forest coverage. Around 30 percent of its geographical area is covered with forest. Several international and national initiatives are presently being undertaken to arrest increasing deforestation in this state. More than 4 percent of Odisha's geographical area and nearly 12 percent of its forest cover has been earmarked as protected. Most of these areas were declared protected in the 1980s without any consultation with the local-level institutions and communities and even without informing them. There are 18 more protected areas in the planning stage. Displacement and dispossession have become major contentious issues for these protected areas. Forest communities are traditionally dependent on nonwood forest products for their sustenance.

Thus, there is strong opposition to the eviction of villagers within the protected areas. Another major problem is wildlife from protected areas coming into human settlement and causing damage to property, crops, and human life.[38] In the last two decades, Odisha has witnessed many struggles by the poor to protect their life and livelihood.[39]

There are several ongoing protests in Odisha against the management of these protected areas. Much of the early activism follows the Gandhian tradition: fasts, mass rallies, and long marches. But gradually some movements adopt violent activities into their repertoire of action to disrupt protected area projects.[40] Odisha is presently experiencing an increase in the number of these protest movements despite its "backwardness." Notwithstanding Odisha's low literacy rate, poverty, industrial backwardness, the fact that it has the largest proportion of "untouchable" and tribal groups in the country, and the very weak record of popular movements in the pre- and post-independence period, people have recently started organizing protests against protected area policies. In the post-independence period, the coastal region of Odisha has been politically and economically more powerful than the inland part. In the inland part, more than 35 percent of the population consists of indigenous tribals. In addition to the regional disparity in the development, there is a difference in the number and outcome of movements in these two regions of Odisha.

While nearly 40 percent of Odisha's inland area is forest area, the share of the coastal region is only 10 percent.[41] Till recently in the inland part, in spite of larger areas declared as protected areas, affected people have not organized protest as often as their coastal counterparts. Both the inland and coastal regions of Odisha have nine protected areas each, but their size is much larger in the inland region. Moreover, in the inland region, affected people have generally failed to mobilize opposition against these projects. And when they have opposed them in the past, they have not been able to expand their support base and sustain their protest for a long period of time.

Some researchers provide the customary class analysis to illustrate the difference between the peoples of the coastal and inland regions over their protesting competence against unfavorable government policies. As Fernandes and Raj argue, powerful middle-class farmers in the coastal areas due to their possession of superior economic resources have been successful in organizing more effective protests against displacement than their poor landless counterparts in the inland area.[42]

If the class factor is the key to the success of the movements in Odisha, then how were the landless lower-caste fishermen of the Chilika Lake able to successfully protest against the state power and big industrial houses? The coastal region has witnessed a successful protest movement in the 1990s against proposed intensive prawn farming in the Chilika Lake, the largest saltwater lake in the country. The affected people of the area took up the opposition against the planned large aquaculture projects in the lake. They carried on the agitation for a longer duration and successfully resisted its implementation. The organization and success of the Chilika movement in the coastal part of Odisha display the weakness in the class-based explanation about the success or failure of the movement. If the lack of formal land ownership has been the reason for nonparticipation of landless people in the protest movements in the inland region, then how could the fishermen in Chilika, who had no land and fishing rights, put up a united fight to protect their fishing ground? Coastal Odisha has also witnessed another successful movement against a proposed mega steel plant at Gopalpur. This movement was spearheaded by the local villagers who have been traditionally the Kewra flower growers. Most of the Kewra plants (the essence of its flower is used as a food flavoring agent) in the region are grown on the government land and villagers had no legal rights over the land, which was planned to be acquired for the steel plant. However, the lack of formal rights over the land had not adversely affected the successful mobilization of affected farmers against the project. Common property resource, like inland fishing, forests and pastures, is extremely important for the livelihood of the tribal population in the inland region as well. They stand to lose access to these resources due to the encroachment by large development projects or protected areas. However, the tribal population in the inland area has not been able to protect these livelihood resources as successfully as the people in the coastal region by failing to organize effective protest movements.

There was a clear distinction between the two areas in the late 1990s. It clearly shows that the protests in the inland part of Odisha had failed to attract larger participation from the affected and sympathetic population. Every movement develops a participation scheme. The patterns, levels, and types of participation define to a certain extent its strength and success. For enhanced participation, an extended group network is needed. The diffusion of the protest is necessary in order to keep it alive when its initial spark begins to sputter. Protest marches and demonstrations need to be translated into an organized popular movement. A successful movement needs to transcend the "volcanic" stage of collective action. It is also easy for the authority to control a movement if it is localized in nature. The spread of the movement to larger areas makes it difficult for the authorities to use their forces to maintain order and that may help to receive concessions from the state.

The transformation of occasional protests to an organized movement against any protected area would be impossible in the absence of flexible and inclusive networks within the broad framework of collective action. The disadvantage that results from the smallness of a group can be overcome through flexible coordination. Flexible coordination can be achieved through a network of small groups, which complement one another to achieve collective efficiency for the popular action. The existence of weak ties in society can set the pace for networking among the groups and thus can help the movement to spread and sustain in order to achieve its aim and objective. If we go to Granovetter's classification of interpersonal ties, strong ties exist when people with close and intimate relationships see one another frequently over long periods of time.[43] On the other hand, weak ties exist among individuals who see one another infrequently and their relationships are casual rather than close.

The more local, social, and cultural interactions are mutually reinforcing, the more likely people are to engage in collective defiance.[44] The strong intragroup ties create high trust within the group but at the same time bring suspicion about the others. By bringing exclusiveness to the group, strong ties may hinder the group's initiative or participation in a broader movement. These exclusive ties need to take the backseat thereby leaving room for secondary associations in the society in order to strengthen the movement. Unlike the strong ties, the weak ties within and outside the own group may bring inclusiveness (at least not exclusiveness) and help to build and spread the movement successfully.

The tribal population in Odisha in the last century was far from being a homogenous group. It was divided into various clans and subclan groups. Sixty-two tribal communities of Odisha have been designated as Scheduled Tribes, of which 13 have been recognized as primitive tribal groups.[45] These tribal communities have different cultural traditions, histories, and languages. They have strong group identities, have very close relationships within the group, and tend to be suspicious of other groups.

The people in the inland part of Odisha, with strong ties to one another within their own caste/clan group, had lesser interaction with outside groups. The associational life in this region was also based on protecting and/or promoting the interest of their own groups, and did not facilitate the cross community engagements. This we believe had significantly impeded the coordination and limited the popular participation in the protest movement in the region.

Influence of External Factors and Actors

Recently, the movement scenario in inland Odisha has gone through a noticeable change. The tribal population of the inland part has started taking up coordinated and large-scale protest movements against various government policies, including protected areas. Several organized protests have come up against the proposed projects, which can potentially lead to population displacement. Even those who have been displaced by development projects in the 1960s and 1970s have revived their protests in a more coordinated manner. There is a significant change being observed about the enhanced cooperation among the various tribal groups of Odisha in each other's issues.

Widespread resistance by tribal people to projects causing displacement has forced the Odisha government now to plan a comprehensive relief and rehabilitation policy. The opposition of the tribal people to new industries because of displacement in the Kalinga Nagar area has resulted in deaths of 12 protesters in police firing in January 2006. Under the banner of the *Visthapan Virodhi Janmanch* (People's Committee against Displacement), they have organized a series of programs to continue their agitation. A main highway of the state had been blocked for traffic for more than one year since the firing incident. The movement has spread to other parts of inland Odisha. Those who have been already displaced by the various projects have started demanding better and further compensation. Several projects, which are under construction or at the planning stage, are also being targeted to demand minimization of displacement and provision of a comprehensive compensation package. These movements are coordinating among themselves and are drawing strength from the support that is being extended by some civil society organizations and prominent citizens from different parts of the state and outside.

The new change in the process of protest mobilization in the inland region has come thanks to the recent entry of both the ultra-Left and ultra-Right groups to the arena and their impact on the network structure in the society. The spectacular rise of the Maoist movement in the inland part in the last couple of years has managed to break the traditional clan barrier among various groups to a large extent and create one tribal identity. In the past, the people in this region were clearly divided along their own clan identity and the interaction among these subgroups was limited, and if there was any in most cases it was hostile. The impact of the Maoist movement has helped to bring these subgroups together and they now see themselves as tribals in their struggle against the "oppressive" state and outsiders.

Moreover, the conversion and reconversion competition between Christian missionaries and Hindu radical organizations have also facilitated the tribal groups' ability and possibility toward networking with each other as well as with the outside support groups. The tribal population in the inland region, which was keeping itself isolated from any interaction with outsiders before, has thanks to the relentless endeavor of these religious activists got rid of that reluctance to a large extent.

A radical communist group had led a violent uprising in 1967 in the neighboring state of West Bengal. The government crushed the movement and subsequently, during the 1970s, it was fragmented into various factions. Some of these groups remained active in parts of Andhra Pradesh and Bihar and their influence was localized in nature. However, with the economic liberalization and the gradual withdrawal of the state from its commitment to welfare policies, the support for these radical groups got a boost in recent years. The Maoist Communist Centre of India (MCC), which was active in parts of Bihar, and the Communist Party of India (Marxist-Leninist) and the People's War Group or PWG, which had its base in Andhra Pradesh, merged to form a new entity, the Communist Party of India-Maoist (CPI-Maoist) on September 21, 2004. The new entity has reaffirmed its commitment to the classical Maoist strategy of "protracted armed struggle," which defines its objectives not in terms of the seizure of lands, crops, or other immediate goals, but the seizure of state authority. Within this perspective, participation in elections and engagement with the prevailing "bourgeois democracy" are rejected, and all efforts and attention are firmly focused on "revolutionary activities" to undermine the state and seize power. The unity move of the Indian Maoists has provided the ultras a bigger base and larger influence and strength. The rebels are forcefully and violently demanding for the creation of a communist state comprising tribal areas of the states of Andhra Pradesh, Maharashtra, Odisha, Bihar, and Chhattisgarh. Their struggle's demand is to improve the economic and social rights of the poor and of indigenous tribes in the region.

Most part of inland Odisha has presently come under the influence of this radical left, that is, becoming part of the Maoist corridor. Prior to their merger in 2004, the Peoples War Group (PWG) had a token presence in the southern districts of Odisha (Koraput, Malkangiri, Nabarangapur, Rayagada, Gajapati, and Ganjam) whereas the Maoist Communist Centre (MCC) had its influence in the western part (Sundargarh, Mayurbhanj, and Keonjhar). After the formation of the Communist Party of India (Maoist), the radical left movement has spread to the remaining parts of inland Odisha (Sambalpur, Kandhamal, Deogarh, Jharsuguda, Jajpur, and

Angul). The rapid upsurge of Maoists in inland Odisha has been possible due to a continuous process of underdevelopment in the region. Taking advantage of the acute poverty, rampant corruption, and regional disparity in the inland parts of Odisha, Maoists have shown the tribal population the dream of a revolution. This armed rebellion has been instrumental in bringing self-confidence among the tribal population and has organized them into a single group. Thanks to this, the protest movement of any tribal group does not remain localized in nature anymore. In any antigovernment demonstration, there is participation also from groups that are not directly affected by the development or conservation projects.[46]

Besides the impact of Maoist rebellion, the politics of religion has also brought changes to the interpersonal ties and therefore the structure of the tribal society of inland Odisha, by encouraging interaction among various subgroups. For many years now, the Christian missionary agencies have been active in the inland part of Odisha to support and promote the conversion of the tribal population to Christianity. However, in the past few years, particularly after 2000 (when a regional party came to power in the state of Odisha in coalition with the pro-Hindu party BJP), a "re-conversion" campaign has given rise to active opposition of the Christian missionary activities. The state government has brought in a controversial Odisha Freedom of Religion Act, which prohibits "the use of force or inducement or fraudulent means" to convert someone from one religion to another. The Christian missionaries are not only facing the obstacle from the state, but also violent opposition from Hindu extremist organizations. According to Biswamoy Pati, religious communalism and divisive politics have entered Odisha and have attained a high level of aggressiveness.[47] Several Hindu fundamentalist groups have become active in the region. The competition between the two religious groups in recruiting support has also exposed the tribal population of inland Odisha to interaction with outside religious organizations and their mobilizing agents.

The conversion and re-conversion moves have facilitated the tribal groups to interact with other clans and castes as well as the nontribal population and organizations. As it is presently experienced in the inland part of Odisha, the social network structure is not a stagnant one and the intervention of external factors and actors can have a significant contribution to its evolution. Ideology and religion have been able to create larger identities in the society, in the process overriding the strong and exclusive bonding within smaller groups. This process has been facilitated by a large number of political and religious entrepreneurs. Weak and bridging ties developed through ideological and religious mobilizations have led to efficient sharing of information and concern cutting across social and geographical clusters, in the process creating a strong network capital.

The outside actors are not exactly the same as "movement entrepreneurs,"[48] as they do not directly act as agents or brokers to help the process of moving a localized action to a coordinated and broader movement. The mobilization processes of the protests in the inland part have taken advantage of a changing social network structure, which has materialized only due to outside intervention. However, these outside actors are involved in this social churning process not with the express interest of helping the mobilization of these protests but because of their own selfish political and religious agendas.

External actors' contributions to manufacturing of diffused and weak ties among the tribal population have led to a denser social network in helping them to identify with a larger social group structure. As a result of that, local protests against the protected areas, which were earlier limited in size and short in duration, have now been able to link with others and initiate coordinated actions. These broadening contacts have helped to mobilize larger and sustained opposition against various projects in the inland part of Odisha. Several protest movements, like demanding better compensation for the people who have been displaced by Hirakud Dam in the 1960s, which had met with an untimely demise in the past, have now even taken rebirth to put pressure on government agencies to gain back the lost rights and better compensation.

Concluding Remarks

The proper and effective implementation of policies on forest conservation is difficult to achieve, if they are not understood, accepted, and supported by the populations that are directly and indirectly affected by them. The idea of protecting the forest without forest people is located primarily in the developed world and is often seen as an imposition on a skeptical third world. The leadership of many developing countries—due to outside pressure—adopts this policy, but rather than protecting the forest this policy in many cases leads to social conflicts.

Often protected area management fails to meet the needs and demands of forest populations that traditionally utilize resources within or adjacent to designated protected areas. Local demands for resource use in several cases conflict with other goals to conserve resources for visitor enjoyment or biological integrity. In the developing countries, the state agencies have mostly failed to include the local communities actively in forest management, particularly in the management of protected areas. Forest conservation priorities, when taken in isolation from local people and broader patterns of natural resource use, are bound to lead to grievances and anger.

This Indian experience shows that a state may force through the implementation of a protected area policy disregarding the interest of the affected population. However, the changing social environment may provide affected people with possibilities to counter the nonconsultative policies of the state. Affected people and their supporters might be able to organize and mobilize large-scale opposition and highly motivated social actions even after the implementation of the policies. The suppressive and manipulative actions of a state may force reconciliation of the popular protest at the surface level, but it neither addresses the root of the problem nor helps to secure lasting benefits for the project.

The trajectory of protest movements in the inland part of the Indian state of Odisha also shows a novel type of contribution of external agents to the mobilization process. The Left radicals and religious preachers have come into contact recently with the affected population of Odisha for their own self-seeking interests, but inadvertently have helped them to strengthen their movements against protected area policies. These outside actors have entered the region to motivate the tribals to join their political party or religion. In the process, they have been able to trigger the interaction among various subgroups of the tribal population, and that has helped the tribals to create a larger social network and successful mobilization against the protected area policies of the state.

Notes

1. Borrie et al., "Protected Area," 133–54.
2. Ghimire and Pimbert, *Social Change.*
3. Ghimire, "Parks and People," 195–232.
4. Nathan and Kelkar, "Case for Local," 2835–45.
5. Geisler, "Endangered Humans," 80.
6. Cernea and Scott, *Anthropological*; World Bank, *Recent Experiences.*
7. Ibarra, *Social Movement.*
8. Grindle, *Challenging the State.*
9. Wignaraja, "Rethinking Development," 6.
10. Snow, Zurcher, and Ekland-Olson, "Social Networks," 787–801.
11. Diani, *Green Networks.*
12. Paxton, "Is Social Capital," 88–127.

13. Ostrom, *Governing.*
14. Rose, "Social Capital."
15. Neidhardt and Rucht, "The Analysis."
16. McAdam and Paulsen, "Specifying the Relationship," 640.
17. Putnam, *Bowling Alone,* 23.
18. Kossinets and Watts, "Empirical Analysis."
19. Piven and Cloward, *Poor People's.*
20. Buechler, "Beyond Resource"; Tilly, "From Interactions."
21. Kothari, "The Indian."
22. Blomkvist and Swain, "Investigating Democracy"; Swain, *Struggle.*
23. Augustine and Sharma, "Gandhi and the Contemporary Challenges."
24. WWF, *Forests.*
25. Gadgil and Guha, *This Fissured Land.*
26. Gadgil and Guha, "Ecological Conflict."
27. Gadgil and Guha, *This Fissured Land.*
28. Arora, "From State."
29. Bandyopadhyay et al., *India's Environment.*
30. Madsen, "Recent Changes in India's Forest Policy."
31. Mehta, "People's Participation."
32. Arora, "From State."
33. Human and Pattanaik, *Community Forest Management.*
34. Swain, *Struggle.*
35. Behar, *People's Social Movement.*
36. Swain, "Social Capital."
37. Council for Social Development, *India.*
38. Pattanaik, "How Protected."
39. Khatua and Stanley, "Ecological Debt."
40. Human and Pattanaik, *Community Forest Management.*
41. Wild Orissa, *Untamed Orissa.*
42. Fernandes and Raj, *Development.*
43. Granovetter, "The Strength of the Weak Ties."
44. Eckstein, *Power and Popular Protest,* 33.
45. Misra, *Tribal Development.*
46. Swain, *Struggle.*
47. Pati, *Identity, Hegemony, Resistance.*
48. McAdam, Tarrow, and Tilly, *Dynamics of Contention.*

Bibliography

Arora, Dolly. "From State Regulation to People's Participation: Case of Forest Management in India." *Economic and Political Weekly* 29 (1994): 691–98.

Augustine, Clymes, and A. K. Sharma. "Gandhi and the Contemporary Challenges: The Emergence of New Social Movements." *Gandhi Marg* 16 (1995): 437–51.

Bandyopadhyay, J., N. D. Jayal, U. Schoettli, and C. Singh, eds. *India's Environment: Crises and Responses.* Dehra Dun: Natraj Publishers, 1985.

Behar, Amitabh. *People's Social Movement: An Alternative Perspective on Forest Management in India.* London: Overseas Development Institute Draft Working Paper, 2002.

Blomkvist, Hans, and Ashok Swain. "Investigating Democracy and Social Capital in India." *Economic and Political Weekly* 36 (2001): 639–43.

Borrie, W. T., S. F. McCool, and G. H. Stankey. "Protected Area Planning Principles and Strategies." In *Ecotourism: A Guide for Planners and Managers,* edited by K. Lindberg, M. E. Wood and D. Engeldrum, vol. 2, 133–54. North Bennignton: The Ecotourism Society, 1998.

Cernea, M., and G. Scott, eds. *Anthropological Approaches to Resettlement: Policy, Practice and Theory.* Boulder, CO: Westview Press, 1993.

Council for Social Development. *India: Social Development Report.* New Delhi: Oxford, 2006.

Diani, Mario. *Green Networks: A Structural Analysis of the Italian Environmental Movement.* Edinburgh: Edinburgh University Press, 1995.

Eckstein, Susan. *Power and Popular Protest: Latin American Social Movements.* Berkeley: University of California Press, 1989.

Fernandes, Walter, and S. Anthony Raj. *Development, Displacement and Rehabilitation in the Tribal Areas of Orissa.* New Delhi: Indian Social Institute, 1992.

Gadgil, Madhav, and Ramchandra Guha. "Ecological Conflict and the Environmental Movement in India." *Development and Change* 25 (1994): 101–36.

Gadgil, Madhav, and Ramchandra Guha. *This Fissured Land: An Ecological History of India.* Delhi: Oxford University Press, 1993.

Geisler, Charles C. "Endangered Humans." *Foreign Policy* 130 (2002): 80–81.

Ghimire, K. B. "Parks and People: Livelihood Issues in National Parks Management in Thailand and Madagascar." In *Development and Environment: Sustaining People and Nature,* edited by D. Ghai, 195–232. Oxford: Blackwell, 1994.

Ghimire, K. B., and M. P. Pimbert, eds. *Social Change and Conservation.* London: Earthscan, 1997.

Granovetter, M. "The Strenth of the Weak Ties." *American Journal of Sociology* 78 (1972): 1360–80.

Grindle, Merilee S. *Challenging the State: Crisis and Innovation in Latin America and Africa.* Cambridge: Cambridge University Press, 1996.

Human, Joe, and Manoj Pattanaik. *Community Forest Management: A Casebook from India.* London: Oxfam, 2000.

Ibarra, Pedro, ed. *Social Movement and Democracy.* New York: Palgrave, 2003.

Khatua, Sanjay, and Stanley, William. "Ecological Debt: A Case Study from Orissa, India." In *Ecological Debt: The Peoples of the South Are the Creditors: Cases from Ecuador, Mozambique, Brazil and India,* edited by Athena K. Peralt, 125–68. Geneva: World Council of Churches, 2006.

Kossinets, Gueorgi, and Duncan J. Watts. "Empirical Analysis of an Evolving Social Network." *Science* 311 (2006): 88–90.

Kothari, Rajni. "The Indian Enterprise Today." *Daedalus* 118 (1989): 51–68.

Madsen, Stig Toft. "Recent Changes in India's Forest Policy." Paper presented at the fifth annual conference of the International Association for the Study of Common Property, "Reinventing the Commons." Bodoe, Norway, 1995.

McAdam, Doug, and Ronnelle Paulsen. "Specifying the Relationship between Social Ties and Activism." *American Journal of Sociology* 99 (1993): 640–67.

McAdam, Doug, Sidney Tarrow, and Charles Tilly. *Dynamics of Contention.* Cambridge: Cambridge University Press, 2001.

Mehta, Ajay S. "People's Participation." *Seminar* 406 (1993).

Misra, Rabi N., ed. *Tribal Development: Post Globalization.* New Delhi: Discovery Publication, 2006.

Nathan, Dev, and Govind Kelkar. "Case for Local Forest Management: Environmental Services, Internalization of Costs and Markets." *Economic and Political Weekly* 36 (2001): 2835–45.

Neidhardt, F., and D. Rucht. "The Analysis of Social Movements: The State of the Art and Some Perspectives for Further Research." In *Research on Social Movements,* edited by D. Rucht, 421–64. Frankfurt & Boulder: Campus & Westview Press, 1991.

Ostrom, Elinor. *Governing the Commons: The Evolution of Institutions for Collective Action.* New York: Cambridge University Press, 1990.

Pati, Biswamoy. *Identity, Hegemony, Resistance: Conversions in Orissa, 1800–2000.* London: Asia Research Centre Working Paper 7, 2001.

Pattanaik, Manoj. "How Protected Are the 'Protected Areas'?" *Community Forest* 1 (2002): 4–9.

Paxton, Pamela. "Is Social Capital Declining in the United States? A Multiple Indicator Assessment." *American Journal of Sociology* 105 (1999): 88–127.

Piven, Frances Fox, and Richard A. Cloward. *Poor People's Movements: Why They Succeed, How They Fail.* New York: Vintage Books, 1979.

Putnam, Robert D. *Bowling Alone: The Collapse and Revival of American Community.* New York: Simon & Schuster, 2000.

Rose, Richard. "Social Capital: Definition, Measure, Implications." Paper presented at the World Bank Workshop on Social Capital, Washington D.C., April 16–17, 1996.

Snow, David A., Louis A. Zurcher, and Sheldon Ekland-Olson. "Social Networks and Social Movements: A Microstructural Approach to Differential Recruitment." *American Sociological Review* 45 (1980): 787–801.

Swain, Ashok. "Social Capital and Popular Mobilization: Environmental Protests in an Indian State." *Asian Journal of Political Science* 8 (2000): 33–46.

Swain, Ashok. *Struggle against the State: Social Network and Protest Mobilization in India.* Franham: Ashgate, 2010.

Tilly, Charles. "From Interactions to Outcomes in Social Movements." In *How Social Movements Matter?,* edited by Marco Giugni, Doug McAdam and Charles Tilly, 253–70. Minneapolis: University of Minnesota Press, 1999.

Wignaraja, P. "Rethinking Development and Democracy." In *New Social Movements in the South,* edited by P. Wignaraja, 4–35. London: Zed Books, 1993.

Wild Orissa. *Untamed Orissa: A Journey into the Wilds of Orissa.* Bhubaneswar: Wild Orissa, 2000.

World Bank. *Recent Experiences with Involuntary Settlement.* Washington, D.C.: The World Bank, Operation Evaluations Department, 1998.

WWF India. *Forests 2009.* New Delhi: WWF, 2009.

Chapter 9

The Environmental Movement in Sri Lanka

Sarath W. Kotagama

Introduction

The changing global interest in the environment reached Sri Lanka in the late 1960s and early 1970s as a result of the interest in the Human Environment Conference scheduled for 1972. The Environment Protection Society under the presidency of the late V. Ambalavanar was established in 1972 prior to the World Environmental Conference. The historic protest against the logging of the Sinharaja Rainforest in the early 1970s may be considered as the modern turning point in the interests of the environmental issues in the country.

Discussions on the environment in Sri Lanka have come a long way since then. They have penetrated almost every discipline of today, and have also become a major concern for all sectors of the government machinery, though some try hard to ignore their consequences. This infusion into different disciplines and sectors of governance has invariably resulted in the environment being interpreted differently with relevance to each such discipline and sector.

The "use" of nature and the environment for the well-being of the human species may be defined here as the basic element of "development." Such use of the natural resources and the environment must be based on the basic principles of ecology and environmental science. The social phenomena of economics, politics, and social dynamics influence this use. This leads to a complicated entanglement that results in the environment becoming important in the new sustainable development debates.

Among the people there is an evident lack of clarity regarding the understanding of ecology and environmental science—its applications in resource use, protection and conservation, pollution and damage, development and sustainability, and so forth. The objectives of environmentalism emanating from the advocacy to protect, conserve, and promote sustainable use of the "natural environment" have themselves lost direction and purpose as a result of this confusion. It has also led to a state of tension within and among people resulting in the rising importance of the issue of environment.

The understanding of the "environment issue" in Sri Lanka therefore requires initially an understanding of the past trends in environmental management and the current state of the environment that has resulted from development.

Background

Sri Lanka is an island in the Indian Ocean; it is 65,610 square kilometers in area. It is pear-shaped; its central area is hilly, rising up to 3,000 meters. This area is about one-third of the total land area. The rest is rolling terrain with a couple of mountain islands and rock outcrops. Radiating out from the central hills are 103 river systems.

The climate is determined primarily by the monsoons that develop in the northeast and southwest seas.

Sri Lanka's history spans over 2,500 years or more. It has had its heydays and still can be proud of a very important contribution to civilization.

Looking back it could be concluded that 2,000 years ago an estimated population of 18 million (today's population is reaching 20 million) had been sustained in the country at a time when Sri Lanka was at its peak development state. The background to this success is attributed to the hydraulic technology, based on some of the best water management practices promoted today. This technology was clearly based on ecological and environmental principles professed today for attaining sustainable development.

The collapse of this heritage has been attributed to the factor of foreign invasions that led to a series of interconnected events. The invasions resulted in the destruction of the systems. Insecurity and mismanagement sets in along with such invasions that led to the neglect of the systems, which is believed to have increased the prevalence of vector-borne diseases such as malaria. The compounded effect of all of these was the downfall of a glorious civilization. There lies here a lesson from the past on how human security can impact environmental conditions and the survival of civilizations.

The modern "Ceylon" commences with the British capturing the country in 1815. They set about "developing" the country to meet the needs of the British Empire. Concern for the changing environment was first documented in 1873, when Hooker and Gregory commented on the destruction of forests in the hill country by the then ruling British government. Forests in the hill country of Sri Lanka were opened up for tea cultivation in a large way. This voice of concern was responded to by the government by establishing a forest department (1889) and also declaring the forests above 5,000 meters as protected. In 1986, all remaining forests in the country were declared as "forest reserves," "proposed forest reserves," and "other state forests." Thus almost the entire country's natural forest habitat came under protection. So began the forest protection legacy of Sri Lanka.

During the same period, in 1889, Colonel Clark, the then conservator of forests, noted the "appalling slaughter which was being carried out on behalf of traders who actually supplied villagers with powder and shot for the destruction of those animals whose bodies furnished the best hides for export." Arising out of these concerns and the subsequent regulations of the practice of hunting wild animals came a series of legislations to regulate animal hunting. The resulting dialogue gave rise to the establishment of the Game Protection Society in 1896. The Game Protection Society was permitted the management of "intermediate areas" set aside by the government for the purpose of licensed hunting. This was subsequently abolished, but the areas ended up becoming some of the earliest protected areas for wildlife. Thus began "wildlife conservation."

In the 1930s Sri Lanka (then called Ceylon) developed fast through numerous settlements with the restoration of some of the ancient irrigation systems. Following independence in 1948, the rate of settlements increased and more land was cleared for agriculture. These developments were not seen to have been the topics of any environmental concerns then. By the middle of the 1960s, with the global emphasis on socialism, Sri Lanka too moved in this direction and also commenced modernizing, with industries brought into the country.

Sri Lanka has been ruled and governed by democratically elected governments under different constitutions over the years. When independence from the British was granted in 1948, Sri Lanka was already enjoying the benefits of universal franchise and voting was permitted above the age of 18 for both men and women. The Republican Constitution that broke the bonds with the British government system happened in 1972. Through numerous transitions of governments there occurred a major change in 1977, when for the first time a government was elected with a landslide two-thirds majority. The government system changed and the presidency was introduced; the economic picture changed and Sri Lanka endorsed the open economic model. It is also during this time in 1983 that the darkest era of Sri Lanka's darkest history began. The ethnic issues/civil war raised its head and over the years engulfed the entire country. The issue dragged on for 30 long years, and ended in 2009 under the reign of President Mahinda Rajapaksa and his government. Development, however, continued at different levels during this period, with implications for the environment.

This chapter discusses the issues relevant to the historic changes that have taken place in the country since the 1960s. The environmental movement in Sri Lanka has evolved very positively through the years. Space constraints make it difficult to discuss all the events that have taken place in the last 40 years in this chapter. However, the most significant events have been listed and some have been elaborated on to drive home the point of the role of environmental NGOs.

State of the Environment in Sri Lanka

Sri Lanka's environmental concerns have been extensively documented over the years in numerous publications. Standing out among them are the *National Conservation Strategy of Sri Lanka* (1988), followed by a series of *National Environmental Action Plans* (1992–1996 and 1998–2001), followed by *Caring for the Environment* (2003–2007), followed by a review of the process by the Institute of Policy Studies (February 2007). This review was the basis for the *Sri Lanka Strategy for Sustainable Development* (February 2007). The country's vision for sustainable development—*Achieving sustained economic growth that is socially equitable and ecologically sound, with peace and stability*—was spelt out in this document. The document thus addressed the issues of poverty eradication, ensuring the competitiveness of the economy, improving social development, ensuring good governance, and ensuring a clean and healthy environment. Following this the National Council for Sustainable Development was established under the chairmanship of President Mahinda Rajapakse in 2009. The minister of environment Champika Ranawaka, taking the lead, developed the National Action Plan for Haritha Lanka Programme (January 2009). This document according to its objective "will lay down the foundation necessary for sustainable development," spanning the period 2009–2016.

The document recognizes 10 "missions" and provides strategies to achieve these missions. The content could in a broad sense be considered as the present issues of the environment. The 10 missions are clean air—everywhere; saving the fauna, flora, and ecosystems; meeting the challenges of climate change; wise use of the coastal belt and the sea responsible use of the land resources; doing away with the dumps; water for all and always; green cities for health and prosperity; greening the industries; and knowledge for right choice.

Parallel to this, since 2001, the environmental situation has been monitored and documented through the publications *State of the Environment* (2001, 2002) and *Sri Lanka Environment Outlook* (2009). Broadly, the summary issues contained in these documents reflect the current state of the environment in the country.

Land Degradation

Improper land use resulting in soil erosion causes land degradation. This along with high application of fertilizer has accelerated the degradation process. The soil condition in the mountain wet zone is considered as quite acute. The production potential of the soil in the agricultural lands of the rest of the country is generally degraded due to the extensive use of agrochemicals. These have in most parts destroyed the soil micro- and macro-organism base that is essential for healthy soil generation and has affected productivity.

Water Pollution

Water has been the major resource base around which our ancient civilization thrived. The technology developed around the use of water for irrigation is unparalleled in the world—the hydraulic civilization of the ancient kings is still the backbone of the present agriculture. At present more than 650,000 hectares of paddy lands is irrigated by diverting and/or augmenting water in basins. To this are added the new needs to obtain hydroelectricity, and 45 percent of the electricity needs are met by water. The groundwater is used heavily by households, factories, and agriculture.

The heavy extraction, use of chemicals in agriculture, sediment loads, and industrial and household waste have in most instances ended up in the waterways and thus they are highly polluted. There exist in the country more than 30,000 water bodies—primarily human-made. Many are yet to be restored. The ones that have been restored are overgrown with aquatic plants, a sign of eutrophication. In certain urban areas they are totally polluted, for example, Beria Lake in Colombo, Lunawa Lake in Moratuwa, Kandy Lake, and Lake Gregory in Nuwara Eliya to name a few.

The groundwater has also been shown to be heavily contaminated with agrochemicals and its use restricted. These have led to problems of drinking water. A thriving bottled water industry has emerged in the recent past resulting from the lack of quality drinking water on tap and in the traditional wells.

Sri Lanka has about 300 high-polluting industries that tap the water sources for their industrial needs and also put out waste into water systems. Many of these still need to develop good industrial water-use practices to ensure that the water is used economically and without polluting it.

Air Pollution

With increased urbanization and industrialization the concern for air pollution has also increased. The main areas of concern remain in and around Colombo and a few other urban centers. A contributory factor to air pollution is vehicle emissions. The vehicle fleet has been growing at 7 percent annually. Vehicle emission testing has been introduced with good results. However, corrupt practices by testing officials have been reported to be an obstacle in some locations. Even though the conditions are not so bad, some places within the city of Colombo exceed World Health Organization (WHO) standards with respect to particulate matter.

Solid and Hazardous Waste

With the growing middle class the increase in solid waste has become a problem that requires urgent attention. It is estimated that Colombo produces around 1,000 tons of solid waste per day. The western province with the highest human density has around 58 waste dumps. All of these are classed as open dumps with absolutely little or no sanitary considerations. At the dumps burning, leaching, and purification lead to many health hazards that are presently taken for granted. With the increasing demand for land and the accompanying increase in waste production the country is faced with a real waste problem. Many suggestions for waste management have yet to take effect.

Loss of Biological Diversity

The prehistoric, historic, and recent events have had a profound effect on the flora and fauna of the country. Amid these large-scale changes the natural forest cover has been reduced to around 23 percent. Many of the other natural systems have also degraded. Sri Lanka has been recognized as one of 35 biodiversity hotspots, which indicates the threat to and the value of the biodiversity. Coupled with the biodiversity we are proud to inherit a rich culture that gives greater value to the use of biodiversity.

Despite current concerns 13 percent of the land area is under high protection as protected areas declared by the Fauna and Flora Protection Ordinance (1937 and 1994). A further 10 percent of the forests come under forest protection, declared under the Forest Conservation Act. Among the major issues is the human-elephant conflict resulting from elephants causing damage to crops and property and occasional loss of human life. Elephants are also killed or injured in the ensuing conflict. Concerted action through various programs to ensure the conservation of biodiversity is yet to receive full support from the authorities.

Unsustainable Use of Natural Resources: Forestry, Fisheries, and Minerals

The current status of all these resources is that they seem not to be sustainably harvested.

The forestry sector has with the Second Forestry Sector Master Plan taken a positive step of management with the involvement of the private and public sector. There is a greater emphasis on participatory management with the people and it has opened new vistas for conservation. These new approaches to become effective are a strong challenge for a 100-year forestry history of "command and control" to be changed to a participatory equal-status situation with users and other stakeholders. Conflicts on boundaries and resource allocations are still present.

The situation of minerals is even more precarious. With no real control over much of the resource exploitation, the current minerals are exploited with little concern for the impacts on the environment. Approved legislation is still to become effective in most spheres of mining. The biggest challenge for the country was with the proposed Eppawala phosphate mining contract. Effective legislative action and public outcry enabled the project to be shelved. The rapid development has resulted in greater demand for soils and granite rubble. The sand mined from riverbeds had led to numerous issues for the environment and remains a major problem with few positive controls.

The fishery resource has seen a continuous downward trend in availability, indicating a nonsustainable harvest strategy. The "near-shore" fishery has been further affected by land pollution, forcing fishermen to go out to the "off-shore" resources. These too seem to be on the pathway of depletion. An emphasis on the "deep sea" fishery has been the answer to all these unsustainable and environmentally unsound activities. This industry has been capital-intensive and is driving the average fish prices higher and thus affecting consumption patterns. One may add here the continuous extraction of ornamental freshwater and marine fish that has been of real concern to many ecologists. Without doubt the resource has been affected severely, especially in the freshwater sector. Even with an average mortality of 35 percent per shipment, the practice continues. The loss of many freshwater sources can be directly attributed to the exploitation, aggravated by the high levels of leaching of agrochemicals into the water sources.

Environmental Health

Sri Lanka has, like most countries, paid more attention to curative interventions and less to preventive interventions. This is considered as the major issue even though our life expectancy has reached an average of 71 years. The occupational health situation in relation to agriculture, industries, and other occupations remains in a serious state. Pesticide poisoning among the youth remains at a very significant level in the country. Similarly the health conditions arising out of water and sanitary conditions is acute in the urban low-income-level dwellers. To this must be added the continuing impact of vector borne diseases.

Socio-environmental Changes

Physical-chemical impacts on the environment caused by the changing lifestyle, development process, and the changes in the sociocultural environment must also be taken into account. Some of the changes are of concern to the well-being of the society. Without debate one has to accept that the society has become more violent, intolerant, abusive, and disrespectful. Whether these are direct results of development or not is questionable, yet the facts of change are evident everywhere. It appears that the social security and respect for children and females that arose out of deep social ethical values have changed.

Many environmental security issues have also arisen over the years. The nature of these security concerns of the present and future need to be understood in line with the trends of development and the needs of the people.

Future Trends, Developments, and Requirements

Sri Lanka has had an average economic growth rate of around 3 to 7 percent per annum during the last 10 to 15 years, and it is predicted that this will continue to remain around this figure in the coming years. With the settlement of the 30-year conflict one could say that the growth rate could increase further. Add to this a population growth of around 1.2 to 1.3 percent of the current population standing around 19 million. A further complication to this demographic picture is that Sri Lanka is gradually becoming an aged population.

The population settlement areas continue to be increasing in the wet zone in the western region. The social structure is changing gradually, as fewer people are taking to the soil as farmers, while the younger generation is moving to employment in the trade and industrial sector. This would inevitably affect the nature of agriculture and its productivity. The impressive economic growth has led to a surge of massive infrastructure development programs. At present Sri Lanka is classified as a "middle income generating country" and placed in the "newly industrial country" category too. All this means a greater concern for the environment is needed if we are to ensure sustainable development as envisaged by the Haritha Lanka project.

Presently the population remains unevenly distributed with over 75 percent still in the rural areas, while the urban population is expected to rise to 40 percent of the total population in 2020. The per capita consumption of natural resources is expected to rise with increasing income levels, resulting in energy demand expected to continue to increase at around 10 percent per annum, with coal contributing largely to its production. Consequently the Urban Development Authority recognizes that environmental degradation will be the major problem of the urban sector in the future, with vehicle traffic growing at 7 percent per annum, and solid waste from households increasing and changing into more nonbiodegradable material.

The per capita income is to reach $4,000 in 2016, but will not be evenly distributed, thus poverty will continue to be a prevalent factor, though clear signs of decrease have been noted.

Industrial growth will be a major driving force for employment and production through private industry and investment. Rapid and large-scale development projects have taken off in the country both in the north and in the south.

It will be quite evident that these aspects are hotbeds for evolving issues within the country. A very significant contributory factor to the environmental issue will be the aspect of devolution. The existing 13th Amendment to the Constitution has itself led to some conflicts. These have not been very significant as the devolved authorities are yet to assert their full rights under the constitutional changes.

The Environmental Movement in Sri Lanka

As indicated in the introduction and background, the environmental movement within Sri Lanka can be broadly categorized under two areas. The historical concerns for forest and wildlife had been the only issues in the latter part of the 19th century and the early part of the 20th century. The redistribution of people and the surge toward industrialization and to "develop" in the modern sense indicate the current thinking of environmental concerns and action. This commences from around the mid-1960s.

The following environmental issues discussed since the mid-1960s indicate the major trends of concern and involvement that both the government and nongovernmental organizations (NGOs) have had with respect to the environmental issues arising out of "development." They have been carefully selected from among many to illustrate a trend. Some have overlapped in time but the issues concerned are significant to indicate the subtle directions of the development and environmental concerns.

Establishment of the Sinharaja World Heritage Reserves (1965–1989)

The mid-1960s saw the emergence of plywood production as a means to overcome wood waste through timber utilization in the conventional saw-milling process. Sri Lanka followed suit and decided to establish a plywood factory. The timber resource for this had to come from the only unlogged largest wet-zone low-mid-country forest patch that was declared as a forest reserve in 1886. This forest reserve of around 22,000 hectares today was targeted for logging. The government departments responsible for promoting the project were the forest department and Plywood Corporation. The forest department had to release the forest for logging. This was challenged by the environmental groups on the basis that Sinharaja was the last pristine patch of low country wet zone forest and needed to be conserved as a sample ecosystem. Further its hydrological importance and its significance in terms of community dependence and other aspects were highlighted. The most important factor was its value in terms of the fauna and flora. These were the days when "biodiversity" was not in the vocabulary, thus there was no collective word for the value.

Spearheading the campaign against logging was the WildLife and Nature Protection Society (WLNPS). Its president Mr Thilo Hoffman and members joined hands with other organizations that were present at that time.

Environmental organizations, academics, students, and the public joined the protest. Despite the protests, logging commenced in 1972. However, this was done with the setting aside of around 3,000 square hectares as a biosphere reserve. Further, a small committee chaired by Member of Parliament George Rajapakse along with some scientists was appointed to inquire into the protest and recommend future strategy. The report came out in 1974. It had strange recommendations that left the decision for the future "hanging in the air." They said that there was inadequate scientific evidence for or against logging. Thus nothing happened till 1977. In response the MAB reserve area was increased and elevated to an international level.

In 1977, the prevailing government changed and a new government came into power. The new government was immediately approached and thus in 1977 logging was suspended and in 1978 completely halted. Sinharaja was saved from the logging process by a political act.

This act was welcome but since political decisions can change with the tide, the need to ensure long-term protection had to be based on sound science. Thus research efforts commenced at Sinharaja. The National Science Council of Sri Lanka, University of Peradeniya, and March for Conservation from University of Colombo researched into the flora and fauna. The results enabled the designation of Sinharaja as the nation's first natural World Heritage Site in October 1989. It also resulted in the enactment of a new conservation law, The Wilderness Heritage Act, for Sri Lanka. This completed a long tussle with government organizations, NGOs, academics and the public to conserve the last largest pristine forest patch in the wet zone of Sri Lanka.

Samanalawewa Reservoir (1968–1998)

Still remaining as one of the major disasters of modern reservoir construction is the Samanalawewa Reservoir. The initial feasibility and construction was started without any environmental impact assessment (EIA) process. Despite protests from environmental groups and other concerned citizens, with the project having gone through almost all the donor agencies to Sri Lanka, it ended up with Japanese government assistance for construction. The Japanese government assistance required an EIA and this was hurriedly conducted within a short period of around three to five months. The reservoir construction was ongoing when the process was taking place. Today it stands as an iconic reminder to all about not heeding to environmental concerns raised with respect to its geological characteristics. A major leak persists despite numerous technological applications with high costs. Is it incapable of fully realizing its planned potential?

The Accelerated Mahaweli Project (1980–1990)

Way back in the 1950s, the government planned to harness the waters of the largest river in Sri Lanka. An ambitious plan to divert the water toward the north central and the north was devised. The project included the development of many new reservoirs and townships. Part of the project commenced in 1976 when the river was diverted to provide power and irrigation water to the southern portion of the north central province. Resettlement of persons was undertaken but no questions were asked about the environmental impacts. The next development phase began in 1977 when the government decided to accelerate the program and achieve what was planned to be implemented in 30 years within six years. This was a major challenge and for the first time in the country's developmental history an EIA was done. Its recommendations included the establishment of an environmental action committee that met regularly and monitored the progress and mitigations. The program involved the resettlement of many persons from the hill country in the planned agricultural areas. As a member of the action committee, in retrospect, there were many environmental security issues that arose in the process. At that time these problems were not seen as environmental security issues. The resettling of persons in cleared areas with assistance to overcome the initial stages was accompanied by uncertainty and trauma. Today these could be interpreted as environmental insecurity. It was not so then. The shock of the harsh new habitats that were completely different from their previous areas had mixed blessings. Those who were able to overcome these conditions of insecurity and alienation have become successful in their new habitats while the others have failed.

One environmental situation that remains to this day is the impact on the resettlement of the traditional "Veddah people" (presently referred to as Wannialoatto). Taking over their settlement areas for a protected area, they had to be resettled and brought in to the "mainstream." This was resented and opposed by a sector of the Veddah people including their leader. Everybody except seven families closely related to the chieftain left for the new agricultural settlement. Their life has changed but at a very high cost. It involves the loss of traditions that were very environment-friendly. The seven families have fought their way to regain some of their lost glory today. But this was through sheer dedication and grit on the part of the chieftain and his close family members. This classic situation could be interpreted as a good example of environmental insecurity created by the so-called development.

Muthurajawela Ecological Profile and Development Plan (1982–1986)

In 1982 the government decided to develop the marsh area north of the Colombo city referred to as Muthurajawela. The environmental NGOs protested at the very outset. In response the government took a very constructive step. For the first time development planning proceeded with an extensive ecological profile preparation. The establishment of a task force under the aegis of the Greater Colombo Economic Commission, consisting of members from all the conservation departments and scientists for diverse disciplines, initiated a series of research studies and data-gathering on the marsh. The result was over 30 scientific reports and technical reports on the ecology of the marsh. Based on these documents a development plan with the declaration of a large sanctuary area and a small development area was recognized. The rational use of resources based on a scientific socio-ecological profile set the stage for a positive future in harmonizing environmental concerns and development. The program also for the first time established a community-based management establishment, which is functional to this day with respect to the fishery use of the lagoon system. Recreational use with the establishment of a wetland facility named the "marsh" functioned for some years very effectively. The drop in tourism affected its functioning along with the Katunayake highway tracing, forcing the facility to be relocated.

There have been many encroachments and development incursions since the last 15 years. Some of these have been diffused but some have affected the marsh severely. The area is under immense pressure due to the growing Colombo metropolitan population.

The biggest damage to the area has come from the ill-planned highway to the airport from the city of Colombo. Skirting the eastern boundary of the wetland with a landfill-raised highway has resulted in many cities to the west of the road going under water after the slightest rain. Bigger culverts and drainage points have been proposed but the damage has now become an annual event with a high cost. Further evidence of sedimentation and severe land-use changes have been reported by the use of remote-sensing satellite pictures. The large land-based changes around the marsh are causing the system to fill with sediments, which is not a visible threat but a major one. Thus the NGO effort, though it resulted in a very positive and novel approach to development, has presently been diffused. The sanctuary coming under the department of wildlife conservation has enabled the marsh to be protected to some degree.

Forestry Sector Master Plan and Its Environmental Concerns (1985–1990)

The early 1980s saw another important environmental issue gripping the country. This was the proposed Forestry Sector Master Plan (FSMP). As soon as the contents came to light the conservationists and environmentalists tore it apart. Prepared by the state agency, the ministry of lands and land development, along with forest department personnel and with technical assistance from the consulting company Jaakko Poyry International under a FINNIDA grant, it was thus at the receiving end of the criticisms. The notable feature in the protest was the involvement of the academic population. They brought to light scientific aspects that were very strongly against the reports. Along with NGOs the direction of the project was changed. When the implementation came about in 1990, the National Conservation Review was made mandatory for the forests over 200 hectares. Thus, a new chapter in forest conservation commenced in Sri Lanka. The biodiversity interest building up in the international arena further influenced the future directions of both the forest and wildlife sectors during these years.

Eppawala Phosphate Mining Issue (1987–1999)

The Eppawala mineral deposit is primarily one of rock phosphate of very high quality. In addition it is believed that the deposit contains rare earths. The extent and quantity of rock phosphate is yet to be properly ascertained. But on crude estimates it covers an area of about 55 square kilometers. This area has many traditional villages and archeological sites. The main opposition to the project is the loss of traditional homelands to a private company, which is foreign, and the immense impact mining will have on the environment during and after mining. In addition the economics of the project were also considered very suspicious and unattractive to Sri Lanka.

The agitation resulted in big protests and threats of bloodshed. The residents had threatened "sit-in-to-death," violence, and trouble. These I believe are issues of security. The people's lives were now insecure. In 1999 in a historic Fundamentals Rights case the court gave judgment in favor of the people and the exploitation proposal by the foreign company was abandoned. The petition in the Supreme Court was heard and the people won the battle. According to the judgment, the right of the land is in the hands of the people and the state had no right to bulldoze development without the consent of the people of the area. The judgment was unprecedented in Sri Lanka before. If similar judgments had been made in the past, many projects would have been stalled.

The present situation is that geological explorations continue and state extraction continues at a very low scale by a state-sponsored company. The people have secured their rights and defused a likely environmental and social catastrophe.

The Coal Power Plant Program (1989–2005)

Since 1989 Sri Lanka has unsuccessfully tried to promote the construction of a coal-powered electricity-generating unit. The first attempt at Trincomale failed to meet the technical criteria of a good EIA and thus it was rejected. A second attempt to put the same in the south was unsuccessful for political reasons. The people of the south very clearly denounced the setting up of a coal-powered thermal power plant; they sent a clear message of antagonism to the ruling party. The ruling party defused the situation by promising the people that the project will not go forward. Subsequently a third attempt was made by planning a plant along the west coast at a place called Norochcholai, close to Kalpitiya. The project was opposed by environmental NGOs with the help of religious organizations. The protests resulted in the loss of one life—when police opened fire against protesters. *This was the first life sacrifice in any environmental protest campaign in the history of Sri Lanka.* Despite these protests the developers went ahead and an EIA was done and received approval. The communities and environmental NGOs continued to oppose the project. The religious leadership was able to get the president of the country to say that the project will not proceed if the people disapprove of the project. This said, the minister for power was always promoting the project. The situation remained volatile with the project virtually suspended; state officials were not welcome to the Norchcholai area. This is a good example of environmental security breaches.

Against this background the present government that was first elected in 2004 recommenced negotiation of the power plant establishment. Having had the Wayamba Provincial Council approve the EIA, funds were raised through a Chinese consortium. The major "peoples concerns" were settled through intense negotiations primarily with the religious institutions.

Construction commenced and the first phase with a 300 megawatt capacity was commissioned in 2010. The background concerns raised by the people were taken into account, adequate compensation and enhanced development incentives seem to have worked.

The Kandalama Hotel Construction (1991–1995)

Destroying a 2,500-year cultural tradition (i.e., of protecting the immediate catchment of a reservoir from any construction), the Kandalama Hotel was constructed in the immediate catchments of a reservoir. The project was by a private company. An EIA that was highly criticized was done. But it had not taken into account the cultural dimension. The project was also mooted with a lot of political power. It also had violated the existing irrigation ordinance. Against this backdrop there was a massive protest organized by environmentalists, religious organizations, and concerned citizens. The protests were confronted by state power that was totally backing the project. The situation was such that even antireligious feelings were raised to squash the opposing camps. During the project development phase the area was totally closed to the public. Amid this uneasy situation the hotel was constructed. The situation that prevailed during the period of the campaign was definitely a national security concern. It could have resulted in inter-religion group confrontation as some Christian clergy and Buddhist monks came out openly against the project while there were those in the same faiths promoting the project. Animosity and bad taste were prevalent everywhere.

The project was completed, and a monstrous concrete structure built at immense cost stands as the memory of the protest. The hotel is spending a lot of money to portray itself as an "eco-resort"—which it never can be. At the same time, the hotel has become one of the best environmentally well-managed hotels in the country.

Expressway Development and the Environment (1970–2011)

Road construction opens the world to development opportunities. But when such construction is done with no heed to sociological and environmental concerns, the cost escalates and it becomes a white elephant. Sri Lanka launched into expressway planning around the 1970s. At that time very few environmental concerns were raised about development. The project proposed to connect the central city of Colombo with that of the south through an expressway running through the country. For economic reasons the project did not commence till around 1992. The Road Development Authority commenced to fill paddy fields for the purpose without any EIA for the development process. On challenging this act the process was stalled till 1994 when the Road Development Authority recommenced the activities. Again due to environmental pressures the activities were abandoned. The program finally recommenced with assistance from the Asian Development Bank and the completion of an EIA.

Despite numerous shortcomings in the EIA, the project went ahead with major hiccups to partially complete the task in 2012. Cost escalation and post-development lessons have resulted in major changes in the thinking of any future expressway development. The environmental concerns have become a major factor in the new decision making. The role of the environmental NGOs was significant in the southern expressway that resulted in this attitudinal change by the developers. The issue of expressways has not finished, but the attitudes have changed and more environment-friendly approaches are being considered in the plans for the future.

Diversion of Menik Ganga to Augment Lunugamwehera Reservoir (1998–2010)

Sri Lanka is abundant with built reservoirs of various sizes. More than 90 percent of these are ancient structures renovated in the present times. Few have been built recently. Among them the largest was built across the Kirindi Oya in the southern area of the country—referred to as the Lunugamwehera Reservoir. Traditionally *oyas* are considered to be flowing rivers that dry up or have little water during most part of the year. Across these *oyas* the traditional practice was to build *amunas* or weirs to divert excess water to the fields. Despite the traditional knowledge on such systems the large reservoir constructed resulted in not having adequate water and thus had to be augmented. The decision was to divert water from the Uma Oya that flows to the west of the country. The cost was considered exorbitant and thus the idea of diverting Kelani Ganga was mooted. But the environmental movement opposed the project, which resulted in it being shelved due to its impact on the environment. Many years later around 1992 the idea was resurrected with a new dimension to first stock the rain water in a reservoir and then permit it to flow across the Lunugamwehera National Park to the reservoir. The need for water to enhance the Lunugamwehera National Park was also a major factor that enabled this modified version of the diversion to be environmentally acceptable. The construction of a new reservoir enabled the enhancement of the protected area system and provides water and improves the PA system in the process. Thus an interesting use of an EIA process to enhance the environmental protective activities and also meet the needs of agriculture was struck. The project was titled "The Diversion of the Menik Ganga to Enhance the Quality of the Lunugamwehera National Park." This enabled the irrigation engineering attitude to be changed to make use of natural landscape drainages for transporting water across land, thus benefitting the wildlife as well.

This recognizes the advantage of EIAs and the new environment-friendly development thinking that prevailed among the new technical and administrative personnel. It also reflects the advanced knowledge of environment-developmental relationships.

Hambantota Harbour (2002–2011)

The final example that I would like to document is the first inland harbor built in the country. Hambantota lies at the southern end of the country. Its strategic position was recognized many years ago with a large number of ships travelling between East Asia and Europe via the Middle East. Initial plans were to build the harbor in the usual manner by the construction of a breakwater out into the sea. Due to prevailing currents and other factors this was abandoned and the plans for an inland harbor commenced. The project involved an EIA, and with investment from China the work was approaching completion in 2011. From the very beginning there was much controversy regarding the construction. The plan included the digging up of a trough and sealing it with concrete in a wetland sanctuary. By making use of the "no-net-loss" principle, this reservation was overcome. Yet, the concerns of salt water being brought inland were going to remain. The construction was completed and the harbor is open to activities. The first few months were also clouded with an unseen event of nature that resulted in the surfacing of a rock at the entrance due to sand movement. This now has been blasted to clear the pathway to the harbor. It resulted in a variation to the cost that was not foreseen during the EIA. This unique experiment would be an ultimate test of project-based EIAs, and if properly monitored, a lesson to be learnt.

Conclusion and Institutional Changes

The selected examples given in this chapter have shown that the response of people to environmental issues have changed over the years. After 40 years in the environmental movement, I can clearly see that environmental issues are definitely more important today and they are on the verge of being more major security-oriented issues than in the past. Protests in the past have been mostly on technical matters through conventional appeals and protests. Today they have become high social mobilizers that end with violence or threats of violence. Some have even extended to self-inflicted violent expressions.

Many more projects and programs have been implemented with and without much concern for the environment over the years. On the whole one could very clearly state that the incidence of concerns over the environmental impacts when development is to be undertaken has risen. This is a positive sign from all sides. *But what is still lacking is the true commitment to monitor the projects to learn of consequences.* Nobody seems to feel that it is the most important aspect of any EIA-based project implementation to avoid any impact in future activities, which will inevitably be needed with increasing populations and demands on resources. The many donor- or loan-driven project supporters are insisting today for the mandatory requirement of EIAs but not the monitoring. We have still a long way to go!

Since the concern for the environment in its present form began around the early 1960s, states and NGOs have grown accordingly. With no state agency to represent the concerns of environment in the 1960s and 1970s, today Sri Lanka has a ministry for the environment and a regulatory body, the Central Environmental Authority (CEA). The Constitution of Sri Lanka, and the judiciary and legal system too have recognized the need for environmental protection and conservation beyond the conventional wildlife/forest agendas. The growth of environmental NGOs in the country very broadly has followed the international change in the realization of the significance of the environment. Commencing in the early 1970s in preparation for the Stockholm Conference of 1972, the Environment Protection Society was established with the participation of the conventional wildlife/forest NGOs. This thus recognized the different aspects and the new dimensions of the environmental concerns. Since then, the number of environmental organizations has increased.

Their concerns have been as diverse as the environment and environmental concerns themselves. Many of these NGOs could easily be considered as "adversarial" or "antidevelopment." But that was the time of showing the negatives of development to a development-biased process that did not have a mechanism to either recognize and or even prevent such negative concerns when demonstrated.

As the significance of the environment built up, and funding became more available, the state started looking for NGO partners, and the development NGOs were able to capitalize on their available infrastructure and goodwill to join and benefit from the flowing funds. The vociferous environmental NGOs found themselves even more marginalized and sidelined. It took some bold initiatives by a few strong-headed NGOs and persons to keep the momentum up and not succumb to the easy path forward. Establishment of legal systems enabled the environmental NGOs to take the state and development agencies to court and obtain some of the very important positive judgments. This happened with the establishment of the Environmental Foundation Ltd in 1988. It also enabled the public to become more aware of the powers they possessed under the systems to force environment-friendly development. Despite many positive trends and victories there have also been negatives.

In 1980 the March for Conservation (MfC) was established from within the University of Colombo with the objectives of enhancing the knowledge-base of environment conservation among the public and researching into the impact of deforestation resulting from logging at Sinharaja. MfC physically marched every year for 15 years in a row. One such march was from Colombo to Kandy. The awareness built up through these gimmicks and numerous schools and public activities were clearly evident in the responses of the public to projects mooted by the governments. The programs even extended to government officers. MfC's research in Sinharaja was instrumental in providing the faunal information for the successful nomination of the forest as a World Heritage Site. The research also brought out the largest database on mixed-species feeding flocks in the world and continues to do so to date. Its role in involving the community in accepting the world heritage status for Sinharaja was one of the first rural education workshops held in the area. Its success has enabled the tour guides to be part of the rural youth employment program since 1989. In 2000 MfC wound up its role and dissolved the institution having successfully achieved its objective.

During the mid-1980s, through a conference held to discuss environmental media roles, the Sri Lanka Environmental Journalist Forum was born. It functioned very effectively to professionalize environmental journalism, and facilitate regional activities too. A parallel outcome of this Panos-sponsored workshop was also the Sri Lanka Environmental Congress. It brought together every environmental NGO in the country except one. The successful functioning strengthened the NGO capacities by providing education and working knowledge of organizations. Because they were mutually assisting one another, the 1990s could be considered the best time for environmental NGOs in Sri Lanka. The information dissemination role was taken over by another organization named Mihikatha, later renamed Mihikatha Foundation. The first local language (Sinhala/Tamil) environmental newspaper was published and distributed by Mihikatha. The environmental voice was strongest during these years (1988–2000).

In a strange twist of events, with civil/ethnic conflict and politics, the environmental NGO movement became the victim of circumstances and the united front fell apart. Each of the major NGOs shrunk in its visibility and is far from being effective today. This was partly a result of the situation of conflict that prevailed in the country that prevented any group from gathering and protesting. The risk was too much, and NGOs, recognizing the constraints, reduced their visibility. The NGOs have not recovered since. Numerous new NGOS have emerged during the last couple of years. But sadly their role has been nothing but the endorsement of state activities. The reporting has also become more personal, lacking in focus and often engaging in vicious mudslinging. This has weakened the movement considerably. This appears to be a common situation in journalism across the field today. The end of the conflict also resulted in a burst of developmental activities large both in magnitude and in impact. With no proper opposition to muster the support of the people and fight back against the potential damage, conditions have turned out a bit negative.

The battle for balanced development has always been a challenge. The changing donor perceptions have however helped in making the environment a major concern and one that must be taken into account at all costs at the beginning. Despite the weak NGOs this factor has to some extent balance the situation. Today no development assistance is granted without environmental clearance. The problem however is that those environmental reports/assessments are far from good in quality and content. Export-oriented large development projects are found to be environmentally more responsible as a result of the global environmental concerns. The construction and industrial developments have also become "green." Thus despite this bleak present situation of environmental NGOs, Sri Lanka can still boast of an environmentally literate society and look positively to the future as the situation in the country slowly becomes normal.

Chapter 10

An Overview of the Green Movement in China

Da Mao

Introduction

China, as a historical, geographical, political, and cultural entity, has its uniqueness in the world. Such uniqueness is also reflected in the contradictory relationship between Chinese people and their natural environment. For example, China is rich in land and natural resources, but the continuous population and economic growth has degraded its environment to a vulnerable status; Chinese intellectuals specialize in portraying nature romantically, but the mainstream social attitude toward nonhuman beings actually has typically been materialistic and pragmatic; the tradition of centralizing political and economic power seems to be a gospel to solve environmental problems due to market failure and individual selfishness, but a lack of citizen spirit and freedom leaves more room for the state to make bigger mistakes.

Despite its particularity, China, like many other parts of the world, faces the common challenge of balancing development with environmental sustainability. Although such challenges have existed for its long history, it was not so acute until the early 1970s, when Chinese leaders first recognized the seriousness of the country's environmental problems and the wave to protect the environment across the Western world. During that time, a top-down environmental governance system was preliminarily established. And since then, no matter how much China changed politically, economically, and culturally, the ignited green fire, though faint at the beginning, has become brighter and brighter.

This chapter aims to introduce the brief history since the early 1970s on how the major players in the environmental protection field in China evolved, and how environmental problems have been addressed by Chinese environmentalists, particularly in the 1990s and 2000s. The chapter is divided into two parts. The first part focuses on displaying the key players in the environmental movement or green movement of China, which includes national leaders, government bureaucrats, conservation experts, elite intellectuals, grassroots activists, and environmental nongovernmental organizations (ENGOs). It demonstrates that the different driving forces of the environmental movement emerged in different historical periods, and civil society groups gradually played a more active role than government officials and individual activists. The second part uses case studies to illustrate how environmental problems are tackled by Chinese environmentalists. These cases cover topics like wildlife conservation, antidam campaigns, actions calling for Apple to monitor its supply chain, and a public debate on "reverence for nature." They collectively reveal that political atmosphere is an important factor influencing the topic selection and direction of the Chinese green movement; Chinese ENGOs have become more capable and mature in launching environmental campaigns, and have more substantial interaction with governments, industries, and international partners; environmental ethics practically grows from both "battles" on physical environmental problems and conceptual debate on the human-nature relationship.

The Driving Forces of the Green Movement

Environmental Protection Becomes Government Responsibility

In early December, 1970, when a delegation of the Japanese Socialist Party was visiting Beijing, Zhou Enlai, the first premier of the State Council of China, invited one delegate (also a television reporter) to introduce the lessons of environmental hazards in Japan to some ministry-level officials of the Chinese government. After the lecture, Premier Zhou instructed the attendants to discuss what they had learnt and ordered that the result of the discussion should be circulated to other officials who would attend the meeting on national development planning.[1] Given that the country was still mired down in the mud of "the great cultural revolution" and most of its government officers refused to admit that socialist countries had environmental problems, the personal farsightedness of a top leader gave China a chance not to lag too far behind the outside world on environmental protection.

A seemingly coincident pollution event happened right after the collective learning of Japanese pollution lessons. In March 1971, the public health officials reported that many Beijing citizens were poisoned after eating some bad-smelling fish. Premier Zhou immediately ordered the relevant departments to conduct an investigation. Later, the investigators reported that the problematic fish came from Guanting Reservoir, one of the most important drinking-water sources for Beijing City, and it had been highly contaminated by DDT and other toxic chemicals. The follow-up campaign to clean up the reservoir and some other seriously polluted water bodies, such as Dalian Bay, Jiaozhou Bay, and coastal water along Shanghai City and Guangdong Province, testified that China was not an exception in facing environmental challenges while pursuing high-speed industrialization and modernization.[2]

Water pollution was not the only negative result of the dramatic socialist transformations since 1949 when the Communist Party took over the ruling power of China. Directed by the ideology to copy the Soviet Union model, the new regime invested most in large-scale and technologically intensive heavy industries such as steel production, chemical plants, coal extraction, cement works, and electricity production mainly via thermal power stations. This development strategy led to two major negative consequences. Firstly, thirst for energy input stimulated rapid extraction of coal resources and clearing of forests for fuels. However, the low efficiency of coal utilization brought many industrial cities heavy air pollution and wide coverage of acid rains. The cutting of fuel-woods also deteriorated the process of deforestation that had lasted for over 2,000 years in Chinese history. Secondly, industrial development increased the input of other natural resources and the output of waste. Mining activities expanded in high speed, but their efficiency was very low and resulted in waste of resources. The "three wastes" (solid, water, and gas wastes) problem became severe in the 1960s. Moreover, without appropriate control measures, these wastes contained dangerous toxic materials, such as mercury, cadmium, lead, arsenic, and complex organic compounds.

Besides the problems caused by rapid but rather primitive industrialization, the huge and continuously growing population of China laid compound pressure on ecological safety. In the countryside, trees continued to be cut for fuel and farming. Forestry was not sustainable due to indiscriminate cutting. As a result, the loss of forest left people with shortage of wood resources, loss of wildlife habitats, and vulnerable ecosystems that resulted in severe droughts or floods in many regions. The abuse of chemical fertilizers and pesticides also had visible impacts in the 1960s and 1970s. The widespread use of DDT not only caused the aforementioned disaster of Guangting Reservoir, but also led to some rice production crises between 1970 and 1972.[3]

Before the 1970s, some government-oriented mass movements contributed to the environmental degradation in China. For example, "the Backyard Furnace Campaign" during the Great Leap Forward period mobilized nearly 60 million peasants to produce iron and steel with primitive technologies. It is believed that during the campaign, numerous trees were cut as fuel for iron production. Another example is "the Four Pests Campaign" which in the 1950s wrongly listed sparrow with fly, rat, and mosquito as a pest to be eliminated, and consequently many insects harmful to agriculture quickly multiplied and posed a heavier threat to agriculture.[4]

It was under the preceding historical background that the national leaders, such as Premier Zhou Enlai, started to emphasize, though not publicly, the emergency of environmental protection. A clear turning point for the Chinese government to put environmental protection as a national responsibility was the United Nations Conference on Human Environment of 1972. As we know today, China sent a big delegation to the conference, with members from ministries like the national planning committee, the ministry of foreign affairs, the ministry of health, the ministry of agriculture, and the ministry of water resources. Despite the fact that some Chinese leaders stubbornly used the conference as another venue for ideological struggle with the Western world and the Soviet Union block, Premier Zhou Enlai explicitly pointed out that the mission of the delegation was to "be informed of the environmental situation of the whole world and the impacts of environmental problems to the socioeconomic development of other countries, and use the foreign lessons as a mirror to understand the environmental problems of China."[5]

After the delegation came back from Stockholm, Sweden, it was reconfirmed to Chinese leaders that similar environmental problems, such as air pollution, water pollution, and ecological degradation were serious in China too and deserved immediate action. Hence, in August 1973, the state council organized the first National Conference on Environmental Protection, which formally triggered the top-down environmental governance in China.[6]

Pioneering Environmentalists and Activists

Between the early 1970s and early 1990s, with the foundation laid by national leaders like Premier Zhou Enlai, China preliminarily established its official system for environmental governance. In this period, some senior officials of the central government made their contribution, and some individual environmentalists or activists also stepped onto the stage.

Professor Qu Geping, the first chief executive of the State Environmental Protection Administration (SEPA), has witnessed and personally participated in the founding of the government environmental protection system in China since the early 1970s. Qu's contribution can be summarized in three aspects. Firstly, he played a leading role to establish an independent government system for environmental management in China. After the first National Conference on Environmental Protection in 1973, the state council established a Leading Group for Environmental Protection as a unit in charge of environmental management for the whole country. But in the following 15 years, the Leading Group or the later Environmental Protection Administration was supervised by a ministry-level department under the state council. According to Qu, environmental management does not fall into the responsibility of any existing departments, such as the ministries in charge of industrial development, agriculture, or construction, and should independently function on its own. With that goal, Qu and his colleagues made hard efforts to give their work a unique identity and finally made SEPA a ministry-level department directly supervised by the state council in 1988. Secondly, under his coordination, some fundamental national environmental policies were formed. For example, during the second and third national conferences for environmental protection that were held in 1983 and 1988, respectively, Qu and his colleagues included "environmental protection, like 'one child policy,' is a basic national policy" into formal government documents, which later entered into the Constitution of China. They also proposed three principles for environmental governance, namely the precaution and prevention principle, the "polluter pays" principle, and the reinforcement of environmental monitoring.[7] Lastly, besides institutional and theoretical building, Qu and his colleagues set up some working targets for the whole country. For instance, on February 5, 1990, Qu wrote the following words in an article: "by the end of this century (the 20th century), the sky will be more blue and the water will be cleaner (in China)."[8]

As criticized by some observers and even some SEPA officials, Qu is an idealist, and some of his work did not fit in with the reality of China. For instance, his goal to clean up China's environment by the end of the 20th century proved to be a dream, and he also admitted that he and his colleagues "are too anxious to improve the situation and underestimated difficulties." In addition, the seemingly independent SEPA and the later ministry of environmental protection (MEP) do not have enough capacity to fulfill their missions, and the basic policies and principles for environmental protection are always undermined by the power of promoting economic growth and natural exploitation. Nevertheless, this does not mean that the work of Qu and his colleagues are in vain, but implies that a pure top-down approach of environmental movement in China is not enough.

After the late 1970s and early 1980s when the Chinese government adopted the "Reform and Opening" policy, the country transformed dramatically. So did its environment. As local government and individual people attained more freedom to run business, the old environmental problems mainly caused by the centralized industrial development and population pressure were compounded with the eruptive desire to make money, compete in markets, and convert nature into profit. While SEPA was struggling to cope with more and more pollution and ecological degradation, some citizen environmentalists started to take their own actions.

There emerged people dedicated to species conservation. One most famous practitioner is Xu Xiujuan, who is now regarded as an environmental heroine for the protection of the red-crown crane (*Grus japonensis*), a unique crane species in northeast Asia. In 1981, when Xu was 17 years old, she inherited her father's career and became a young crane cultivator in Zhalong Crane Nature Reserve, Heilongjiang Province. Very soon, Xu became an expert and was invited by the managers of the new Yancheng Nature Reserve in Jiangsu Province, a winter migrating habitat of cranes. There, Xu helped cultivate and save more cranes. Unfortunately, in the winter of 1987, when searching for two lost cranes, Xu was drowned in the marsh and lost her life. The death of Xu Xiujuan brought her the official title of Environmental Martyr, which is quite unique in Chinese history. However, it was the exposure of her diary, which recorded her love of the beloved species and the desire of pursuing a meaningful life, that moved thousands of Chinese people and gave them a chance to connect their lives to mother nature.[9]

There emerged people who started to investigate their motherland and mother rivers. Wu Dengming, born in 1940, lives in Chongqing, the second-largest city along the Yangtze River. As early as the late 1970s and 1980s, Wu, as a university security officer, voluntarily surveyed the upper reaches of the Yangtze River and some of its major branches. He found that, compared with the situation in the 1960s, deforestation in the Yangtze watershed had been accelerated due to economic development and lumbering-technology upgrade. Back to Chongqing, Wu wrote a survey report and started making petitions to the government and organizing the youth to promote the idea of protecting their mother river.[10]

There emerged people specialized in thinking and writing about the environment. Tang Xiyang, an old journalist who survived from cruel political movements since the 1950s, created a new journal, *Da Zi Ran* (Nature), at the beginning of his fifties. During the whole 1980s and 1990s, the focus of Tang and his journal was reporting natural conservation in China and all over the world. Tang wrote several books too, in which he elaborately recorded the beloved wild animals, plants, and landscapes that he had seen, and people dedicated to nature preservation, including the aforementioned environmental heroine, Xu Xiujuan. He also raised some basic questions on natural conservation, such as why we should care about nature, what is the practical way to do so, and what we can learn from other countries.

Tang Xiyang once wrote in his book: "my only writing creed is 'stick to the facts.'"[11] Such a creed is also pursued by Xu Gang, a professional writer famous for his ecological reportage and later respected as the pioneer of environmental literature in China. Compared with Tang's work, Xu's writing is more critical of the reality, and full of anger on the exploitation of nature that took place everywhere in China. Such a mood is clearly reflected in the chapter titles of his first environmental literature book: "Woodcutter, Wake Up!," "Rivers Do Not Flow Forever," "The Degrading Land," and "The Sand Storm Lines of China."[12] Without doubt, the unchanged theme of Xu's work has always been that China is in ecological crises and people can get rid of the crises only by rethinking their role in nature. He expressed these views not only based on his personal investigation and ecological knowledge but also in literary words. For example, while explaining the importance of the diversity and health of the ecological system, he said: "only when all beings on earth are safe, the human being is safe."[13]

The Reemergence of Civil Society

As one famous environmental reporter, Feng Yongfeng, described, there are three basic ways for Chinese citizens to do something for their environment. Firstly, they can try their best to behave environmentally in daily life. Secondly, they can report environmental destruction or crimes to the government and push government officials to take action. Lastly, they can establish environmental organizations, and then, under the name of "civil society," lead more people to consistently intervene in environmental issues.[14] Among these ways, the last one is potentially most powerful.

Civil society is not a new thing in China. Before 1949, civil society organizations, such as chambers of commerce, charity groups, academic institutions and citizens' clubs, were very common across the country. After 1949 when the ruling Communist Party declared that it represented the interest of all Chinese people, civil society groups, except some that were strictly supervised by government, lost their space to exist. Thirty years later, as the Reform and Opening policy was adopted, civil society has come back again in China.

There are many reasons for people to organize themselves. Perhaps the most self-evident one is that people "need" to connect with one another. Furthermore, self-organized people can do something good for themselves and for the whole society, or even correct the failures of government and the market. So can civil society organizations focused on environmental issues, or, in another name, environmental nongovernmental organizations. In China, the emergence of ENGOs in the 1990s supplemented the shortcomings of government and individual activists in the field of environmental protection.

As mentioned before, despite the fact that China has had an independent environmental governance system since the late 1980s, which included SEPA at the national level and over 3,000 environmental protection bureaus (EPBs) at the provincial, municipal and county levels, SEPA or EPBs actually have been the weakest government departments, because, compared with economic growth, protecting the environment is not a political priority. Moreover, considering that governments at different levels in China do not receive sufficient supervision by people or "people's representatives," it is understandable that both SEPA and EPBs have little pressure from the bottom to deliver their responsibilities. Some of these agencies even became "protecting umbrellas" for polluters and environmental criminals.

Individual environmentalists like the four figures mentioned earlier also met their obstacles. The head of SEPA might be able to promote some environmental legislation, but was not able to mobilize thousands of lazy officers to deliver their duties or to challenge environmentally unfriendly industries that have spokesmen within government. When some wildlife conservationists or environmental investigators were able to save lives and raised public concerns in a small area, it was hard for them to curb the bigger trend of destruction and pollution. Environmental literature could bring fresh environmentalism to thousands of people's minds, but the way to bridge knowledge and the willingness to act might go beyond the environmental writers' ability. Therefore, when time entered into the early 1990s, all these obstacles became acute, and it seems that they could be overcome only by well-organized environmental actions. In another word, without civil society groups, a true environmental movement would not emerge in China.

It should be noted that individual actions were the basis of the emergence of environmental groups. In fact, it was the recognition of the necessity to convert individual actions to collective ones that triggered some pioneering environmentalists to initiate ENGOs. For example, the Yangtze River investigator, Wu Dengming, created a local ENGO called Chongqing Green Volunteer Association (CGV) in the mid-1990s. Tang Xiyang, the journalist and environmental writer, launched Green Camp in 1996, a program later transformed into an ENGO focused on training young environmental activists. Even Qu Geping, a ministerial-level government official, also established a semi-governmental organization called China Environmental Protection Foundation, after his retirement from SEPA.

Interestingly, before Chinese environmental organizations emerged, international ENGOs landed first and left some heritage to the Chinese green movement. The first two organizations that started to work in China are the International Crane Foundation (ICF) and the World Wildlife Fund (WWF, now known as World Wide Fund for Nature). In the early 1980s, WWF, known to Chinese people for selecting the panda as its logo image, "naturally" invested first in protecting the panda in China. Later, it further helped the Chinese establish a number of natural reserves such as the Black-necked Crane Reserve around Caohai Lake in Guizhou Province. ICF worked at Caohai too. After getting familiar with the local conditions, ICF intervened in the conflict between local residents and natural reserve staff, and gave them some constructive advice for solving problems.[15] In addition, both WWF and ICF have trained a lot of Chinese conservation experts. Their working spots, such as Caohai Lake, were succeeded by Chinese NGOs later.

Two Types of Environmental Movements

With conditions such as the harsh reality of environmental degradation, looser government control of people's networking, and pioneering works initiated by domestic and international environmentalists, the green movement in China, represented by the formation of environmental groups, emerged in the early 1990s.

In terms of the motivation of the establishment, environmental organizations can be divided into two types: private-interest organizations and public-interest organizations. The former is created based on protecting the members' own environmental rights such as breathing clean air, drinking clean water, and living with sustainable natural resources. By contrast, the members of the latter are not directly threatened by environmental degradation. What they are dedicated to is the "public" rather than a specific group of people. However, despite this dichotomy, private-interest organizations will sometimes go beyond their members' personal interests, and the public-interest organizations may ally with private-interest groups to fulfill their goals.

In China, the emergence of public-interest environmental organizations far preceded private-interest groups. Such a phenomenon may be different from many other countries. A most likely reason is that the earliest ENGOs were established by elite intellectuals who had more capacity to call for people to work with them. By contrast, though industrial laborers or farmers were more vulnerable to environmental problems, they were less capable to act because their citizens' rights were repressed by big companies and local governments.

The first Chinese ENGO, the Saunders Gull Conservation Society of Panjin City (SGCSPJ), was established in 1991. According to the preceding taxonomy, it falls into the category of public-interest organization, because its work is all about the conservation of one wildlife species that benefits the whole society. In addition, its founder, Liu Detian, is an experienced local newspaper journalist, a typical intellectual in Chinese modern society. Three years later, a more influential ENGO, Friends of Nature (FON), was founded in Beijing, and all of its four founders were well-known scholars of the time. These founding scholars had sensitive insights into China's environmental problems and took a brave step to work together as a group. Very soon, many common citizens, including school teachers, journalists, and students became the earliest volunteers of FON. In the same year of 1994, Li Lailai, a sociologist, established the Institute of Environment and Development (IED) and began work by conducting environmental research, personnel training, and information distribution.

In the following 10 years after the three earliest ENGOs were founded, there have been over 50 ENGOs (not including university student groups and "NGOs" with government background) established in China. In general, these ENGOs have three common features. Firstly, the majority of the founders are professionals in the humanities and social sciences. This implies that intellectuals of natural science in China were less sensitive to social problems or had less courage to get involved in the activities questioning governments and industries. Secondly, few of these groups were initiated by environmental victims, and as a result their activities were more about the interests of the general public. Thirdly, in terms of geographic coverage, these ENGOs concentrate in Beijing City and southwest China, such as the Yunnan, Sichuan, and Guizhou Provinces.[16] This is because Beijing is a relatively safe political harbor and a good information center for civil society organizations, and the southwest region is a hotspot of biological and cultural diversity conservation, which is what attracted financial resources from international foundations in the first place.

Some ENGOs like FON and Global Village of Beijing (GVB) (established in 1996) even did not know what to do at the beginning. They started from what they could do and what might be needed from society, such as natural observation, mountain climbing, garbage separation, documentary production, and environmental education. But gradually, ENGOs found their own niches. For examples, FON cares more about wildlife conservation and environmental education; GVB selects energy, municipal solid waste, and media communication as its strength; Green Earth Volunteers (GEV) (also established in 1996) forges itself as a home for environmental journalists and spent all of its energy to oppose dam construction in southwest China; Green Hanjiang and Huai River Guard are local ENGOs located in Hubei and Henan provinces, respectively, and focus more on the protection of their mother rivers. This trend of specialization has continued until the present, with more and more Chinese ENGOs having their clear working field, geographical area, target group, and action methodology.

Would environmental campaigns led by ENGOs really change something in reality, for example, stopping a factory's pollution, suspending a dam project, or saving a plot of forest? The answer is "yes" and "no." In general, the scale of ENGOs is still small and their voices can always be ignored. One or two visible successes, such as the Tibetan Antelope protection, were achieved because ENGOs attracted wider public attention, especially support from the media. Given this, it is not difficult to understand that grassroots mass movements, though they emerge later than the activities led by elite intellectuals, attain some immediate victories.

The campaigns opposing the construction of paraxylene (PX) chemical plants and municipal solid waste (MSW) incinerators are the two typical grassroots mass environmental movements in China. In 2006, the central government approved the plan of building a PX chemical plant in Xiamen City, Fujian Province. Following some open opposition by local politicians and public intellectuals, thousands of Xiamen citizens networked with one another via the Internet and mobile phones, and finally walked the streets on June 1 (also Children's Day in China) and demonstrated their anger. Five months later, the municipal government decided to quit the original plan and move the plant to another city of the same province. Following the Xiamen incident, on August 14, 2011, the same movement happened again in Dalian City, Liaoning Province, and the scale was even bigger, with the participation of over 10,000 people. On the evening of the demonstration day, the city leaders decided to meet all the demands of the citizens.[17]

As for the anti-incineration movement, it concerned many middle-class people who lived in the largest Chinese cities, such as Beijing, Shanghai, Nanjing, and Guangzhou. After these people "suddenly" found that MSW incinerators would be built in their "backyards," they learned by themselves the risk of incineration and then launched campaigns to stop the projects. Between 2007 and 2009, incinerator opponents organized street demonstrations in different cities, some of which effectively changed government policies.

During the preceding grassroots mass movements, citizen groups were also formed. For example, the residents opposing incineration in Beijing and Guangzhou established informal volunteer groups by themselves. These groups had a good internal cooperation and responsibility division. However, compared with professional ones, these citizen groups were temporary and had their limitations. For one thing, once their initial goal was reached, the organizations virtually collapsed. Moreover, because the relevant campaigns were based on self-interests, their victory might pose harm to people in other places. As seen from the anti-PX and incineration cases, the citizens' action did not stop or mitigate pollution, but rather moved them elsewhere.[18]

Nevertheless, private-interest environmental action is necessary and should not be ignored by professional ENGOs. Some Chinese ENGOs, such as the Institute of Public and Environmental Affairs (IPE) (established in 2006) and Green Beagle (established in 2009), have already emphasized facilitating grassroots actions. Additionally, despite the fact that mass movement may vanish at some point when the original self-interest goals are reached, some activists have extended their concerns beyond their own problems and area. For example, some residents in Beijing and Guangzhou who used to call for "not in my backyard" are trying to establish formal ENGOs and turning their focus to the general municipal waste management issues.

In sum, before the second decade of the 21st century, different kinds of people like national leaders, government bureaucrats, conservation experts, and elite intellectuals have played active roles for cultivating the green movement in China. Later, these people and their individual activities transformed into organized groups and campaigns with clear focuses to tackle actual environmental problems of the country.

The Battlefields of the Green Movement

"Pleading on Behalf of Wretched Nature": The Wildlife Conservation Movement in China

A remarkable phenomenon of the Chinese green movement is that most environmentalists turned their eyes to wildlife conservation first, instead of pollution. For example, the earliest International ENGOs working in China (ICF and WWF) and the oldest ENGO established by the Chinese (SGCSPJ) all started working on protecting wild animals, such as the panda, the crane, and the gull. This is not a coincidence. On the one hand, as mentioned earlier, the first generation of environmentalists are either governmental officials or intellectual elites particularly with a social science and humanities background, and almost all of them lived in the urban areas. Emotionally, these people have a romantic affection with "wilderness," and rationally, they have the impulse to think about the delicate relationship between human beings and the other species. Such a social background and mental characteristics impelled them to do something for wildlife protection. On the other hand, the biodiversity of China is rapidly being lost due to commercial hunting, habitat destruction, pollution, and climate change. As official reports described, in the recent 100 to 200 years, China has lost over 10 mammal species, and over 20 animal species are close to extinction; 4,000 to 5,000 higher plant species are in severely endangered status.[19]

Personal affection and the dangerous situation however cannot fully explain why environmentalists acted first in the wildlife conservation field. Another important reason behind this is that environmentalists would have least conflict with the government in this field, because natural conservation, compared with pollution, is less contradictory with the goal of industrializing and "modernizing" the country. It should also be noted that the relatively stringent political atmosphere in the 1990s in China forced the newly born ENGOs to carefully ally with pollution victims from the grassroots. As a result, biological conservation is a practical start for the newly sprouting green movement in China.

The slogan "pleading on behalf of wretched nature" for wildlife conservation comes from an article by Liang Congjie, one of the four founders of FON:

> (In 1996), tens of volunteers traveled over one thousand miles to the heart of the Hengduan Mountains in Northwest Yunnan Province. That was not an ordinary summer camp nor scientific expedition. Instead, their goal was to protect the living rights of a plot of primary forest and a population of wild monkey, and to plead on behalf of the wretched relic of nature.

The common name of the monkey that Liang referred to is Yunnan snub-nose monkey (*Rhinopithecus bieti*), an endemic species in China like red-crown crane and Saunders gull. The campaign of protecting this animal emerged in the early 1990s.

Xi Zhinong is now the most famous ecological photographer in China. His career has been closely connected with the Yunnan snub-nose monkey. From 1992, Xi started to be involved in a three-year research scheme in the northwest region of Yunnan Province, which the monkey inhabits. Three years later, he made an outstanding documentary, *Exploring Yunnan Snub-Nose Monkey,* and received an international reputation. In 1995, Xi was shocked by news that the local government of Deqin County had decided to cut 1 million square kilometers of primary forest, which would cause Yunnan snub-nose monkeys to lose their home. After some vain efforts within Yunnan Province, Xi asked for help from outsiders. He wrote a letter to Tang Xiyang, who later passed his request to many media in Beijing and ENGOs like FON. Then, in December 1995, FON launched a campaign to lobby the central government to stop the tree-cutting. In 1996, Tang Xiyang organized some university students to go to the threatened forest and further attracted the public's attention (also mentioned in the quote shown earlier in the chapter, by Liang Congjie). After that, the seesaw match between loggers and environmentalists did not stop until August 2, 1998. On that day, a news program of China Central Television (CCTV) heavily criticized the man-made deforestation in Yunnan, which triggered a strong intervention from the national leaders to stop illegal logging.[20]

Soon after the campaign for protecting the natural reserves in Yunnan, FON was engaged in a tougher war against the illegal hunting of the Tibetan Antelope (*Pantholops hodgsoni*). Like the Yunnan snub-nose monkey, the Tibetan Antelope is an endemic species in China. Because its fur is the raw material for Shahtoosh wool, a luxury fur product in markets of Western Europe and North America, since the 1980s, a large number of Tibetan Antelope were slaughtered and soon the species was close to extinction.

In general, there were three questions that environmentalists needed to address for protecting the Tibetan Antelope, especially when the government was not willing to take responsibility. The first one was: who is able to enforce the law? Fortunately, in 1992, a semi-government military force, called the Wild Yak Troupe, was established and took the responsibility to patrol Kekexili, Qinghai Province, one of the most important habitats of the Tibetan Antelope. Its existence was a deterrent to illegal hunters and slowed down the slaughtering. The second question was: who pays for the protection? At the beginning, volunteers and ENGOs contributed a lot to the patrol. Yang Xin is a natural explorer from Shenzhen, the most economically developed city in China. After he heard the stories of the Wild Yak Troupe, he decided to raise funds to help the troupe build a permanent conservation station in the reserve. With this station, the patrol time would be longer. In 1997, the station was finished, and it is the first non-government-invested conservation facility in China. In addition, in 1997 and 1998, FON, collaborating with WWF and the International Fund for Animal Welfare (IFAW), launched a mass communication campaign and appealed to the central government for intervention. As a result, after 1999, the ministry of forestry, a department responsible for natural reserve administration, and the Qinghai provincial government invested much more in the anti-illegal hunting campaigns.[21] The last question was: who can stop the illegal trade? For this, ENGOs also played a key role by raising the issue on the international stage. In October 1998, Liang Congjie wrote a letter to Tony Blair, the prime minister of the United Kingdom of the time, on the evils of Tibetan Antelope–smuggling. Blair replied and expressed his commitment to stop the fur smuggling within Great Britain and the whole European Union.[22] In the meantime, some Indian ENGOs also helped lobby their government to control wildlife fur import, and encouraged the Indian craftsmen to make other products to substitute Shahtoosh.[23] In sum, with contributions from governments, local people, volunteers, domestic and international ENGOs, the illegal hunting of the Tibetan Antelope gradually vanished after 1999. This is perhaps the first successful natural conservation campaign in China.

Let Rivers Run Freely: The Campaigns against Dam Construction

Compared with wildlife conservation, the dam issue poses bigger challenges for environmentalists, because the cost-benefit analysis for hydropower development looks much more complicated.

Since the early 20th century, harnessing rivers and exploiting potential hydropower has been a dream of Chinese leaders. To many ordinary people, dams are also a symbol of modernization. However, criticism of dam building never stops, even in the era of Chairman Mao Zedong. In 1957, Huang Wanli, a professor of Tsinghua University who taught hydrological engineering, openly opposed the building of the Sanmenxia dam at the Yellow River. He predicted that the dam would cause silting-up and create disaster for people in the region. Less than 10 years later, his unwelcome words became true. In the mid-1980s, Huang wrote letters to Chinese leaders again and opposed the building of the Three Gorges dam on the Yangtze River.[24] This time, Huang failed again, but differently from the Mao era, he was not alone anymore.

Compared with their predecessors, the new Chinese leaders were less dictatorial on decision making on large-scale public construction projects.

Before 1989, the central government invited a wide range of experts including environmental and ecological scholars to evaluate the Three Gorges dam proposal. According to some experts, the dam would result in severe environmental impacts, such as blocking migrating routes of wild fish, damaging the habitat of some aquatic animals, changing the local climate, and creating a huge number of ecological refugees. Most of these counter-opinions were collected in a book, *Should the Three Gorges Dam Be Constructed: A Collection of Scholars Debate*, edited by Dai Qing, a famous journalist during that time.[25] Unfortunately, the democracy within the community of scientists did not last long, and soon the experts supporting the project outweighed the opponents. In 1992, the state council decided to put the project scheme to the vote by the National People's Congress, but the result showed that only two-thirds of the representatives endorsed the bill, an unprecedented low percentage in the history of the People's Republic of China.[26]

The approval of the Three Gorges dam project was followed by a bigger wave of exploiting hydropower resources in the southwest part of China, including Sichuan, Guizhou, and Yunnan provinces and Chongqing City. The government and hydropower developers declared that hydropower was a green energy, enabling the country to be less dependent on dirty coal, and would improve the life quality of the local people and the local economy by providing them cheap electricity. These arguments have their attractiveness, but, differently from the previous debate of the Three Gorges dam, which is narrowed within politicians, techno-bureaucrats, and scientists, civil society groups and ENGOs joined in the new round of antidam campaigning and gradually played a critical role.

In August 2003, the central government gathered experts to discuss the scheme of 13-ladder hydropower exploitation along the middle and lower reaches of the Nujiang River. This news soon raised serious concerns from the public. Considering that the Nujiang River is one of the two Chinese rivers (the other one is Yarlung Zangbo River) that maintained their natural virginity, and an integral part of the World Natural Heritage, "the Three Parallel Rivers of Yunnan Protected Areas," enlisted by the United Nations in July 2003, many scholars, artists, writers, journalists, and environmental activists signed an open letter to call for protecting the river from hydropower exploitation. In February 2004, Premier Wen Jiabao stated in a written instruction that the Nujiang River project must be "cautiously researched and the decision making must be scientific."[27] Although this instruction did not entirely freeze the original construction scheme, it actually has suspended the project for more than eight years already.

After the Nujiang River debate in 2003, ENGOs gradually became the driving force for the antidam movement. Among them, Green Earth Volunteers and its founder Wang Yongchen are the most well known. Wang is a journalist of China Central Radio Station and many volunteers of GEV are journalists too. Influenced by the debate in 2003, Wang and her colleagues dedicated themselves entirely to the fate of rivers in China. In the following eight years, led by GEV, Chinese environmental activists continuously investigated dam-building sites, wrote research reports, organized media salons, and consulted with scientists on the issue of dam building. Despite many failures, their efforts brought some positive changes. For example, on December 27, 2008, a dam construction company formally invited ENGOs to participate in a meeting on the environmental impact assessment (EIA) report of Ahai Hydropower Station at Jinshajiang River. Two days later, two representatives of ENGOs sat in the meeting in Beijing, and their written statements were included in the meeting conclusion.[28]

Dam building is a topic more controversial than wildlife conservation, and given it hydropower indeed has some benefits, it is necessary for environmentalists to present their reasons against decisions made by governments and developers. At the beginning, these arguments sounded rather general and transcendental. Some common statements that they made include: a freely run river has its intrinsic value and embodies ancient Chinese philosophical wisdom such as Taoism; a "wild" river has high ecological and aesthetic values, which may be damaged by dam building; there are alternative ways to get energy and reduce poverty; the decision making of any large-scale public construction project must be open and just.

Apparently, the preceding arguments are useful in awareness-raising but impractical for challenging specific dam projects. ENGOs also realized that without substantially questioning some ongoing project, the overall strategy of the central government to exploit hydropower in China cannot be overturned either. Hence, as ENGOs practically intervened in more and more dam-building cases, their arguments became more concrete. For example, in the latest campaign against Xiaonanhai Hydropower Station at the upper reach of the Yangtze River and in Chongqing City, ENGOs provided stronger evidence to oppose the project: the project is illegal because it is located in the natural reserve for protecting endangered fish in the upper reach of the Yangtze River; the EIA process violates relevant laws; the project is not economical because its average capital investment per unit kilowatt reaches 16,000 Chinese Yuan, much higher than the investment of the Three Gorges dam, which is 4,950 Chinese Yuan; the other existing hydropower stations of the same region have enough capacity and responsibility to transmit electricity to Chongqing City, which is supposedly to benefit from the Xiaonanhai project.[29]

Faced with the above pressure by ENGOs, one EPB official of Chongqing City who endorsed the project responded that "following the extinction of some species, there will be a new species born. This is what is called 'survival of the fittest in natural selection.'"[30] Such absurd sophism actually sets off the progress made by environmentalists over the past few years. As the antidam campaign further develops, it will not be very long before a new river ethic prevails across the country.

"Take Your Green Choice": The Campaign Calling for Apple to Change Its Supply Chain

Although ENGOs received little success in stopping dam projects, they won in some antipollution cases. Among these cases, the anti-Apple campaign led by the Institute of Public and Environmental Affairs since 2010 is the most remarkable one.

IPE, though a relatively young Chinese ENGO, applied new techniques and tactics to tackle environmental problems. In 2006, also the year of its establishment, IPE introduced the "China Water Pollution Map" to the public, and immediately attracted much attention. Briefly speaking, the map is an Internet-based database combining the digital map and information of water pollution around the country. Its creativeness lies in the fact that all the data showed on the map is from government publications, including water quality monitoring information and legal violation records of factories. As Ma Jun, the leader of IPE, said, "when these data are silently scattered in innumerous corners, people do not recognize their power. Once they are congregated and well presented, a 'regular armed force' tackling pollution is organized overnight."[31]

Almost at the same time that the IPE publicized its Water Pollution Map, companies listed on the map, especially those with good reputations, approached IPE and tried to convince the latter that they would change.[32] Despite the success of the map, Ma Jun and IPE were not satisfied yet. They found that those companies responding to the map were usually big foreign corporations investing in China, while so many Chinese state-owned and private enterprises did not care much about the public's views on their environmental performance. Hence, in 2007, IPE collaborated with another 20 ENGOs to launch a "Green Choice" campaign. They set the goal of the campaign as to provide full information for consumers to select environmentally friendly producers and products and to press big foreign companies to turn their supply chain greener.[33]

Considering that many multinational companies contract their business out to original equipment manufacturers (OEMs) in developing countries, especially China, the ENGOs' emphasis on the responsibility of big foreign companies to monitor their supply chain is an effective tactic to tackle Chinese companies. Between April and October 2010, the ENGOs' network, Green Choice Alliance, released three reports on the heavy metal pollution caused by information technology (IT) companies and their supply chain in China.[34] Right after that, IPE and its partners started to reveal the dark side of Apple, one of the most popular IT companies in the world.

On January 20, 2011, IPE, FON, and Green Beagle publicized their first report on the environmental problem of Apple's supply chain in China. The report exposed that since May 2008, Wintek, a Taiwanese company that manufactured Apple's products in Suzhou industrial park in Eastern China, had illegally used a toxic chemical called hexyl hydride to clean the screen of iPhone, and this process caused poisoning of many workers. The report also emphasized that Wintek's fault has direct relation with Apple's marketing style, because the rush to meet consumers' demand indirectly forced OEMs to use un-environmental techniques to accelerate the production process. Considering that Apple's engineers were directly involved in its OEMs' daily operation, ENGOs believed that Apple should not be innocent at all.[35]

The exposure of Apple's bad management of its supply chain called the public's attention soon, but what outraged people most was not the misdeed of Wintek but the arrogant attitude of Apple toward its responsibility. In the following seven months, Apple refused to respond to any questioning by victims, ENGOs, and the media. In that situation, the Green Choice Alliance published the second report in August, which included more solid evidence on the pollution caused by Apple's OEMs. This time, Apple agreed to talk with ENGOs.[36]

However, the informal dialogue between Apple and the Green Choice Alliance in September did not virtually change the IT giant's policy of keeping its supply chain as a business secret, including its environmental records. During this moment, the involvement of a U.S. NGO, the Natural Resource Defence Council (NRDC), brought a breakthrough to the campaign. On September 19, the executive director of NRDC wrote a letter to Apple, requiring the company to substantially respond to the request of Chinese ENGOs. On October 31, Ma Jun and Linda Greer, the environmental health program director of NRDC, visited Apple's headquarter in San Francisco and met some senior officers of the company. As stated by Ma Jun, during the meeting, Apple mentioned the necessity to increase the transparency of its supply chain, a stand quite different from its previous one. Then, on November 15, representatives of five Chinese ENGOs formally negotiated with Apple's representatives in Beijing. As described by the attendants and media, the meeting was full of the smell of gunpowder, but both sides saw it as a good start for positive cooperation.[37]

It must be admitted that when Chinese ENGOs initiated their campaign targeting Apple, they did not plan to be part of a bigger movement calling for Apple to take its corporate social responsibility internationally. A seemingly coincident event happened in February 2012. On that day, two U.S. labor rights organizations made an accusation against Apple that one of its major OEMs in China, Foxconn, treated workers inhumanely. This event later caused a big wave of anti-Apple demonstrations in the United States and Apple was soon named a "sweatshop" by many environmentalists and social activists globally.[38] Obviously, the exposure of the labor problem of Apple's supply chain, from another angle, strengthens the legitimacy of the Chinese ENGOs' action. Conversely, the Chinese ENGOs' campaign unintentionally became part of an international movement dedicated to curbing the negative effects of economic globalization.

Reverence for Nature: Building Environmental Ethics through a Public Debate

The three cases discussed in foregoing section are all directly related to some physical environmental problems in China, such as biodiversity loss, degradation of water systems, pollutant discharge, and the use of toxic chemicals. When environmental activists were engaged in the relevant campaigns, their focus was more on the facts, laws, policies, and people's actual activities. Yet, the emphasis on the material level of the environmental issue does not mean the absence of ethical struggle between environmentalists and their opponents. As mentioned in the dam-building case, the local government official openly devalued the survival of endangered fish, actually reflecting a "cornucopian" environmental attitude, which believes that nature is unlimited and even if it is limited, other species should be sacrificed for human benefit. Obviously, environmentalists do not agree with this attitude at all, their slogan of "pleading on behalf of wretched nature" and "let the rivers freely run" contain strong beliefs that the nonhuman parts of nature have their intrinsic value, which cannot be purely judged by the pragmatic demands of human beings.

After all, the direct aim of each environmental campaign is more about the change of environmental policy, law enforcement, the level of public participation, and eventually the physical status of an environmental problem, and less the ethical implication behind it. However, in January 2005, Chinese environmentalists participated in a public debate on whether human beings should have reverence for nature, a peculiar event that introduced the idea of environmental ethics into the discourses of the masses.

The debate originated from an article, "Humans Do Not Have to Have Reverence for Nature," which was published in a Chinese magazine called *Globle* on its first issue of 2005. In the article, the author, Professor He Zuoxiu, who is a famous physical scientist and academician of the Chinese Academy of Science, criticized the saying that "man should have reverence for nature," which was quite popular among the media during the disasters of Severe Acute Respiratory Syndrome (SARS) in 2003 and the Indian Ocean Tsunami at the end of 2004. He argued that the ultimate goal for environmental protection is humanity's benefit; therefore, some environmental "damage" such as building dams is worthy. He also explained that "reverence for nature" was a slogan that characterized human-nature relationship morally, not scientifically, and therefore was a typical antiscience notion.[39]

The dissemination of He Zuoxiu's article soon triggered criticisms. In the meantime, some mass media acutely caught the market value of a potential social debate. From January 11, 2005, the *Beijing News,* the most influential newspaper in the capital of China, set aside a special column to publish articles and comments on "whether human beings should have reverence for nature." In the end, it collected 31 papers. Seen from these published opinions, the people involved in the debate were extremely divided. The debaters standing by He Zuoxiu were some well-known writers or scholars who often criticized environmentalists and believed science is the ultimate power to protect human welfare. On the contrary, people challenging He and his supporters included the leaders of ENGOs, such as Liang Congjie (FON), Wang Yongchen (GEV), and Liao Xiaoyi (GVB), and some other environmental activists. It is not surprising that the opinions of the two sides were quite conflicting with each other. This can be seen in the topics of their articles: "'Reverence for Nature' Is an Anti-Scientific Notion," "'Reverence for Nature' Is a Common Scientific Knowledge," "What Fault for 'Reverence for Nature'?."

Although the *Beijing News* ended the column on February 25, the fire of the debate did not stop but extended to other media, including some newspapers, magazines, and influential websites. In addition, more people like scholars studying the philosophy of science joined in and brought wider perspectives to the debate. Consequently, the discussion on "reverence for nature" became a hot cultural phenomenon in the spring of 2005 in China.

This chapter does not intend to review the debate in detail and make judgments on which side is the winner. Nevertheless, it can conclude that such a debate gave a good chance for many ordinary Chinese people to think about the relationship between human beings and the rest of nature. Or, it can at least highlight the harvest of environmentalists from the debate, namely the answers to some key questions for building environmental ethics for Chinese society.

The first question is: is science the only medium for people to build their relation with nature? The debate told the public that having some moral connection with nature falls into the "nonscientific" realm, which may not necessarily be "antiscientific." Additionally, the nonscientific connection, such as the religious, spiritual, and aesthetic perceptions of nature cannot be devalued by science. The second question is: can science solve all environmental problems? In the debate, it was well presented by environmental activists and philosophers that if science represents people's best knowledge about nature, it is still too limited and its application has already posed many untold impacts to nature and ultimately to human beings. By contrast, being respectful to nature implies a humble attitude toward the power of human science and technology.[40] The third question is: does nonhuman nature have "value"? This question can also be converted to a more academic discussion: which principle is the base for environmental protection, the anthropocentric view or the biocentric/ecocentric view? From the debate, the audience can have a chance to think about whether the human being, as the ethical subject, should give the nonhuman nature similar rights as human beings, and whether human beings should adapt to the natural system or manipulate it.[41]

Conclusion

Generally speaking, "environmental movement" can be defined as a coordinated group action focused on environmental issues. In China, between 1949 and 1976, the governments or the ruling Communist Party used to be the driving force of many political and social movements, but after that, they gradually lost the interest and power to mobilize citizens to do something extra to what is requested by law. Yet, in the new era after the late 1970s, when people have more room to pursue their private interests, the deteriorating environmental problems have demanded more of people's attention and participation. Who can be the driving force for the new movement of environmental protection? The answer is obviously the people themselves. As shown by this chapter, national leaders, government bureaucrats, conservation experts, and elite intellectuals in China have played important roles in different historical stages, and eventually, with the emergence of ENGOs and grassroots citizens' groups, the separated green powers in China are more likely to congregate together.

Between the early 1990s and the beginning of the 2010s, the burgeoning green movement in China has covered a wide spectrum, including natural conservation, pollution control, environmental health, fair trade, environmental justice, environmental education, and climate change. Among a great deal of work done by ENGOs and individual activists, the four cases introduced by this chapter may give readers a general picture of the characteristics of the contemporary environmental movement in China. Firstly, as shown in the wildlife conservation and antidam campaigns, national politics is an important factor influencing the topic selection and direction of the movement. Secondly, all cases exemplify that ENGOs become more capable and mature in initiating campaigns. Their substantial interaction with citizens, governments, and industries is rapidly increasing. The anti-Apple campaign further reflects that the Chinese green movement has a closer connection with global environmental and social movements. Lastly, "battles" on physical environmental problems always imply the different or even contradictory understanding of environmental ethics, and sometimes the difference would evolve into conceptual battles like the one on "reverence for nature" in the spring of 2005.

By and large, for the past 20 years, the green movement in China has "found" its driving forces and demonstrated its capacity of tackling environmental problems. Although the movement still lacks successful cases, it is indeed changing more and more Chinese people's minds and their living conditions, as well as the way in which they participate in public affairs in China.

Notes

1. Qu, *Mengxiang yu qidai: Zhongguo huanjing baohu de guoqu yu weilai* [*Dream and Wish: the Past and Future of Environmental Protection in China*], 36–37.
2. Qu, *Mengxiang yu qidai*, 37–38, 40.
3. Murray and Cook, *The Greening of China*, 34.
4. Murray and Cook, *The Greening of China*, 33–34.
5. *Huanjing juexing: renlei huanjing huiyi he Zhongguo diyici huanjing baohu huiyi* [*Environmental Awakening: The Conference on Human Environment and the First Environmental Protection Conference of China*], 206.

6. Qu, *Mengxiang yu qidai,* 39–40.
7. Ibid., 71–79.
8. Ibid., 55.
9. Feng, *Jiaoni ruhe zuo huanbao: Zhongguo minjian huanbao chuanqi* [*How to Protect Environment: Chinese Folk Legend of Environmental Protection*], 100.
10. Ding, *Gongdao keng yousheng: Ding Pin huanjing xinwen ji shici zuopin xuan* [*Clatter of Bow and Sword: A Selection of Environmental Reports and Poems of Ding Pin*], 144–53.
11. See Tang and Marks, *Huanqiu lüse xing* [*A Green World Tour*]
12. See Xu, *Famuzhe, xinglai!* [*Woodcutter, Wake Up!*].
13. Cited from Zhou and Zhang, "Yige zhizhe de youlü: fang zhuming huanjing wenxue wenxue zuojia, shiren Xu Gang [A Sage's Worry: An Interview with Environmental Literature Writer and Poet, Xu Gang]."
14. Feng, *Wei minjian huanbao liliang nahan* [*Whoop for Civil Environmental Protection Power*], 145.
15. Futrell, "The Evolution of International NGOs in China: Broadening Environmental Collaboration and Shifting Priorities," 159–84.
16. China Development Brief. "NGO List." http://www.chinadevelopmentbrief.org.cn/ngo_info.php?newclass = 101000000. (accessed March 10, 2012).
17. Johnson, Thomas. "Extending Environmental Governance: China's Environmental State and Civil Society," 171–76; College of Journalism and Communication, Jinan University. "Dalian PX Shijian [Dalian PX Event]." Accessed September 12, 2012. http://yqjd.jnu.edu.cn/multimedia/49.html.
18. Johnson, "Extending Environmental Governance," 174; Global Times, "Haidian Plans to Build New Incinerator," http://beijing.globaltimes.cn/society/2010–11/593002.html (accessed November 18, 2010).
19. Lü, "Biological Diversity: Double Meanings for Climate Change," 185; Li, "Environmental and Ecological Protection: Nation's Responsibility and Citizen's Rights," 25.
20. Baidu Baike, "Xi Zhinong," http://baike.baidu.com/view/281716.htm (accessed March 2, 2012); Friends of Nature, "*Zhichi Xi Zhinong tongzhi baohu dianxibei yuanshi senlin* [Supports to the Efforts by Comrade Xi Zhinong on the Protection of Yunnan Sunb-Nose Monkey]," 329–31.
21. Wang, "Xuere changjiang yuan: lüse xingdong [Blood Hot at the Source of Yangtze River]," 46–57; Friends of Nature. "Guanyu baohu zanglingyang wenti de baogao he jianyi [A Report and Recommendations on the Protection of Tibetan Antelope]," 342–47.
22. Friends of Nature. "Wei baohu zanglingyang jinzhi zanglingrong feifa maoyi 'ziran zhiyou' zhihan Yingguo shouxiang Bulaier bing huo fuhan [Friends of Nature's Letter to the British Prime Minister, Tony Blair, on the Banning of Illegal Trade of the Fine Hair of Tibetan Antelope for Protecting Tibetan Antelope and the Reply of Tony Blair]," 203–206.
23. See Wright and Kumar, *Fashion for Extinction: An Exposé of the Shahtoosh Trade.*
24. Baidu Baike, "Huang Wanli," http://baike.baidu.com/view/20750.htm (accessed March 5, 2012).

25. See Dai, *Changjiang sanxia gongcheng yingfou xingjian: xuezhe zhenglun wenji* [*Should the Three Gorges Dam Be Constructed: A Collection of Scholars Debate*].

26. Hu, "Xingjian Changjiang sanxia gongcheng yi'an biaojue qinli ji [Witness of the Vote on the Bill of Constructing the Three Gorges Dam]," 4–8.

27. Chen, "Nujiang shuidian kaifa zao huanbao renshi dizhi tiaozheng fang'an zanwu houwen [The Exploitation of the Hydropower along Nu Jiang River Faces Opposition from Environmentalists, There Is No Follow-Up on the Adjusted Scheme Yet]."

28. Feng, *Jiaoni ruhe zuo huanbao,* 106–9.

29. 18 Chinese ENGOs and some scholars and environmental volunteers, "An Open Letter to Premier Wen Jiabao and Vice Premier Li Keqiang on Stopping the Unworthy Xiaonanhai Hydropower Project," http://www.fon.org.cn/content.php?aid=15176 (accessed April 5, 2012).

30. Liu, "Wei Xiaonanhai dianzhan touzhu nide yan'guang [Pay Your Attention to the Xiaonanhai Hydropower Station]," http://news.sina.com.cn/pl/2012–03–28/110124187489.shtml (accessed March 28, 2012).

31. Feng, *Jiaoni ruhe zuo huanbao,* 147–49.

32. Ibid., 149.

33. *Zhongguo huanjing baohu changdao zhinan* [*Manual for Law-Based Environmental Advocacy in China*], 166–68.

34. Ibid.

35. NetEase, "Pingguo zai Zhonguo de 'xuehan gongchang' [Apple's 'Sweatshop' in China]," http://discover.news.163.com/special/apple/ (accessed March 12, 2012; Global Entrepreneur Magazine, "Pingguo de qianxian muou [Apple's Marionette]," http://www.gemag.com.cn/html/2012/chinese_company_0112/28149.html (accessed January 12, 2012).

36. Cao, "Pingguo shoudu chengnuo rang gongyinglian touming [Apple Promised the First Time to Make Its Supply Chain Transparent]," http://www.dfdaily.com/html/113/2011/11/4/691237.shtml (accessed March 11, 2012).

37. Cao, "Pingguo shoudu chengnuo rang gongyinglian touming"; Sohu IT, "Pingguo mihui Zhongguo huanbao zuzhi: shencha jiaqiang dan toumingdu bugou [Apple Secretly Met Chinese Environmental Groups: Better at Examination but Still Lack of Transparency]," http://it.sohu.com/20111116/n325734353.shtml (accessed March 16, 2011).

38. Chen, "Meiguo liang tuanti kangyi Pingguo 'xuehan gongchang' yali huo zhuanyi zhi gongyinglian [Two American NGOs Protested against Apple's Sweatshop, the Pressure May Transferred to Its Supply Chain]," http://www.nbd.com.cn/articles/2012–02–10/632922.html (accessed March 10, 2012).

39. See He, "Renlei wuxu jingwei daziran [Human Do Not Have to Have Reverence for Nature]."

40. Gong, "Yi kexue de taidu duidai kexue shi xuezhe de tianzhi: xu *Jingwei Ziran* [Foreword of *Reverence for Nature:* Treating Science with Scientific Attitude is Scholars' Bounden Duty]," 5–11.

41. Liang, "Renlei wuxu jingwei daziran ma? [Does Human Not Have to Have Reverence for Nature?]."

Bibliography

Baidu Baike. "Huang Wanli." http://baike.baidu.com/view/20750.htm (Accessed March 5, 2012).

Baidu Baike. "Xi Zhinong." http://baike.baidu.com/view/281716.htm (Accessed March 2, 2012).

Cao, Hong. "Pingguo shoudu chengnuo rang gongyinglian touming [Apple Promised the First Time to Make Its Supply Chain Transparent]." *Dongfang Daily,* November 4, 2011. http://www.dfdaily.com/html/113/2011/11/4/691237.shtml (Accessed March 11, 2012).

Chen, Hongwei. "Nujiang shuidian kaifa zao huanbao renshi dizhi tiaozheng fangan zanwu houwen [The Exploitation of the Hydropower along Nu Jiang River Faces Opposition from Environmentalists, There Is No Follow-Up on the Adjusted Scheme Yet]." *China Economic News,* June 21, 2006.

Chen, Shijun. "Meiguo liang tuanti kangyi Pingguo 'xuehan gongchang' yali huo zhuanyi zhi gongyinglian [Two American NGOs Protested against Apple's Sweatshop, the Pressure May Transferred to Its Supply Chain]." *National Business Daily,* February 10, 2012. http://www.nbd.com.cn/articles/2012–02–10/632922.html (Accessed March 10, 2012).

China Development Brief. "NGO List". http://www.chinadevelopmentbrief.org.cn/ngo_info.php?newclass=101000000 (Accessed March 10, 2012).

Dai, Qing. *Changjiang sanxia gongcheng yingfou xingjian: xuezhe zhenglun wenji* [*Should the Three Gorges Dam Be Constructed: A Collection of Scholars Debate*]. Hong Kong: China Book Press, 1989.

Ding, Pin. *Gongdao keng yousheng: Ding Pin huanjing xinwen ji shici zuopin xuan* [*Clatter of Bow and Sword: A Selection of Environmental Reports and Poems of Ding Pin*]. Beijing: Public Literature and Art Publishing House, 2008.

18 Chinese ENGOs and some scholars and environmental volunteers. "An Open Letter to Premier Wen Jiabao and Vice Premier Li Keqiang on Stopping the Unworthy Xiaonanhai Hydropower Project." http://www.fon.org.cn/content.php?aid=15176 (Accessed April 5, 2012).

Feng, Yongfeng. *Jiaoni ruhe zuo huanbao: Zhongguo minjian huanbao chuanqi* [*How to Protect Environment: Chinese Folk Legend of Environmental Protection*]. Beijing: World Affairs Press, 2011.

Feng, Yongfeng. *Wei minjian huanbao liliang nahan* [*Whoop for Civil Environmental Protection Power*]. Beijing: Intellectual Property Publishing House, 2010.

Friends of Nature. "Guanyu baohu zanglingyang wenti de baogao he jianyi [A Report and Recommendations on the Protection of Tibetan Antelope]." In *Wei wugao de daziran,* 342–47.

Friends of Nature. "Wei baohu zanglingyang jinzhi zanglingrong feifa maoyi 'ziran zhiyou' zhihan Yingguo shouxiang Bulaier bing huo fuhan [Friends of Nature's Letter to the British Prime Minister, Tony Blair, on the Banning of Illegal Trade of the Fine Hair of Tibetan Antelope for Protecting Tibetan Antelope and the Reply of Tony Blair]." In *Wei wugao de daziran* [*Pleading on Behalf *of the Wretched Nature*], edited by Congjie Liang, and Xiaoyan Liang, 203–6. Tianjin: Baihua Literature and Art Publishing House, 2000.

Friends of Nature. "Zhichi Xi Zhinong tongzhi baohu dianxibei yuanshi senlin [Supports to the Efforts by Comrade Xi Zhinong on the Protection of Yunnan Sunb-Nose Monkey]." In *Wei wugao de daziran,* 329–31.

Friends of Nature, and Dongping Yang, eds. *Annual Report on Environment Development of China (2010).* Beijing: Social Science Academic Press, 2010.

Futrell, Chad. "The Evolution of International NGOs in China: Broadening Environmental Collaboration and Shifting Priorities." In *The China Environment Yearbook* Volume two, 159–84.

Gong, Jingcai. "Yi kexue de taidu duidai kexue shi xuezhe de tianzhi: xu Jingwei Ziran [Foreword of Reverence for Nature: Treating Science with Scientific Attitude Is Scholars' Bounden Duty]." In *Jingwei Ziran* [*Reverence for Nature*], edited by Xiangui Su, Song Tian, and Huajie Liu, 5–11. Baoding: Hebei University Press, 2005.

"Haidian Plans to Build New Incinerator." *Global Times.* http://beijing.globaltimes.cn/society/2010–11/593002.html (Accessed November 18, 2010).

He, Zuoxiu. "Renlei wuxu jingwei daziran [Human Do Not Have to Have Reverence for Nature]." *Globe,* 2005 (1).

Hu, Changmin. "Xingjian Changjiang sanxia gongcheng yi'an biaojue qinli ji [Witness of the Vote on the Bill of Constructing the Three Gorges Dam]." *Wuhan Cultural and Historical Materials* 1, 2007: 4–8.

Johnson, Thomas. "Extending Environmental Governance: China's Environmental State and Civil Society." PhD diss., University of Glasgow, 2009.

Li, Dun. "Environmental and Ecological Protection: Nation's Responsibility and Citizen's Rights." In *Annual Report on Environment Development of China (2010),* 1–40.

Liang, Congjie. "Renlei wuxu jingwei daziran ma? [Does Human Not Have to Have Reverence for Nature?]." *The Beijing News,* January 25, 2005.

Liang, Congjie, and Xiaoyan Liang, eds. *Wei wugao de daziran* [*Pleading on Behalf of the Wretched Nature*]. Tianjin: Baihua Literature and Art Publishing House, 2000.

Liu, Jianqiang. "Wei Xiaonanhai dianzhan touzhu nide yan'guang [Pay Your Attention to the Xiaonanhai Hydropower Station]." http://news.sina.com.cn/pl/2012–03–28/110124187489.shtml (Accessed March 28, 2012).

Lü, Zhi. "Biological Diversity: Double Meanings for Climate Change." In *Annual Report on Environment Development of China (2010),* 181–88.

Murray, Geoffrey, and Ian G. Cook. *The Greening of China.* Beijing: China International Press, 2004.

NetEase. "Pingguo zai Zhonguo de 'xuehan gongchang' [Apple's 'Sweatshop' in China]." http://discover.news.163.com/special/apple/ (Accessed March 12, 2012).

"Pingguo de qianxian muou [Apple's Marionette]. *Global Entrepreneur Magazine.* http://www.gemag.com.cn/html/2012/chinese_company_0112/28149.html (Accessed January 12, 2012).

Qu, Geping. *Mengxiang yu qidai: Zhongguo huanjing baohu de guoqu yu weilai* [*Dream and Wish: the Past and Future of Environmental Protection in China*]. Beijing: China Environmental Science Press, 2000.

Qu, Geping, and Jinxin Peng, eds. *Huanjing juexing: renlei huanjing huiyi he Zhongguo diyici huanjing baohu huiyi* [*Environmental Awakening: The Conference on Human Environment and the First Environmental Protection Conference of China*]. Beijing: China Environmental Science Press, 2010.

Sohu IT. "Pingguo mihui Zhongguo huanbao zuzhi: shencha jiaqiang dan toumingdu bugou [Apple Secretly Met Chinese Environmental Groups: Better at Examination but Still Lack of Transparency]." http://it.sohu.com/20111116/n325734353.shtml (Accessed March 16, 2011).

Su, Xiangui, Song Tian, and Huajie Liu, eds. *Jingwei Ziran* [*Reverence for Nature*]. Baoding: Hebei University Press, 2005.

Tang, Xiyang, and Marcia Bliss Marks. *Huanqiu lüse xing* [*A Green World Tour*]. Huhehot: Inner-Mongolia People's Publishing House, 1993.

Wang, Weiqun. "Xuere changjiang yuan: lüse xingdong [Blood Hot at the Source of Yangtze River]." In *Wei wugao de daziran*, 46–57.

Wright, Belinda, and Ashok Kumar. *Fashion for Extinction: An Exposé of the Shahtoosh Trade.* New Delhi: Wildlife Protection Society of India, 1998.

Xu, Gang. *Famuzhe, xinglai!* [*Woodcutter, Wake Up!*]. Changchun: Jilin People's Publishing House, 1997.

Ma, Tiannan, ed. *Zhongguo huanjing baohu changdao zhinan* [*Manual for Law-Based Environmental Advocacy in China*]. Beijing: Intellectual Property Publishing House, 2011.

Yang, Dongping, ed. *The China Environment Yearbook* Volume two. Leiden: Brill Academic Pub, 2008.

Zhou, Shiping, and Shutong Zhang. "Yige zhizhe de youlü: fang zhuming huanjing wenxue wenxue zuojia, shiren xugang [A Sage's Worry: An Interview with Environmental Literature Writer and Poet, Xu Gang]." *Environmental Education* 2010 (7): 12–16.

Chapter 11

The Environmental Movement in South Korea: The Discourses of Anti-developmentalism and of Saengmyung (Life)

Woojoo Chang

Introduction

> . . . Cruelly, the Saengmyung (Life) of Saemanguem has been massacred . . . Central government has distorted the value of Saengmyung (Life) . . . We cannot say Saengmyung (Life) when we destroy Saemanguem because we are Saemanguem and it is ourselves . . . For the Saengmyung (Life) of Saemanguem mud-flats and a better future, we should struggle. . . .
>
> — *The Manifesto of Solidarity against the Reclamation Project in Saemanguem,* March 2006.

The preceding quotation is an extract from *The Manifesto of Solidarity against the Reclamation Project in Saemanguem,* an area of mudflats on South Korea's west coast, where the government began a reclamation project in 1989 to create agricultural land by building a huge dam. In response to the project, fishermen, environmental activists, religious and local community groups came together to campaign against the project and to try to protect the wetland ecosystem—home to various bird species, animals and other living creatures, as well as fishing communities.

The *Manifesto* repeatedly uses the expression, "Saengmyung (Life)" of Saemanguem. In English, Saengmyung may be translated as "life" or "living beings." In this manifesto, however, Saengmyung designates not only living beings but also the whole ecosystem and the place of Saemanguem itself. Most declarations and environmental activist manifestos use rhetoric to justify and encourage environmental activities. But interestingly, instead of using language advocating the "protection of the ecosystem and the environment," the expression of protecting or respecting "Saengmyung" has been used within the environmental movement and its campaigns since the 1990s. The term Saengmyung in this context refers not only to the life of human beings, but also to the environment, embracing nonliving things, the ecosystem itself, living creatures of the mudflats, such as shellfish, crabs, and birds, and also the local fishing communities. In the South Korean context, respecting life has become the symbolic rhetoric used by environmental activists since 1990.

South Korea, together with Hong Kong, Singapore, and Taiwan, is characterized as one of the "Four Little Dragons"[1] after decades of sustained rapid economic growth since the 1970s. South Korea was regarded as one of the least promising countries in the world after the devastation of the Korean War of 1950–1953. Postwar reconstruction was facilitated by foreign aid and intensive government economic development programs. In 1960 the GDP per capita was $67. Following a military coup in 1961 and under the Park Chung hee dictatorship, the government instituted economic reforms to implement rapid development. By the end of the second military dictatorship under Chun Do Hwan, GDP per capita had risen to $5,000 in 1988.

Rapid economic growth during this period has resulted in industrialization and urbanization, and the environment has been of little consideration. The emergence of the environmental movement in South Korea is inextricably linked with the processes of industrialization that occurred during the 1970s. The state-led development activities in South Korea during the 1970s and 1980s resulted in rapid economic growth that brought economic affluence but also generated and exacerbated environmental pollution.[2] Since the 1970s, environmental concerns and the environmental movement have problematised this development process and the focus of environmentalism in South Korea has been largely about protecting life from the ravages of rapid and aggressive industrialization pursued by the government.

This chapter examines the history of environmental movements and environmentalisms in South Korea since the 1970s. It also examines the main distinctive discourses emerging from the Korean environmental movement—the discourse on anti-developmentalism and the discourse of Saengmyung (Life), based on Korean traditional ecological values. I will examine how the South Korean environmental movements have evolved in terms of the strategies, main agents, themes, and discourses. I will also analyze the characteristics of Korean environmental discourses in the context of the two case studies of the Nakdong River phenol-contamination incident and Hansalim's cooperative activities. In particular, I will focus on the role of women within Hansalim as one of the main actors within environmental movements in South Korea.

The History of the Korean Environmental Movement

The emergence of the Korean environmental movement can be divided into three stages, each of which is identified with particular issues and with specific organizational features: the 1970s (the early period), the 1980s (the period of significant public pollution: *Gonghae* in Korean), and post-1987 (the period of the proliferation of the movement) to the mid-1990s. A significant turning point in the environmental movement came in 1987 when the democratization movement[3]succeeded in ending the long-term military dictatorship and in which various civil groups were established, including a number of environmental organizations.[4] In this chapter, I will divide the history of environmental movements into three periods—pre-1987, between the mid-1980s and the mid-1990s, and from the mid-1990s to the present—according to the political changes and influences.

The Environmental Movement Prior to 1987

It is necessary to begin by considering the "stage of early period" of the environmental movement in the 1970s.[5] The Park regime's economic development plan, announced in 1962, led to massive pollution in and around industrial sites such as Ulsan, Onsan, and Yeonchun, causing serious health problems in industrialized areas. Environmental issues were raised by victims of pollution in these areas and young activists organized small investigative groups. However, these activities were on a small scale and were not well organized. Environmental groups such as the Pollution Research Group (*Gonghoe Munje Yeongu Hoe* in Korean) were formed but lacked the resources to support the victims' movements.[6]

While in the 1970s the environmental movement was not effective and not established, the 1980s saw the emergence of an "antipollution" movement. In 1982, some Protestant pastors, Catholic priests, and social activists established the Pollution Research Institute (PRI, *Gonghoe Munje Yeonguso* in Korean), which sought to offer support to pollution victims by researching pollution problems and by providing education for the public about the environment. In 1983 Onsan residents reported symptoms of neuralgia, first of the lower back and later of the whole body, which they linked with pollution from local metal foundries. Some college students and researchers investigated this and noticed that even teenagers showed these symptoms, which could develop rapidly (in two to three years) to include general paralysis, dermatitis, and respiratory disease. This environmental group suggested that up to 500 residents suffered from these symptoms as a result of the harmful effects of pollution from factories. After debate with the government as to whether or not these illnesses were caused by pollution, in 1985 the South Korean government proposed a relocation plan, moving about 10,000 residents (around 2,000 families) to another region. During this period, the environmental movement undertook investigations of polluted areas and tried to gain compensation for victims.[7]

The concept of "public nuisance" (*gonghae*) is important for those concerned with understanding the characteristics of the Korean environmental movement between the 1970s and the mid-1980s.[8] This concept was adopted to refer to pollution that affects the public. Environmentalists assumed that pollution issues had significance for the public and they regarded it not as the result of any one company's action, but rather as the result of the government's policy of rapid industrialization, which they considered to have ignored the harmful consequences of such development. Prior to 1987, environmental movements were pitted against the state, which tried to cover up the negative aspects of development. Environmental activists campaigned against the government, launching campaigns to expose the suffering of peasant farmers and urban residents, which they attributed to industrial pollution.[9] The term *environmental rights* had wide circulation during this period, suggesting that people were keen to protect their living environment, homes, fields, and health against the negative impact of economic and industrial development.

Most environmental activists of this period had been involved in the student movement that had campaigned against the dictatorship. In the 1980s the Korean environmental movement might therefore be said to have focused not only on environmental problems but also on social structures that they identified as related to the causes of pollution.

The Emergence of the South Korean Environmental Movements between 1987 and the mid-1990s

Successive military dictatorships between 1961 and 1987, from the Park regime to that of Doo-whan Chun, were challenged by the democratization movement. This movement aimed to overthrow authoritarian South

Korean governments, which severely repressed and restricted basic democratic rights and freedoms.[10] The government of this period maintained that it was in citizens' interests that democratic rights be restricted to enable economic growth. Many protesters from student and labor organizations were arrested during the period of the military dictatorships. The labor movement, the student movement, and middle-class citizens who were against the authoritarian regimes came together to try to bring about political change. This culminated in a massive demonstration in 1987. The protests of this period led to the calling of a presidential election and to the establishment of a more democratic regime.

After the transition to a democratic government occurred in 1987,[11] new social movements involving groups that emphasized "citizens' rights" emerged within the new civil society. The South Korean version of civil society can be analyzed with reference to Jurgen Habermas's concept of the "public sphere." Habermas's framework posits that society can be divided into various sectors and domains—the state, the market, and civil society. The state sector is associated with the range of activities organized and legitimated by formalized and centrally coordinated coercive power, which governs the nation and regulates various other sectors. The market sector consists of private enterprise, which is mainly engaged in business activities for profit through the production and exchange of goods and services.[12] The civil society sector is comprised of the residual domain outside the realms of the state and the market, sometimes referred to as the "independent sector," the "third sector," or the "voluntary sector."[13] This domain involves a wide range of self-governing and self-generating private organizations in which citizens can exercise their initiatives in their own right for the common good.

The characteristic of civil society in the South Korean context was that the key actors were industrial laborers and middle-class citizens. Since 1987, citizen movements have rapidly expanded and have constituted a new set of organizations, thereby creating space for the association of diverse interest groups seeking the extension of citizens' rights and the enlargement of the public sphere. The citizen movement groups led to the establishment of several key organizations, such as Citizens' Coalition for Economic Justice (CCEJ) and the People's Solidarity for Participatory Democracy (PSPD), which motivated others to organize other nongovernmental organizations (NGOs).

Environmental groups and organizations flourished with the blossoming of civil society. For example, the Korean Anti-Pollution Movement Association (KAPMA) was established in 1988 by merging two existing environmental organizations (the Korea Anti-Pollution Citizens' Movement

Council and the Korean Anti-Pollution Movement Youth Council). KAPMA focused on pollution control and raising public awareness of pollution and it also started the debate on nuclear issues, launching the first major campaign against the construction of nuclear power plants in South Korea.

In 1991 a major industrial accident resulted in the release of phenol into the Nakdong River from local beer factories in Daegu, located in the southern province of South Korea. This accident raised public environmental awareness about quality-of-life issues as members of the public began considering the relationship between drinking water and water pollution from factories. This incident captured mass media attention and became a symbolic issue in South Korea.

Also, the 1992 United Nations Conference on the Environment and Development (UNCED) marked another turning point for the South Korean environmental movement. At the nongovernmental forum of UNCED, South Korean environmental activists exchanged accounts of their experiences with representatives from Western countries. After UNCED, environmentalists and South Korean government reformers began to address global environmental issues and to formulate a South Korean environmental agenda.[14] Since the early 1990s, many of the theories and concepts of Western environmental discourses have been introduced and discussed both in the NGO community and by academics in South Korea.

After 1995, women-led community environmental organizations grew, undertaking a range of local activities and campaigns: against local waste-burning plants, antidevelopment projects, and advocating recycling. Campaigns included attempts to prevent the construction of incineration plants in the Sanggye and Guro areas in Seoul, and to prohibit the construction of golf courses in Gangnam and Woojang Mountain. Some women's activities focused on developing green consumption patterns through organic cooperatives that connected directly with organic agricultural production. Local women also launched the *Noeksaek Gagae* (The Green Store) initiative, a community-based project in Gwachun, a second-hand shop where local people could exchange goods, focusing on the quality-of-life issues. So, in this period, the environmental movement had been constructed and structured as the growth of civil society in South Korea.

The Third Stage: From the Mid-1990s to the Present

From the mid-1990s onward, environmental movements have flourished and become varied. Several developments have been contested and halted as a result of campaigning efforts for natural conservation, for example, in the case of the Anti-Dong River Dam Campaign and the Anti-Saemangeum Reclamation Project Campaign, which were grand governmental projects of development. Secondly, various cooperatives between consumers and producers for organic food materials have been established. Religious leaders have become involved in the environmental movement in conducting the Anti-Saemangeum Reclamation Project Campaign and Protecting Salamander of Chungsung Mountain Campaign.

Plans for the Dong River dam project were announced by the government to prevent flooding in Yeongwol, Gangwon Province, in 1991.[15] According to the plan several small villages would be submerged for the dam. In April 1998, Korea Federation of Environmental Movement (KFEM) and local people's groups campaigned to cancel the project so that the natural ecosystem would remain protected. The disputes of this project were between the necessity of the dam to protect from floods and the natural and cultural values of the submerged areas. The protesters and media showed the clear and beautiful scenery and unique species of the Dong River areas. These beautiful scenes moved people to act in defense of the environment. Through the citizen-governmental investigation panel to examine the environmental impact of the dam construction, Kim Dae-jung's government finally announced the cancellation of the project in 2000.[16] This was a landmark case in the history of the environmental movement in South Korea. It was the first time that ecological values had been placed above the necessities of human beings in decision making. The natural beauty and ecosystem were regarded to be protected. Also, the main activists were varied—there were professional photographers who were involved in producing pictures of natural beauty in this area; there were local people, and experts of ecology and environment; media journalists were significant and helped to broaden the environmental concerns to the public. Finally, the campaign was successful in influencing policies in the public sphere.

Organic cooperatives, such as Hansalim, Doorae organic cooperatives, ico-op, and other localized cooperatives, have been set up and have flourished. The cooperatives' members have established direct exchange of organic products between consumers and organic farmers. The cooperatives have also had an active role in local environmental issues, such as the anti–Sungmi Mountain water tank project and the antiwater fluoridation campaign, which were related to local people's everyday lives.

The main reason these cooperatives have been successful is because of consumers' increasing consideration of the relationship between environmental concerns and health. The majority of consumers are mothers, who care for family members' health. According to recent newspaper reports, about 40 percent of Korean children suffer from atopic dermatitis, 32 percent of them have allergic rhinitis, and 24 percent of them have asthma.[17] The cause of atopic dermatitis has not been confirmed medically, but the main causes are suspected to be genetic heredity, environmental contamination, or a combination of both these factors. Symptomatic treatment is available at dermatology clinics but some mothers found treating these skin problems with steroid creams ineffective. Organic foods are helpful to decrease the atopic symptoms. So mothers became members of consumers' cooperatives. Furthermore, the consumers' groups became concerned with local environmental issues and organized local campaigns such as the antiwater fluoridation campaign in Gwachun, which was initiated by the Gwachun branch of Hansalim between 2000 and 2003.

The environmental movement has also seen the active involvement of religious leaders. In 1992, the plan for the construction of an express railroad was announced between Daegu and Kyungasju in Kyungsang Province of the southern part of South Korea. These plans included the construction of a tunnel through Chungsung Mountain. Ji-yul, a Buddhist nun and a leader of a temple of the mountain, claimed to stop the project to protect the salamanders, an indigenous species that live specifically on the mountain. When the case went to the court, the case was called the "Salamanders' Suit." The Buddhist leader went on hunger strike four times between 2003 and 2004, the longest one for 100 days, which threatened her life. The hunger strike for the protection of the salamanders of Chungsung Mountain was symbolic and evoked people's empathy for salamanders and the ecosystem. Owing to her hunger strike, the case went to the Supreme Court, though the court ruled in favor of the project in 2005. The hunger strike was a new turn in the environmental movement in campaigning against governmental development projects.

In the case of the Anti-Saemanguem Reclamation Project Campaign, it was led by religious leaders, such as Buddhist monks, pastors, priests, and others. They led the campaigns in a way of "three steps and one bow" all through the South Korean peninsula from March 28 to May 31 in 2003. During the day protesters walked and bowed peacefully and slowly through the land and in the evening, they organized discussions with local people, with whom they stayed. Although the campaign was not successful in stopping the construction, this new symbolic way of protesting impacted other social movements and activities, which represent a peaceful way. The three steps and one bow originated from a way of Buddhist practices. The protesters were showing their sacrifices and peaceful attitudes to present their claims.[18] During this period, the environmental movements have begun to flourish, as social changes have seen an increased participation in civil society.

Two Discourses of South Korean Environmentalisms

An Australian political-ecological scholar, John Dryzek, explains Foucault's concept of discourse as meaning a shared way of apprehending the world. Discourses can be embedded in language, which can enable the narration of coherent stories and accounts.[19] In his book, *The Politics of the Earth: Environmental Discourses,* Dryzek emphasizes Foucault's notion of discourses as constructed with "signs of power that actors can get the discourse to which they subscribe accepted by others. Discourses can themselves embody power in the way they condition the perception and values of those subject to them, such that some interests are advanced, others suppressed."[20] According to Dryzek's view, environmental discourses can be identified with (1) basic entities recognized or constructed, (2) assumptions about natural relationships, (3) agents and their motives, and (4) key metaphors and other rhetorical devices.[21]

In following Dryzek's idea of environmental discourses, I suggest that South Korean approaches to the environmental movement may be organized under two categories of discourses—that of anti-developmentalism and of Saengmyung (life). Each discourse may be characterized as having different values, conceptualizations of nature, main environmental issues, types of activities, main actors, and strategies of activities. Those environmental discourses may produce critical rhetoric or alternative solutions to environmental issues.[22]

The Nakdong River Phenol-Contamination Incident and the Discourse of Anti-developmentalism

On March 21, 1991, Doosan Electro-materials company, located in Gumi in the southern part of South Korea, unintentionally leaked 30 tons of phenol extracts into the Nakdong River, a source of tap water for the citizens in the city of Daegu.[23] Environmental groups criticized the company and organized the boycotting of Doosan's products. The citizens of Daegu refused to pay their water bills. A month later and despite strong campaigns against them, the government and Doosan announced that they would resume operations as normal. However, within a week, the leaked phenol contaminated the Nakdong River again.

The incident became known as the Nakdong river incident and was a key moment in the South Korean environmental movement in empowering civil society through citizens' concerns and participation. The issue was covered widely by the media and there was widespread outcry by citizens whose everyday lives had become affected by this environmental problem.

An association was formed by women who suffered miscarriages in the aftermath of the Nakdong River spill. The formation of this association marked a turning point in the involvement of women in the South Korean environmental movement because the campaign was initiated by women, who had miscarried, suggesting a relationship between women's health—in this case, that of pregnant women and mothers—and a harmful environment.[24] However, these women's concerns were not given much coverage in the mass media. When these women raised their health concerns to the government and to the company responsible for the pollution, they sought scientific evidence linking polluted drinking water and miscarriage. However, they did not find it easy to establish a connection between pollution and miscarriage. Some government officials thought that these women were seeking illegitimate ways to get compensation. In 1991, these women could not garner support from other women's groups because they did not acknowledge the importance of environmental issues at that point. Without support from women's groups, experts, or the mass media, these women's complaints were largely ignored. Several years later, some women environmental activists and ecofeminist scholars did reevaluate the activities of these women who had suffered miscarriages in the wake of this industrial accident as a crucial moment in the history of women's environmental activism in South Korea.

Following the Nakdong river contamination incident, there was widespread criticism of the Korean governmental policy of prioritizing development and industrialization and disregarding environmental concerns. Anti-developmental sentiments were widespread throughout the environmental movement. With the support of the public, the organization of the environmental movement was expanded and institutionalized. More citizens were able to become members of environmental organizations and support the organization. The incident was critical in establishing a civil society in which public concerns could be voiced for the first time.

Since the mid-1990s, environmental movements in South Korea have successfully campaigned for the protection of nature and ecosystems over human-centered development. In the contemporary context, environmental discourses in South Korea not only address the environment *per se*, but also consider the broader social context in which the environment is situated. Many South Korean environmentalists have pointed out that South Korea's main environmental problems stem from rapid industrialization since the 1970s, claiming that the state and corporate sectors have not considered the environment in prioritizing economic growth. The main debate on environmental issues has generated two opposing perspectives, which can be labeled as *environmentalism* and *developmentalism.*[25] Environmentalists are concerned with the conservation of the environment, as problems emerge as a result of development and construction projects. Conversely, developmentalism asserts the priority of economic growth and development, suggesting that environmental problems may be tackled through the deployment of advanced technology.[26]

Sung-tae Hong points out the relationship between the national development project, led by the Park Jung-hee regime, and the ecological crisis. He claims that the development projects were initiated with industrialization and constructions of land planning. The result of compressed industrialization had influenced to cause pollutions and destroy ecosystem.[27] So, the discourse of anti-developmentalism has constructed the contradictory discourses between developmentalism and environmentalism.[28] The discourse of anti-developmentalism criticizes the negative aspects of market-led and capitalist economic growth. South Korean environmental activism and discourses have thus placed an emphasis on seeking alternative social systems that favor "green" values such as environmental preservation, respect for life, and grassroots democracy to reform the social system and which thereby create an ecological value system. The South Korean environmental movement has grown exponentially in a very short time: extending from its initial passive approach that focused on environmental-pollution victims in the 1970s and 1980s. South Korean environmental activists went on to propose full-blown environmental agendas, which questioned the prioritization of development and industrialization.

The discourse of anti-developmentalism has primarily been politically motivated and has emphasized the need for change to social and political structures and systems. The goals of the movement are to change social and political structures from ones that prioritize economic development progress to ones that emphasize the process of democratization. Specifically, the case of phenol contamination in Nakdong River inspired the public to campaign for the environment and with the establishment of civil society, various professional environmental organizations were established with the opening of new political opportunities.

Myung-rae Cho notes that there was a change of focus in environmental consciousness in the early 1990s, as attention shifted from the concept of "public nuisance" to a wider, prevention-oriented notion of "the living environment as a whole."[29] Hence, the Korean environmental movement has not confined itself only to issues pertaining to the environment, but rather, it has positioned itself as a leading political force for a broad "green politics," which has involved pursuing the preservation of the environment, life, and extending democracy more broadly. Civic consciousness about the environment has evolved, shifting from a more passive approach, involving activities such as seeking compensation for specific pollution cases, to more proactive strategies, which entail trying to prevent damage to the environment as a whole and arguing for civil and environmental rights.[30] So the South Korean environment movement has gained political influence. Though the anti-developmentalist approach has seen many gains it has not been able to change social and political structures based on green politics, establishing the Green Party.

The Discourse of Saengmyung (Life) and the Case Study of Hansalim

The Discourse of Saengmyung (Life)

While the anti-developmentalism discourse was constructed with sociopolitical concerns, reflecting the emergence of civil society and democratic political processes, another discourse that emerged in the South Korean environmental movement was the discourse of Saengmyung (life). The discourse of Saengmyung, however, offers an alternative ecological discourse, one which emphasizes the interrelationship between nature and human beings, respect for life, direct participation in local democratic process, and direct exchange between consumers and producers in the local communities.

The term *Saengmyung* was first used in an environmental context by the ecological thinker and poet Ji-ha Kim in the late 1980s. Ji-ha Kim is also a vocal political leader. In 1964, he protested against South Korea's military dictatorship and was imprisoned several times—for a total of seven years. After his release, he devoted himself to studying both Korean traditional thought, "Donghak,"[31] and Western ecological thought. He also became involved in organizing Hansalim, the biggest organic cooperative in South Korea, established in 1986.

Since Ji-ha Kim began to use this term as a core concept, the slogan "respecting Saengmyung" has become a popular expression amongst ecological and environmental activists. The expression carries emotional appeal, since it is linked to the desire to save the life of the ecosystem. Ji-ha Kim has developed the ecological discourse of Saengmyung around this understanding of the concept. The discourse of Saengmyung has expanded the understanding of life and has inspired many within the South Korean environmental movements to shift away from a human-centered viewpoint toward a more ecological viewpoint.

The discourse of Saengmyung first emerged in the wake of Ji-ha Kim's exploration of the concept in the early 1980s. He argued that the background against which the term Saengmyung in modern Korean culture arises is the "culture of death," that is, the rapid economic development and growth which, in Ji-ha Kim's view, has been paradigmatic in Korean society and which he regards as threatening the value of life.[32] He contends that rapid economic development has led to a Korean society which is willing to take risks with life, resulting in serious disasters such as the collapse of the Sungsoo Bridge on November 21, 1994, and of the Sampoong department store on June 29, 1995.[33] Kim saw these events as precipitating a "total crisis of values" in Korean society.[34] For him and for others sharing similar views to his, respecting life in the sense of Saengmyung has profound implications for transforming the core values of modern industrialized Korean society.

In the discourse of Saengmyung, the core term itself implies a special sense of life, distinguishing it from its literal meaning. The concept of Saengmyung means literally being alive as an organism. According to the *Oxford English Dictionary*,[35] life primarily refers to the condition that distinguishes animals and plants from inorganic matter. Rather than invoking this literal definition, Ji-ha Kim suggests that there is a wider concept of life. He claims that, in general, the concept of "environment" assumes a division between the natural world and human beings. On the other hand, the concept of Saengmyung assumes the interrelatedness between the natural world and human beings.

Ji-ha Kim also suggests that the concept Saengmyung assumes that nonliving matter is a part of the same realm as organic animals and plants and he sees nonliving matter as sustaining the life of animals and plants. He calls inorganic matter a "hidden world."[36] This sense of the concept of life is closely associated with the notions of "vitality, vigor, or energy." Kim assumes that inorganic matter contains vital energy that is crucial in supporting life. Thus, not only living beings but also inorganic matter is seen as a significant component within the whole organic cycle.[37]

From Kim's viewpoint, the concept of Saengmyung construes nature as neither passive nor determined, but rather, as "active" and "alive." This understanding of nature in the discourse of Saengmyung is similar to the view of nature as an organism, which the ecological feminist philosopher Carolyn Merchant traces in *The Death of Nature* (1983). As Merchant explains, prior to the Scientific Revolution in the 16th and 17th centuries in the Western world, nature was regarded as "a living entity", organically alive and vital.[38]

In the Korean context, the term *Saengmyung* is much more popular and has greater resonance than the term *ecology*. Whilst as a scientific term *eco* is not familiar to Korean people, the term *life* evokes "sentimental feelings of familiarity,"[39] which is significant since the concept of life in the Korean context incorporates the implication of respecting or protecting life. The rhetorical slogan of "respecting life" and "protecting life" encourages individuals and groups toward emotional investments that sustain them in taking responsibility for protecting life.

Furthermore, the rhetorical invocation of "respecting Saengmyung" can be extended and translated into protecting "the Earth" as "*oikos*." The advocate of deep ecology, Warwick Fox, suggests that within Arne Naess's "ecocentric" idea, the "eco" implies "*oikos*," meaning home. This metaphor can be expanded to include the notion of "the Earth" as "the home of all beings."[40]

The main ideas of the discourse of Saengmyung are that:

1) human beings and nature have immanent spirits, which make them respect beings of moral concern;
2) human beings should respect others, including matter;
3) according to holistic interrelations, "individuals and parts" are related to the "whole" and wholes should not be reduced to smaller parts;

Thus, this discourse of Saengmyung challenges hierarchical dualistic viewpoints, such as those associated with the construction of divisions between human beings and nature. Furthermore, emphasizing the interrelatedness of humans and nature can lead to the valuing of coexistence and harmony between humans and nature as an ecological precept.

Hansalim

The discourse of Saengmyung asserts the value of interrelationship between human beings and nature as a way of exploring ecological values and green politics. This discourse has inspired the establishment of small, local communities based on ecological values such as the local cooperative Hansalim. Here, local people participate in the process of decision making as members of small cooperatives and groups.

The cooperative movement has its origins in utopian socialism. Cooperatives began in Britain where, in 1844, the idea of cooperatives emerged based on laborers' activities. From this original form, cooperatives have evolved into different forms in different locations. In 1995 the International Co-operative Alliance (ICA) offered the following account of these social and political formations:

> A co-operative is an autonomous association of persons united voluntarily to meet their economic, social, and cultural needs and aspirations through a jointly-owned and democratically-controlled enterprise.[41]

South Korean organic cooperatives were first set up in the late 1980s in order to develop a direct relation between producers and consumers around organic materials and to realize a sustainable environment.

One of the aims of organic cooperatives in South Korea is to develop an alternative to the economic system associated with modern industrial societies. The critics who focused on the by-products of industrialization reflected broadly on the problems of modern culture. The South Korean movement emerged from concern about the crisis of Korean agriculture, which was under pressure with the opening of markets and as a result of the requirements imposed by the World Trade Organization.

Where the exchange between consumers and producers is direct, as in organic cooperatives, women predominate in most of the consumer groups—mainly housewives and mothers. In South Korean society, women still often take the main responsibility within the family for shopping for foods and for products required for daily life. Women's everyday lives are often seen to revolve around domestic work and the responsibilities of caring. Organic cooperative movements can provide appropriate resources given some women's environmental concerns related to everyday matters including eating, health, and consumption. The style and system of organic cooperatives facilitate women's involvement: ordinary citizens can register for membership and buy organic materials and foods from cooperative shops and they can participate in local environmental activities.

It is worth noting that women, who are often marginalized in the public sphere, may become core activists in their local communities through local organic cooperatives. As the Japanese cooperative movement leader Yokota Katsumi points out, the subjects of cooperatives are "housewives, women and people who have been marginalized from modern industrial society."[42] In South Korea, most husbands work in towns and cities, at a distance from their places of residence and most are absent from the home during the daytime, while housewives, children, the disabled (who are not employed), the unemployed, and old people stay at home all day. Hence, these are the people who are most able to participate in cooperative activities. Another characteristic of the organic cooperative movement is that it first emerged in urban areas, meaning that most of its activities occur in cities. The organic cooperatives in South Korea were organized in metropolitan centers and they mainly involved women living in large residential areas.

Hansalim, the biggest farmer/consumer cooperative in South Korea, was started in 1986, originally with the aim of protecting Korean agriculture, which was deeply affected by the initiatives of the World Trade Organization (WTO) and by the increasing globalization of trade. Literally, the meaning of "Han" in Hansalim is "one, the whole, together" and "Salim" means "to save everything" or "to live" in Korean. So Hansalim translates as "saving every life" and has connotations referring to the social implications of living with nature, concern for future generations, and the cooperative coexistence of urban and rural areas.

Hamsalim presents the following as its aims and core values:

to achieve a holistic life, as an integral part of the life of the evolving individual, and to coexist in harmony;
to be ecologically aware of nature, because human beings are responsible for the earth and the eco-system;
to build a social community and rebuild communities we have lost;
to build an alternative life-affirming culture;
to build social activism according to the principles of life by proactively refusing to participate in the industrial culture, which supports the bureaucratic system;
to practice a life-affirming spirituality and recognise the relationship between nature, the individual and society, and
to follow a practice of "united lives," involving everyone in the making of Korean history.[43]

The core emphasis of Hansalim is on the valuing of the "interrelationship" between humans and nature, humans and society, and consumers and producers. Originally, Hansalim was set up by men from the farmers' movement of the 1970s and 1980s, resulting initially in a male-dominated cooperative. The consumers were predominantly city women, while the producers were farmers in agricultural areas. In the early stages of the movement, women's roles were primarily to consume organic foods and products while living in the cities.

The first Hansalim store opened in Seoul in 1986 to promote the direct transfer of organic rice between rural and urban areas, when Korean agriculture faced a severe decrease in the number of farmers, caused by the falling price of rice.[44] In other words, Hansalim was, at its onset, initiated by farmers and producers, rather than by consumers. In April 1988 the consumers' group within the cooperative was officially set up by urban housewives. Hansalim has 16 local branches, eight producer associations, and a research institute for Hansalim activities. It became a nation-wide organization in 2006.

In 2007, Hansalim had about 147,339 families and gross sales to the value of 108,950,000,000 South Korean won (KRW) a year. In 2011, it had 293,442 families and gross sales to the value of 222,581,000,000 KRW.[45] Through facilitating direct exchange around organic foods and products, Hansalim is oriented around social issues relating to the environment and with local agendas in urban areas. For instance, the Seoul branch of Hansalim has organized various campaigns: protesting against the building of an expressway across Bukhan Mountain; a healthy school meals campaign; a campaign to protect Korean rice (from the regulations of the WTO); and a campaign against GM (genetically modified) foods.

Although several women leaders were unhappy with women's status in Hansalim, their concerns were largely ignored in the 1990s.[46] However, female leaders with feminist viewpoints began to appear as consumer leaders and women's leaders began to demand recognition for their activities within Hansalim. Recently, there has been a debate about payment for women's activist work in Hansalim, as such work had previously been unpaid.

Women's concerns within Hansalim have gradually shifted from a main focus on ecological matters and on the crisis of Korean agriculture from a consumers' perspective, toward greater concern for gender issues, such as paying women members for their work and women's status within the organization. With reference to the core values of Hansalim, many women activists in this cooperative consider the phrase "respecting Saengmyung (life)" as important in grounding their personal activities.

The following things may be observed from the increased participation of women at Hansalim. First, women activists become involved in environmental issues through caring for their children and their health-related issues such as atopic dermatitis. In caring for their unwell children, they begin to become conscious of the link between health issues and the environment and begin to get involved in environmental activism. Second, women's interest and engagement with environmental issues have evolved from personal experience to wider local environmental issues. For example, when women become involved in the consumption of organic food products or in helping out in local stores, they are more likely to eventually take an active interest in local environmental issues and eventually become leaders.

Conclusion

This chapter has examined the history of the South Korean environmental movement. Two distinctive discourses, of anti-developmentalism and of Saengmyung, have emerged in reaction to rapid development and its environmental penalties. Anti-developmentalism challenges the current social and political structures while Saengmyung and cooperative activities offer an alternative to the market system.

Anti-developmentalism is a reaction to the prioritization of rapid industrialization and economic growth. The discourse has been constructed with sociopolitical concerns, reflecting the emergence of civil society and democratic political process. In addition, an alternative ecological discourse of Saengmyung suggests the interrelationship between nature and human beings, respect for life, direct participation in local democratic process, and direct exchange between consumers and producers in the local communities (see Table 11.1).

Both the discourse of antidevelopment and the discourse of Saengmyung are opposed to rapid development and construction. They are also similarly critical of the government and corporations who continue to push forward with development projects. The discourses are also aligned in having strong stances toward environmental health, water, air pollution, waste, nature preservation, and nuclear power plant and nuclear waste agendas.

The discourses diverge in their approaches in the following ways. Their approach to environmental activism itself is different. Anti-developmentalism aims to radically influence structural and political changes. It believes that the democratic participation of citizens is an important part of the process. In contrast, the discourse of Saengmyung emphasizes the consideration of the interrelationship between human beings and nature in everyday life and promotes the direct participation of people in local politics. Also, while discourses of anti-developmentalism see the primary agents of change as professional researchers, experts, and activists, the discourse of Saengmyung sees women, farmers, and local citizens as important activists of change.

Although the environmental movement has seen significant gains and successes, as yet neither discourses have been able to effect radical social

Table 11.1. Comparison of the Characteristics of the Discourses of Anti-Developmentalism and Saengmyung (life)

Contents	Discourse of Anti-Developmentalism	Discourse of Saengmyung (life)
Main value	Democratization and participation in political structure Political empowerment for sound environment	Interrelationship between human beings and nature Respect for life Direct participation in local politics
Main issues	Waste contamination of industrial sites Environmental health Water, air pollution and waste Nature preservation Nuclear power and nuclear waste	Quality of food products Environmental health Water, air pollution and waste Nature preservation Nuclear power and nuclear waste
Type of activities	Demonstration Organizing local environmental movement Consciousness direct action, boycotts	Co-operatives: direct exchange and producers and consumers Education Consciousness-rising through the media
Main activists	Professional environmental activists Professional researchers Local environmental groups and ordinary citizens Workers Artistic and cultural groups	Co-operatives including housewives, female members, organic farmers Grassroots ecological groups Religious groups Artistic and cultural groups
Characteristics	Impact to change environmental policies Structural change through movements Nation-wide movement	Changes of lifestyle based on ecological ideas
Social and political goals	Empowerment of civil society in the public sphere Changes to social structure based on social ecology	Local autonomous political system

and political structural change. For example, it was only as late as 2011 that the Green Political Party was established in Korea, as a result of the breakdown of the Fukushima nuclear reactor following the earthquake in Japan. Unfortunately the Green Party was unable to secure any seats during the 2012 parliamentary election.

On the other hand, the cooperatives have offered a space for local grassroots groups, women, scholars, and religious leaders to experience a new form of political practice based on ecological perspectives. These experiments have not reformed any social or political structures, but they have established a subculture of localized political practice.

Notes

1. Vogel, *The Four Little Dragons: The Spread of Industrialization in East Asia.*

2. Gu, "The Korean Environmental Movement: Green Politics through Social Movement," 185–219; Cho, "Emergence and Evolution of Environmental Discourses in South Korea,"139–64; Kim, "The State and Civil Society in Transition," 595–613.

3. The Democratization Movement of the 1980s focused on bringing down the military authoritarian government, which held power in South Korea between 1961 and1989.

4. Kim, "The State and Civil Society in Transition," 595–613; Cho, "Emergence and Evolution of Environmental Discourses in South Korea," 139–64.

5. Moon, *Korean Women's Environmental Movement: The History, Subjects and Types of Activism.*

6. Cho, "Emergence and Evolution of Environmental Discourses in South Korea," 139–64.

7. Gu, "The Korean Environmental Movement: Green Politics through Social Movement," 185–219; Shin, *Friends of Nature: 25 Years History of Environmental Movement.*

8. Cho, "Emergence and Evolution of Environmental Discourses in South Korea," 145.

9. Ibid.,145–46.

10. Kim, "The State and Civil Society in Transition," 599.

11. The uprising of June 1987 involved a symbolic political transition in South Korea, with the mass participation of citizens in peaceful demonstrations designed to change the military government, the Chun Regime. This involved the call for a directly democratic presidential election.

12. Cohen and Arato, *Civil Society and Political Theory.*

13. Habermas, "Further Reflections on the Public Sphere," 453.

14. Shin, *Friends of Nature 2: 25 Years History of Environmental Movement,* 128–74.

15. Gu, "The Korean Environmental Movement: Green Politics through Social Movement,"198.

16. Ibid., 199.

17. *Seoul Newspaper,* June 2, 2006.

18. Hong, *Criticizing Developmentalism: Beyond Park, Jung-hee Regime System to Ecological Welfare Society,* 169.

19. Dryzek, *The Politics of the Earth: Environmental Discourses,* 9.

20. Ibid.

21. Ibid., 19.

22. Dryzek suggests the checklist of items for assessing the effects of discourses; (1) politics associated with the discourse, (2) effect on policies of governments, (3) effect on institutions, (4) social and cultural impact, (5) arguments of critics and (6) flows revealed by evidence and argument. Dryzek, *The Politics of the Earth: Environmental Discourses,* 21.

23. Gu, "The Korean Environmental Movement: Green Politics through Social Movement," 197.

24. Moon, *Korean Women's Environmental Movement: The History, Subjects and Types of Activism,* 57.

25. Cho, "Emergence and Evolution of Environmental Discourses in South Korea," 139–64.

26. Ibid., 160–61.

27. Hong named the state of governmental development projects as "construction state." In his view, construction and land planning could provide the economic benefits and growth. Hong, *Criticizing Developmentalism: Beyond Park, Jung-hee Regime System to Ecological Welfare Society,* 59.

28. Cho, "Emergence and Evolution of Environmental Discourses in South Korea," 152–53.

29. Ibid., 146.

30. Ibid., 147.

31. Donghak literally means Eastern learning in Korean. In the 1860s, Donghak was established, drawing its ideas from Confucianism, Buddhism, and Taoism. Donghak was the catalyst for a peasant revolution (1894), which set out to replace feudalism and to realize social reform, including tax reduction, land distribution, and human rights.

32. Kim, *Life and Autonomy: What Is Saengmyung's Thought and Movement?,* 247.

33. The Sungsoo Bridge, which crosses the Han River, in Seoul, suddenly collapsed in 1994, killing 32 people. The Sampoong Department Store collapsed in 1995, and 501 people died and 6 were declared missing. The main causes of these disasters were rapid and faulty construction work and poor maintenance, linked to fraudulence.

34. Kim, *Life and Autonomy: What Is Saengmyung's Thought and Movement?,* 11.

35. This *Dictionary* offers the following definitions: **1.** the condition that distinguishes animals and plants from inorganic matter, including the capacity for

growth and functional activity **2.** the existence of an individual human being or animal.etc., **3.** a particular type or aspect of people's existence **4.** living things and their activity **5.** a biography **6.** vitality, vigor, or energy **7.** informal sentence of imprisonment for life **8.** (in various games) each of a specified number of chances each player has before being put out **9.** before a noun (in art) based on a living rather than an imagined form. (http://www.askocford.com/concise_oed/life?view=uk)

36. Kim, *Life and Autonomy: What Is Saengmyung's Thought and Movement?,* 45–46.

37. Here, the term *Saengmyung* is similar to Arne Naess's concept of "life" in deep ecology theory. Naess refers to the "intrinsic values of non-human life," indicating that by this he includes what biologists classify as "non-living; rivers (watersheds), landscapes, the ecosystem." Like Naess, Ji-ha Kim also stresses the need to consider inorganic matter as a significant part of the ecological system, with its own intrinsic value. Naess, "The Deep Ecological Movement: Some Philosophical Aspects," 197.

38. Merchant, *The Death of Nature: Women, Ecology and the Scientific Revolution,* 111.

39. Moon, *Discourses on Ecology,* 142.

40. The discourse of Saengmyung is, in some respects, similar to James Lovelock's "Gaia theory," which views living and nonliving parts of the earth as a complex interacting system that can be thought of as a single organism. Fox, "The Deep Ecology-Ecofeminism Debate and its parallels," 8.

41. Kim, *Life and Autonomy: What Is Saengmyung's Thought and Movement?,* 296.

42. Katsumi and Na, *Tender and Strong Citizens in a Foolish Country: Building up Models of Life-oriented Club Movements and Democratic Grassroots' Activism,* 10.

43. www. hansalim.co.kr.

44. Mosim and Salim, *20 Years Old Hansalim Embraces the World.*

45. Cho, "The Present and Future of Korean Co-operatives: Focusing on Organic Co-operatives," 12.

46. Moon, *Korean Women's Environmental Movement: The History, Subjects and Types of Activism,* 130.

Bibliography

Chang, Woojoo. "Women's Collective Caring Practices in the South Korean Environmental Activism." PhD thesis, Lancaster University, 2011.

Cho, Myung-Rae. "Emergence and Evolution of Environmental Discourses in South Korea." *Korea Journal* 44, no. 3 (Autumn 2004): 139–64.

Cho, Wan-hyung. "The Present and Future of Korean Co-operatives: Focusing on Organic Co-operatives." Paper presented at the 2012 Indramang Village School, Seoul, June 14, 2012.

Cohen, Hean L., and Andrew Arato. *Civil Society and Political Theory.* Cambridge, MA: MIT Press, 1992.

Dryzek, John S. *The Politics of the Earth: Environmental Discourses.* Second Edition. Oxford: Oxford University Press, 2005.

Fox, Warwick. "The Deep Ecology-Ecofeminism Debate and Its Parallels." *Environmental Ethics* 11 (Spring 1989): 5–25.

Gu, Do-wan. *Hankuk Whankyung Woondongui Sahuihak (Sociology of Korean Environmental Movement: Towards Justice and Sustainable Society).* Seoul: Moonhakgwa Jisungsa Publispher, 1996.

Gu, Do-wan. "The Korean Environmental Movement: Green Politics through Social Movement." *Korea Journal* 44, no. 3 (Autumn 2004): 185–219.

Habermas, Juergen. *Between Facts and Norms: Contributions to a Discourse Theory of Law and Democracy.* Cambridge, MA: MIT Press, 1996.

Habermas, Juergen. "Further Reflections on the Public Sphere." In *Habermas and the Public Sphere,* edited by Craig Calhoun, 421–61. Cambridge, MA and London: The MIT Press, 1992.

Hong, Sung-tae. *Gaebaljooeuiul Bipanhanda* (*Criticizing Developmentalism: Beyond Park, Jung-hee Regime System to Ecological Welfare Society*). Seoul: Dangdae Publishe, 2007.

Katsumi, Yokota, and Il-kyung Na, trans. *Arisukun Naraui Budruwoomyunsoedo Ganghan Simin* (*Tender and Strong Citizens in a Foolish Country: Building Up Models of Life-Oriented Club Movements and Democratic Grassroots Activism*). Japan: Honnonki Inc.; Seoul: Nonhyung publishing co. 2004.

Kim, Hyuk-Rae. "The State and Civil Society in Transition: The Role of Non-Governmental Organizations in South Korea." *The Pacific Review* 13, no. 4 (2000): 595–613.

Kim, Ji-ha. *Chai (Gap).* Seoul: Sol, 1995.

Kim, Ji-ha. *Saengmyung(Life).* Seoul: Sol, 1992.

Kim, Ji-ha. *Saengmyung ga Jachi (Life and Autonomy: What is Saengmyung's Thought and Movement?).* Seoul: Sol, 1996.

Kim, Ji-ha. *Saengmyung Hak (A Study of Saengmyung).* Seoul: Hwanan Publisher, 2003.

Merchant, Carolyn. *The Death of Nature: Women, Ecology and the Scientific Revolution.* San Francisco: Harper & Row Publishers, 1983.

Moon, Soon-hong. *Chungchi Saetaegak kwa noeksaek gukga (Political Ecology and Greening State).* Seoul: Arche Publisher, 2006.

Moon, Soon-hong, ed. *Hankukui Yosong Whankyung Undong: Gui Yoksa, Joochae grigo Yoohyungdul (Korean Women's Environmental Movement: The History, Subjects and Types of Activism).* Seoul: Sol Publisher, 2001.

Moon, Soon-hong. *Saengtae Damron (Discourses on Ecology).* Seoul: Sol, 1999.

Mosim and Salim. *20 Years Old Hansalim Embraces the World.* Seoul: Guimulco, 2006.

Naess, Arne. "The Deep Ecological Movement: Some Philosophical Aspects." In *Environmental Philosophy: From Animal Rights to Radical Ecology,* edited by M. Zimmermann et al., 193–212. Englewood Cliffs: Prentice Hall, 1993.

Shin, Dong-ho. *Jayeonui Chingoodul 1.2.: whankyung Woondong 25yeonsa (Friends of Nature 1.2: 25 Years History of Environmental Movement).* Seoul: Doyose Publishers, 2007.

Yoon, Hee-jin. "Daedoociaesui Hansalim Woondong (Hansalim Movement in Cities)." *Hansalimdocument* (unpublished), 2003.

Vogel, Ezra. *The Four Little Dragons: The Spread of Industrialization in East Asia.* Cambridge, MA: Harvard University Press, 1991. www.hansalim.co.kr.

Chapter 12

Development as a Double-Edged Sword: Managing an Environmental Movement in Singapore's Urban Environment

Sofiah Jamil

Introduction

Most environmental politics scholars agree that there are three main currents of environmentalism: conservationism, which argues for the protection of wilderness and natural habitats; eco-efficiency, which pushes for economic development that is compatible with conservation and sustainability; and environmental justice, which seeks to address the environmental concerns of poor and marginalized sections of society.[1] The extent to which these currents of environmentalism are able to gain traction in society depends on the approaches taken by civil society organizations (CSOs) in engaging other stakeholders, such as government officials and the private sector, in the context of existing government policies related to economic development. While some CSOs have been more accommodative of government policies, others have forthrightly opposed them, leaving little room for negotiation.

The Singapore environmental movement provides an interesting case study of how the success of these currents of environmentalism are shaped or influenced by the degree of urban development and by extension, government involvement in addressing environmental issues. In this chapter, I will discuss how the processes of urbanization and economic development have both supported and impeded the development of the conservation and eco-efficiency currents of environmentalism in Singapore. I will argue that initiatives leaning toward environmental justice have the potential to be, and should be, pursued further in Singapore because they help to address the challenges that conservation and eco-efficiency groups face—namely deadlock with government bodies and being disconnected from sections of mainstream society—by emphasizing the socioeconomic needs of a broader range of stakeholders in society.

In the first section of the chapter, I give a brief history of Singapore's urbanization and economic development strategies from independence in 1965 to the late 1980s. I then show how existing urbanization policies provide support to cultivating the conservation and eco-efficiency currents in Singapore's environmental movement. The third section provides a discussion of how the process of urbanization has challenged the development of these currents of environmentalism. In the fourth and final section, I suggest potential avenues through which environmental justice may better operate in Singapore in the future.

Background: Singapore's Urbanization and Economic Development Strategies

Singapore is often cited as a successful model of economic development. At the time of gaining independence in 1965, it had a land space of only 581.5 square kilometers,[2] no natural resources, and was categorized as a highly water-stressed nation.[3] Despite these challenges, within 48 years the country has emerged as one of the most industrialized in East Asia,[4] with one of the highest levels of human development in the region. This success is reflected in its strong annual growth in GDP over the years: as high as 13.8 percent in 1970, 11.1 percent in 1988, and 11.5 percent in 1993.[5] The growth rate remains strong at 4.9 percent in 2011.

This exceptional progress is due to the fact that Singapore's centrally planned urbanization process has been both a result and a drive of economic development. To understand this process, it is first important to acknowledge Singapore's political history in the independence period. In a bid to break away from British colonial rule after World War II, Singapore along with other states in British Malaya, as well as Sabah and Sarawak in Borneo, formed the Federation of Malaysia in 1963. However, due to disagreements between political leaders in Kuala Lumpur and Singapore, and a series of racial riots in 1964, Singapore was forced to leave Malaysia in 1965. The separation subsequently put Singapore in a vulnerable position. It was faced with acute challenges such as unemployment in the 1960s and

1970s and a lack of capital to remain competitive in the 1980s. In spite of these adverse circumstances, the Singapore government, under the leadership of Prime Minister Lee Kuan Yew and later his successor Prime Minister Goh Chok Tong, was able to rally a high degree of support and gained the central authority to build Singapore into a newly independent state. This process was facilitated by having an already robust political system, which was established under British colonial rule.

Several aspects of Singapore's urban planning policies have facilitated its development. In order to overcome unemployment and the shortage of capital in the 1960s and 1970s, the government created jobs by developing a manufacturing industry; in the 1980s it focused on extensive built-infrastructure projects to increase the technical skills of the workforce. Infrastructure was such a crucial part of Singapore's economic development in the 1980s that government expenditure on it was three times higher than foreign direct investment during the same period.[6] This resulted in a high degree of land clearance and reclamation in order to make space for industrial and other economic projects to flourish. Over the years since independence in 1965, Singapore's land size increased from 518.5 square kilometers to 714.3 square kilometers.[7]

Planned urbanization has enabled the Singapore government to address a range of environmental problems. For example, urbanization increases the likelihood of runoff waters and therefore the risk of flooding. In view of this, the government improved drainage systems to reduce flood-prone areas from 3,178 hectares in 1970 to 62 hectares as of July 2010.[8] Effective drainage systems are not only a means of flood prevention[9] but are also significant in mitigating the spread of dengue fever, which is a common occurrence during periods of high rainfall when mosquitoes breed in stagnant water. As a means of addressing water-stress issues, urbanization was planned to ensure the preservation and even enlargement of water-catchment areas in Singapore. Another example is the use of a centrally planned housing policy to deal with the lack of sanitation and spread of communicable diseases and to avoid the formation of slums, which are apparent in many other cities.[10] The meticulous planning of public housing in Singapore also ensured easy access to amenities in various neighborhoods, thereby avoiding overcrowding in the city center.

Since the 1980s, an aim of urban planning policies has been to create a clean and green Singapore. Tree planting ceremonies have been a tradition in neighborhood and community events since the early 1960s when Lee Kuan Yew planted a tree annually as part of his vision to make Singapore the Garden City of the tropics.[11] Tree Planting Day was officially launched in 1971 and was timed to occur just before the rainy season in November: a conscious effort to minimize the need for watering the new tree saplings.[12] In addition, enhancing Singapore's greenery was seen as a means for ensuring Singapore's attractiveness as a tourist destination.

So we can see that Singapore's rapid but planned urbanization process was not only designed to ensure its economic development and therefore survival as an independent country, but also over time to make it a livable and attractive place. However, the extent of this urbanization has been at the expense of Singapore's natural environment. To date, about 95 percent of Singapore's natural habitat has been lost as a result of land clearing and reclamation projects.[13] The effects of urbanization can thus be seen as a double- edged sword. On the one hand, it has led to the destruction of natural habitats, but on the other hand it has brought about the conditions for supporting environment-friendly lifestyles and a strong base on which to build public environmental concern. This has in turn led to the emergence of two dominant currents of environmentalism in Singapore: conservationism and eco-efficiency, which have both benefitted and suffered from various urban planning policies.

To examine how Singapore's urban development has supported and impeded the conservation and eco-efficiency currents of environmentalism, the next two sections of the chapter will examine the work of local environmental organizations, which fit into the two categories. For conservation, we will look at the Nature Society of Singapore (NSS) and the Animal Concerns Research and Education Society (ACRES); for the eco-efficiency current, we will look at Green Drinks and ECO-Singapore.

Supporting Environmentalism through Urban Development

Urbanization has often been regarded as the antithesis of environmentalism. Yet, as I will now demonstrate, Singapore's urban policies have actually provided support for the development of its environmental movement. This support has been given primarily by (1) facilitating conservation amidst Singapore's urban landscape; (2) expediting the growth of the eco-efficiency current of environmentalism; and (3) providing opportunities for groups from different environmental currents to work together.

Conservation amid Urbanization

When considering the conservation current of environmentalism in Singapore, it is vital to examine the work of the Nature Society of Singapore (NSS), which is one of the oldest and most influential conservation organizations in the country. While there is evidence of there being conservation enthusiasts in Singapore as far back as 1921,[14] the NSS first started out as the Singapore branch of the Malayan Nature Society in 1954.[15] It broke away in 1991 to become the independent NSS.[16] Members come from a wide range of backgrounds and the NSS committee in particular has often included individuals with a range of expertise and experience with conservation issues. These include academics from the natural and social sciences as well as business entrepreneurs. While the early activities focused on environmental education and an appreciation of nature (driven by various interest groups for flora and fauna and birds), in recent years the NSS has taken a more proactive stance in lobbying the Singapore government on conservation.

The NSS has played a significant role in campaigning for the conservation of Singapore's remaining natural habitats. Two of its most successful campaigns have been to preserve the Sungei Buloh and Chek Jawa wetlands. The former is an important stopover for birds migrating from as far as Europe to Australia, while the latter is a 100 hectares wetland located on the southeastern tip of Pulau Ubin that comprises of seven interdependent ecosystems, namely coastal and mangrove forests, rocky and sandy shores, sand bar, seagrass lagoon, and the coral rubble.[17] What has been particularly notable in these two issues has been the fact that the government was highly amenable to full conservation of these reserves. Chek Jawa in particular stands as a significant milestone for NSS, where strong public opposition toward the land reclamation projects in 2001 successfully halted government development plans that threatened Singapore's scarce biodiversity areas. These areas are now two out of the four protected nature reserves in Singapore, which make up 5 percent of Singapore's total land area.

The conservation movement was given a further boost when the Animal Concerns Research and Education Society (ACRES) was formed in 2001 with the objective of protecting wildlife in Singapore's urban environment. Although a much younger organization than the NSS, ACRES is much more proactive in its work to rescue wildlife that has been stranded, injured, or trafficked. What started as an individual's determination to ensure the protection of three primates has in just over 10 years developed into a nonprofit organization supported by full-time staff and backed by volunteers and supporters numbering around 18,000.[18] ACRES has conducted several campaigns over the past decade, but its most notable achievements have been addressing the consumption of illegally traded wild animals and animal parts, as well as the treatment of animals in Singapore's urban landscape.

While the scope of NSS's work is much broader than that of ACRES, there are common factors that have led to the success of their respective campaigns. The first of these factors is their efforts to get the Singapore government to take conservation issues seriously. Both NSS and ACRES have always been aware that to win government support on the issues, they need to demonstrate that their campaigns would not only be in the interest of conservation groups but also benefit Singapore's development agenda. For example NSS's concerns over natural habitats, such as Chek Jawa and Sungei Buloh, were stipulated comprehensively in its 1990 master plan for the Conservation of Nature in Singapore,[19] which ultimately played a significant role in institutionalizing conservation in the Singapore government's development plans. Conservation principles are embedded in the ministry for environment and water resources annual Singapore Green Plan, which is now in its 11th edition.[20] In subsequent years, the Singapore government took steps to develop these nature reserves into hubs for public environmental awareness. In close consultation with conservation groups and experts, development plans were formulated to incorporate recreational and environmental education activities with minimal adverse impact on natural habitats. Many Singaporean environmentalists have argued that an increased proximity to nature would increase people's appreciation of nature, and that increased appreciation would in turn foster greater awareness and action toward environmental conservation and protection. Their argument has resulted in a notable commitment by government bodies to maximize the potential of nature reserves, not only to increase Singaporeans' environmental awareness, but also to improve their quality of life.

In the case of ACRES, the organization's efforts in tackling the illegal trade of wild animals and animal parts has allowed them to develop a good working relationship with various government departments, beyond the conventional environmental agencies. For instance, ACRES has conducted a series of covert investigations that have led to the exposure of illegal trade in bear parts for traditional Chinese medicine and to the seizure of tiger parts sold in Singapore.[21] ACRES has also actively campaigned to amend sections of the Endangered Species (Import and Export) Act, which was done successfully in 2006, when the amendments made served to strengthen penalties for trading in or owning endangered species and to give more enforcement powers to Singapore's Agri-Food and Veterinary Authority to tackle the illegal wildlife trade. These efforts have been recognized by Singapore authorities who see the value-added role that ACRES plays not just in endangered wildlife conservation but also in how it has complemented government efforts to address other societal issues such as illegal trafficking and food safety.

The second common factor contributing to these organizations' success is the hands-on activities that they conduct to increase environmental awareness and public participation. For instance, the Sungei Buloh wetland has become a space for members of NSS to conduct bird-watching excursions, while the preservation of Chek Jawa's seagrass lagoons allows marine scientists to conduct scientific surveys of marine life. With regard to the latter work, one particular project is sea grass monitoring, which is jointly conducted by the National Parks Board, the National Biodiversity Centre, Team Seagrass, and volunteers from the public. Regular surveys are conducted of the sea grasses in the inter-tidal areas around Singapore as part of a global seagrass assessment and monitoring program known as Seagrass Watch.[22] As such, not only does government support provide conservation groups like Team Seagrass with the opportunity to increase public awareness of the significance of seagrass habitats as highly biodiverse habitats, it also facilitates their work in contributing to scientific research that is used for examining the trends and condition of seagrass habitats at the local, regional, and global scales. ACRES's work is largely facilitated by public participation through its 24-Hour Wildlife Crime Hotline, which is used to report incidents of illegally traded animals. Over time, ACRES has been able to secure a place of its own for the care and rehabilitation of these animals. It has also effectively worked out a volunteer management system, allowing members of the public to volunteer in specific areas of ACRES work they would like to help in, such as taking care of animals in their rescue shed and actual animal-rescue operations. What should be noted about these examples is how the groups have been able to leverage government support to facilitate their activities and thereby to gain greater public visibility.

A third common factor has been the organizations' ability to seize opportunities to highlight issues via the traditional and social media. For instance, when the Singapore government announced plans to redevelop Chek Jawa, NSS's extensive campaigning to conserve the site included a documentary entitled *Remember Chek Jawa* that brought attention to a biodiversity survey of the area and was widely disseminated not just to various stakeholders, such as national planning officials, but also the wider public and private sectors, many of whom may have had very little prior knowledge of natural habitats in Singapore. For ACRES, some of its best campaigns have been conceptualized with the help of advertising companies, such as its "When sharks die, the oceans die" campaign in 2009. The campaign was aimed at highlighting degradation of marine ecosystems as a result of the killing of sharks that are usually for the consumption of shark's fin during Chinese weddings or the Chinese New Year. With increased awareness and pressure on suppliers of shark's fin—particularly in response to an advertisement by a shark's fin supplier that included the message "Screw the divers!"—the campaign has gradually achieved substantial success when major supermarket chains in Singapore decided to stop selling shark's fin in early 2012.[23] The most ingenious use of media engagement however is their "Save the Dolphins" campaign against the mistreatment of dolphins used in entertainment shows in a new resort on the southern island of Sentosa. In addition to engaging members of the public to sign petitions and be part of their social media campaigns, ACRES was very strategic in its engagement with Resorts World. By sending open letters and press releases to the CEO of Resorts World and disseminating footage of the poor conditions in which the dolphins were held, Resorts World officials were trapped into having to answer difficult questions.

These examples demonstrate that channeling efforts into collaborating with the government, providing clear avenues for public participation, and using traditional and social media wisely have allowed the conservation movement in Singapore to institutionalize the first wave of environmentalism in the country while maneuvering around the constraints of a highly urbanized society.

Expediting Eco-Efficiency

As in many other developed countries, eco-efficiency has become a growing theme within Singapore's environmental movement. While there have been government environmental campaigns such as recycling since the early 1990s, it is only in the last decade that the eco-efficiency current has truly taken off. This is largely due to the role of capable individuals in civil society, increased avenues for environmental education and awareness, and the provision of necessary infrastructure and incentives for taking environmental action.

Two individuals have played a significant role in precipitating the eco-efficiency current in recent years. The first is Wilson Ang, founder of ECO-Singapore, an environmental organization advocating sustainable lifestyles for youth and by youth primarily between the ages of 17 and 35. One of ECO-Singapore's notable initiatives has been its Re-Live event, which was essentially a series of roadshows and exhibitions in several Singapore neighborhoods aimed at increasing the level of public awareness on aspects of eco-efficiency.[24] This, and other ECO-Singapore activities, has garnered strong support and collaboration with various government agencies and town councils. In addition, ECO-Singapore has been active in sending youth delegations to represent Singapore at major international climate change and sustainable development conferences. The second key individual is Olivia Choong, who founded the Singapore chapter of Green Drinks[25] in November 2007. This initiative involves monthly discussions that bring together environmentalists and other interested parties to share ideas and network with one another. Given that most of Green Drinks Singapore's initial talks provided a chance for green businesses to discuss eco-efficiency issues as well as products ranging from electric cars to organically produced fashion apparel, several green businesses and start-ups have since become strong supporters of Green Drinks. Over time, however, the range of topics has widened and included other aspects of sustainability, such as food security and nuclear energy.[26] As a result, Green Drinks Singapore has grown to be one of Singapore's most prominent environmental forums and a means of support for other currents of environmentalism such as conservation. These individuals have made important contributions to the environmental scene in Singapore, and have opened up spaces for greater interaction across various sectors and groups interested in the environment.

While the efforts of these individuals are commendable, their success has also been a result of increasing education and infrastructure to support the eco-efficiency current. The Singapore government in particular has made extensive strides in mainstreaming environmental efforts in various ministries. In the ministry for education, for example, while environmental education has conventionally been a means of simply enhancing students' appreciation for nature, the government has gone a step further to introduce tertiary level courses to prepare students for jobs in environment-related industries. It has also invested substantially in research and development, such as in solar energy, despite lacking sufficient land area to meet domestic energy demand. In terms of the transport sector, the government has continued to improve its well-planned public transport system and also introduced cars using alternative fuel sources. While the initial interest in using compressed natural gas (CNG) cars has died down,[27] there is ongoing research and test-bedding of electric cars in Singapore.[28] Green certification and labeling is another growing element in the eco-efficiency current, as it gives companies an incentive to cater to more environmentally conscious consumers. Green Certification in Singapore was given a boost thanks to the government's introduction of the Building and Construction Authority's (BCA) Green Mark Scheme in 2005 and the Singapore Green Building

Product Certification in 2010. Both schemes were established to encourage environmental sustainability in the design of building and development projects, and to contribute to the government's target of "greening" 80 percent of all buildings by 2030. Incentives under the Green Mark Scheme for businesses to adopt environment-friendly practices in development projects include rewarding construction companies with building extra floors if they incorporate green spaces and energy efficient features in the blueprints of future buildings. These measures not only complement the existing Singapore Green Labeling Scheme (introduced in 1995 for consumer products certification under the Singapore Environmental Council) but also expand the opportunities for various sectors to contribute to sustainable development in Singapore.

Breaking Down Barriers between Environmentalist Currents

In recent years there has been a progressive breaking down of barriers between the two main currents of environmentalism through increased collaboration on common concerns. This was particularly evident with the Green Corridor campaign, which represents the land area along the defunct historical railway lines connecting Singapore and Malaysia, which the Singapore government plans to clear for redevelopment. Although now officially known as the Rail Corridor, NSS and other environmentalists have referred to this land area as the Green Corridor during their campaign and proposed to conserve it for sustaining biodiversity[29] (it connects various green spaces in Singapore) and as a means of increasing environmentally sustainable lifestyles.[30] Eco-efficiency groups such as Green Drinks and Eco-Singapore publicly lent their support for these conservation efforts. In fact, a Green Drinks session on the Green Corridor had one of the largest turnouts of all Green Drinks events in Singapore.[31] Participants included not just members of various environmental groups, but also research institutes, think tanks, and the private sector. These developments have thus increased the level of environmental awareness and action in Singapore.

Moreover, the Green Corridor campaign and other similar campaigns have highlighted not only biodiversity issues but also the historical significance of these areas in Singapore. This comes on the back of a growing public concern that rapid changes in the physical landscape are dissolving the nostalgia that Singaporeans have for their homeland and its history. In the case of the Green Corridor, members of NSS organized walks along the defunct tracks as a means of explaining the history of the railway system and increasing people's appreciation of the corridor's natural surroundings. The effects of these efforts have been fruitful. As word on the guided corridor walks spread, members of the public subsequently took the initiative to walk along the corridor, reflecting greater public concern over the issue. The Singapore government has since sought to redevelop the area with greater public consultation by opening a contest for Singaporeans to recommend ideas on how to integrate the Green Corridor's greenery and heritage sites with ways of developing activities to foster community bonding.[32]

With such forward-thinking national policies, Singapore would certainly seem to demonstrate a holistic approach to sustainable urban development in the long run. However, while the government has supported the growth of the environmental movement in Singapore, one could also argue that top-down efforts are based on selective engagement and are primarily a means of enhancing economic interests. The next section will demonstrate that while urbanization and economic development policies may seem to support Singapore's environmental movement, it is also a double-edged sword in that they are also in line with the government's interests—which are often at odds with some environmental goals.

Urban Development *versus* Environmentalism

Despite its role in facilitating the conservation and eco-efficiency currents of environmentalism in Singapore, urban development has also impeded it. Conservation campaigns organized by civil society organizations (CSOs), such as the campaign to save the Bukit Brown Cemetery from being lost to redevelopment, and state-led campaigns, such as for waste reduction, come up against tremendous challenges when urban development is given priority in government decision making and when there are high levels of public apathy.

There are two main ways in which the processes of urban development have limited or diluted environmentalism. First, the Singapore government routinely takes a pragmatic stance toward the country's long-term development. For instance, although the Bukit Brown Cemetery is highly biodiverse and rich in historicity, pragmatism must override romanticism. While the government has helped to preserve existing biodiversity, their efforts are largely dependent on ensuring Singapore's economic attractiveness using the limited land space for other socioeconomic development needs. Given that only 5 percent of Singapore's natural habitats (i.e., the four nature reserves) are gazetted as protected areas, any remaining plot of natural habitat would be subjected to the government's development plans. Prime Minister Lee Hsien Loong has expressed some understanding of these nostalgic sentiments. In his 2012 National Day Rally address, he shared his memories of what different parts of Singapore looked like during his youth. However, he concluded that the dramatic changes have been for the overall benefit of Singapore.[33]

This preference for development over preservation means that despite the best efforts of environmental groups, the government will still have the upper hand, particularly when the government perceives its plans to be popular with powerful stakeholders.[34] In the case of Bukit Brown, some environmentalists have observed that while the government did consult with members of Singapore's environmental movement, they were not given sufficient time to review the redevelopment plans, and thus could not contribute as substantively as they would have wanted to.[35]

The second way in which the processes of urban development have impeded some forms of environmentalism is that it has given rise to apathy at the grassroots level. This has been particularly evident in Singapore's efforts to reduce waste. Its recycling campaign, for instance, which has been in operation for more than a decade, has had limited success due to inherent flaws in some urban planning policies. From 2000 to 2011, the overall rate of recycling has increased from 40 percent to 59 percent. However, an analysis of the breakdown of different types of waste indicates that much of the recycled waste comes from the industrial sector. While plastics and food waste are among the top five wastes generated in Singapore (after paper/cardboard, ferrous metal, and construction debris), their rate of recycling is still low, at only 11 percent and 10 percent, respectively, and make up the top two most disposed waste categories.[36] One could argue that while legislation to regulate recycling in industries has been successful, measures to improve recycling efforts among the mass public has been limited.

The waste problem has a number of causes, but perhaps the most significant is a general lack of public awareness of the specific conditions for recycling. Many people are unaware that items to be recycled cannot be contaminated with food waste, which results in a proportion of items placed in recycling bins not being recyclable.[37] In addition, the design of the rubbish disposal system in public housing areas (which make up about 80 percent of housing in Singapore) is a single rubbish chute, and does not allow for the separation of recyclables from organic waste.[38] While there have been efforts in public housing estates to place more recycling bins in prominent areas, and for even a door-to-door pick up service for recyclable items,[39] there would nevertheless still be the common practice (due to convenience) of using a rubbish chute. Unlike in countries such as Japan and Germany, there has yet to be a system in Singapore that forces people to separate recyclables from organic waste. There are also supply-side issues that need to be addressed. For example, recycling vendors have suggested that plastic waste is not as profitable as other recyclable materials like paper and tin. In the same vein, the lack of traction that Bring-Your-Own-Bag/Receptacles campaigns have had in Singapore is due to the fact that Styrofoam and plastic takeaway cartons are still widely available and are the preferred choice of food vendors.[40] So it is clear that economics and affordability remain significant impediments to concrete environmental action in Singapore.

The waste example illustrates nicely the argument that the eco-efficiency approach to environmentalism in Singapore fails to address the needs of the ordinary Singaporean and remains an elite concern. Despite the obvious benefits of various eco-efficiency products and services in the long run, they are highly costly. The lack of buy-in for environmental certification, by food vendors for instance, is due to the fact that it costs SGD 2,000 for certification of one product and subsequently an annual renewal fee of SGD 1,000.[41] In view of these arrangements, while there is strong top-down support for these eco-efficiency measures, it is often the rich and middle class who are able to be proactive in the environmental movement. There is consequently a lack of resonance with the less-privileged and poorer sections of society, who tend to disregard these environmental measures because they do not meet their immediate socioeconomic concerns, such as livelihood options, healthcare, and rising costs of living. This lack of connection raises the question of the potential of the environmental justice current of environmentalism to address the concerns of Singapore's less-privileged population.

Potential Pathways for Environmental Justice in Singapore

In Western developed countries, currents of environmentalism tend to move from conservation to eco-efficiency and then eventually to environmental justice. Is this the case in Singapore? Before we can answer this question, it is important to note that the term *environmental justice* is highly context-specific. The concept emerged in the United States in the 1980s and was defined as a means of "highlighting the plight of marginalised, ethnic or poor communities experiencing the brunt of environmental health hazards, and is often associated with the term environmental racism."[42] More recently the concept has evolved to reflect the circumstances in other places.

For example, whereas in the U.S. context the focus tends to be on race and ethnicity,[43] the U.K. context tends to concentrate more on socioeconomic status.[44] The narrow American definition of environmental justice is inadequate for several reasons. One is that it lacks a clear definition of the victims of environmental injustice, particularly in terms of according victim status in law. Another reason is that it has tended to account only for localized geographical areas, and therefore to ignore environmental injustices that occur over a larger spectrum nationally or even transnationally.[45] There is a debate within the scholarly literature over how to make the concept more inclusive. David Schlosberg, for instance, argues that understanding environmental justice simply in terms of equitable distribution does not adequately meet the diversity of demands for justice made, such as for recognition of cultural identity and participatory democratic rights.[46]

Given its contested application in developed Western countries, it is not surprising that applying the environmental justice concept to Singapore is not a simple move. For one thing, the development and maintenance of Singapore's urban environment creates a different set of circumstances than those in the West. Thanks to Singapore's public housing scheme that ensures a proportional mix of various ethnic groups in each estate, Singapore has avoided the physical development of urban slums with concentrated ethnic enclaves. One may argue that Singapore's public housing scheme clearly distinguishes the rich from the poor, given that the housing ranges from one-room rental flats to five-room apartments with recreational amenities. However, there are strong antipollution regulations,[47] which the National Environment Agency implements and monitors to protect residents from environmental health hazards regardless of geography or economic status. In this respect, Singapore's urban planning structure has, to some extent, reduced the conditions needed for an environmental justice movement to develop.

Moreover, there are elements of Singapore's political system that have prevented the development of the kind of environmental justice movement that has grown in liberal Western democracies. Recent work by Hobson[48] suggests that environmental justice in Singapore is not overtly visible because of the strong top-down approach to governance. Environmental injustices cannot be articulated with direct reference to citizens' rights and/or justice because doing so would imply the lack of oversight by the ruling government, or may cross over the conventional "OB markers" (short for "out-of-bounds marker"; in other words, topics not permissible for public discussion). Discussing environmental injustice as akin to environmental racism, for instance, would tread into areas beyond conventional OB markers, which include issues relating to race and religion. Moreover, it would not be in the government's interest to be seen as ignoring the needs of its people. Hobson therefore suggests that to the extent that there is an environmental justice current in Singapore, it operates through "everyday actions and representations, which often fall under the radar of environmental justice studies."[49] While one might suggest that there is greater liberty in Singapore's political landscape after the 2011 general elections, which marked the ruling party's lowest performance since independence (winning about 60 percent of total votes), it has not necessarily brought about a change in the modes of engagement that are deemed appropriate by the government. So, rather than voicing overt critical opposition to the government, environmental groups in Singapore still tend to favor the strategy of influencing government officials by contributing solutions that acknowledge existing development constraints and the interests of other stakeholders such as the private sector and concerned members of the public.

In light of these reasons, it is not surprising that the term environmental justice is rarely used in Singapore. However, there are signs that a growing number of activists seek to address some of the socioeconomic or community concerns that are commonly associated with environmental justice in Western literature. For example, one common theme is the need to catalyze intra- and inter-communal bonds across different socioeconomic backgrounds, where justice is seen as something to be achieved for communities rather than just for individuals. An issue of increasing importance is food security because Singapore imports more than 90 percent of its food resources. According to Teng and Escaler,[50] there are various dimensions to examining food security: from macroeconomic issues relating to agricultural productivity and supply chains to household purchase and utilization of food. The last two micro-level issues provide opportunities for environmental justice activism, particularly in relation to food security for people in poverty. It should be noted that while Singapore has high levels of economic development, the level of economic inequality (based on the Gini coefficient) is one of the highest in the world.[51] There is a section of Singapore society whose financial resources fall substantially below subsistence levels. While there is no official poverty line in Singapore, there is evidence of about 200,000 local families living in the bottom fifth of the income scale. The average monthly income in Singapore in 2010 was $2,040, but the poorest 10 percent—about 100 households—earned only $1,400 per month.[52] This statistic does not include the transient workers or the homeless.

There have been efforts by environmentally conscious Singaporeans to provide food to sections of the poor in Singapore. The Soup Kitchen Project, for instance, is a weekly vegetarian soup kitchen project for needy folk living around the Little India precinct. The project was initiated by the founders of the social enterprise Food #03, a vegetarian café that doubles as a space for the arts and the environmental community in Singapore. Beneficiaries of the Soup Kitchen Project range from poor senior citizens to unemployed foreign laborers who are served by volunteers through Food #03 network.[53] Despite facing challenges in maintaining the Food #03 premises, and recently having to close down the café, the Soup Kitchen Project has been able to use the kitchen of other vegetarian restaurants in the vicinity for its project operations.[54] The project has been gaining momentum since 2009 and recent volunteers include groups of school children. The Soup Kitchen Project is thus an interesting example of members of Singapore's environmental movement meeting social justice needs by providing the less privileged with healthy and locally produced foods.

There have also been efforts to highlight the plight of the poor and the elderly in Singapore through a strategy of merging art and environmental awareness. As part of a 2011 community project entitled "Rubbish is (P)art of us!," the Singapore Contemporary Young Artists created two art pieces from unwanted materials that they collected from senior citizens living alone in rented flats. Interaction between the young artists and senior citizens was of benefit to both groups. For the senior citizens, it provided an avenue of social support contributing to their well-being and happiness. For the young artists, the project allowed them to shed light on the growing aging population and the importance of maintaining values such as filial piety, via stories and snippets of oral history that were prompted by the discarded items. More importantly, the friendships formed during the project have been a motivating factor for the artists to conduct further community projects in the future.

Evidence of a growing concern for environmental justice in Singapore can also be found in efforts to advocate a greater public understanding of local environmental history and practices. While the NSS has touched on the importance of history in several of its nature walks, new groups such as the Ground-Up Initiative[55] have sought to provide hands-on activities and formulate innovative ideas that can be implemented at the community level. From conducting workshops to introduce garbage enzymes and composting for household use to providing a space for inventors to create gadgets that enhance sustainable development, the Ground-up Initiative has rekindled a process of enhancing community bonds in Singapore while simultaneously increasing environmental consciousness.

According to Dobson, "no theory of justice can henceforth be regarded as complete if it does not take into account the possibility of extending the community of justice beyond the realm of present generation human beings."[56] A faith-based perspective allows for this process of extending environmental concern beyond the here and now through religious teachings on environmental stewardship. In addition to a growth in scholarship on the role of faith in environmentalism,[57] there are emerging examples of faith-based environmental initiatives worldwide.[58] In Singapore, there have been some efforts driven by religious leaders to inspire their faith communities to become more environmentally aware. These include initiatives such as Project ME: Muslims and the Environment[59] and Catholic group discussions on ecology.[60] There has also been some level of government support for the installation of energy-efficiency features in places of worship such as the Al-Mawaddah mosque[61] and the Poh Ern Shih Buddhist temple.[62] Though still in its formative stages, engaging faith communities in environmental issues in Singapore will potentially increase environmental awareness and action amongst Singaporeans, as well as being a means for faith communities to work together on environmental justice–related charity work. Faith-based environmental initiatives may be an emerging pathway for developing an environmental justice current in Singapore.

These new and emerging initiatives suggest that an environmental justice current may be developing in Singapore, albeit in a way that reflects unique place-specific conditions and that is therefore different from environmental justice movements in the developed West. While Western case studies of struggles for environmental justice often involve overt criticisms of social inequalities and unjust governance, a radical-critical approach is not an option in the Singaporean context. The further development of an environmental justice current in Singapore will depend on environmental activists working within established political boundaries. They will also need to combine their concerns with conservation and eco-efficiency with public expressions of sensitivity toward community concerns for health and well-being. And they will need to demonstrate a capacity to contribute concrete solutions that are for the benefit of building community bonds and societal resilience. Given the strong level of support for welfare and economic development among various sectors and communities in Singapore, there is likely to be a wide range of opportunities through which advocates of environmental justice might advance their cause.

Conclusion

An emphasis on planned urban development has been a defining factor in the emergence and operation of Singapore's environmental movement. There is a constant need for environmentalists to be conscious of, and to accommodate, the pragmatic stance of the Singapore government in addressing environmental concerns. A major challenge for environmentalists of all currents is public apathy toward the environmental movement. This sense of apathy has emerged over the years as a result not only of the dynamics of urbanization but also the perceived lack of connection between environmental issues and the concerns of some sections of society. While there is increasing visibility of the environmental movement in Singapore, few organizations have been able to demonstrate effectively how concern for the environment is relevant to poor and marginalized communities who are hurt by a growing income gap. There is nevertheless ample room for improvement for Singapore's environmental movement to embrace the third current of environmentalism—environmental justice—which is aimed at promoting a sense of responsibility to the common good and enhancing communal bonds. While multi-sectoral engagement via this third current of environmentalism is necessary for more effective community action, it may be perceived by some as being an encroachment into some fairly protected but vital areas, such as social work. It would therefore take a fair amount of time to overcome turf issues and effectively engage local constituencies and community leaders. Nevertheless, small and effective projects such as those mentioned in this chapter would provide commendable models for others to emulate, provided there is support from the appropriate government bodies in Singapore's highly organized political system. But, in light of the government's default preference for economic growth and creating an attractive urban environment for tourists and investors, it is almost certain that all environmental movements in Singapore will need to operate under the shadow of the double-edged sword of development for many years to come.

Notes

1. Martínez-Alier, *The Environmentalism of the Poor.*
2. Tan et al, *The Natural Heritage of Singapore,* 78.
3. Lee, "Water Management Issues in Singapore."
4. Singapore along with South Korea, Hong Kong, and Taiwan were known as the Asian Tigers for their high economic growth rates in the 1980s to 1990s, following suit of Japan.

5. Time Series on GDP at 2005 Market Prices and Real Economic Growth.

6. Ermisch and Huff, "Hypergrowth in an East Asian NIC: Public Policy and Capital Accumulation in Singapore," 21–38.

7. Key Annual Indicators.

8. Hua and Loh, "Managing Storm Water in Urbanised Singapore for Flood Control."

9. Floods in Singapore were particularly damaging for Singapore between the 1960s and 1970s, and the Singapore government responded with extensive efforts to improve drainage systems. See Public Utilities Board Singapore, "Managing Flash Floods."

10. Ooi and Phua, "Urbanization and Slum Formation," 32.

11. The Roots of Singapore's Tree Planting Season, MNDLink, http://www.mndlink.sg/2009/2009_Dec/NParks_article.htm

12. Lee, *From Third World to First: The Singapore Story: 1965–2000* (Singapore: Times Media, 2000), 188–211.

13. Sodhi et al, "Southeast Asian Biodiversity: An Impending Disaster." See also Brook, Sodhi and Ng, "Letters to Nature," and Pickrell, "Singapore Extinctions Spell Doom for Asia?"

14. In 1921, the Singapore Natural History Society (SNHS) was formed to develop "friendly intercourse between local naturalists and the increase and diffusion of knowledge concerning natural history. By 1922, the SNHS had 66 members. It produced the *Singapore Naturalist* publication up to 1928 but then faded away.

15. In 1940, the Malayan Nature Society (MNS) was formed in the British colony of Malaya. The MNS still flourishes today as the Malaysian Nature Society.

16. Nature Society (Singapore).

17. Pulau Ubin, "Chek Jawa on Pulau Ubin."

18. Animal Concerns Research and Education Society (ACRES), "How It All Began."

19. Avadhani et al., "Master Plan for the Conservation of Nature in Singapore."

20. Ibid.

21. Animal Concerns Research and Education Society (ACRES), "Our Achievements."

22. Seagrass-Watch, a global seagrass assessment and monitoring program spanning 18 countries with more than 200 monitoring sites.

23. Sim, "Carrefour Says No to Shark's Fin."

24. See Report on Re-Live @ Heartland Choa Chu Kang."

25. The Green Drinks movement began in London in 1989 as a self-organizing network of individuals discussing green issues over drinks.

26. Green Drinks, Singapore.

27. "CNG Cars in Singapore: Moment in the Sun Over," *The Straits Times*, September 21, 2010.

28. Energy Market Authority of Singapore, "Electric Vehicles Test-Bedding."

29. Nature Society (Singapore), "The Green Corridor: A Proposal to Keep the Railway Lands as a Continuous Green Corridor."

30. Green Solutions, "The Green Corridor."

31. Personal observation and correspondence with Olivia Choong. Also see "Green Drinks: The Need for Conservation of Singapore's Wildlife."

32. Urban Redevelopment Authority of Singapore, "The Rail Corridor: About the Project."

33. Lee, "A Home with Hope and Heart."

34. Personal correspondence with Olivia Choong, Founder of Green Drinks, Singapore, May 23, 2012.

35. The Online Citizen, "Call for Moratorium over Plans for Bukit Brown."

36. National Environment Agency, "Waste Statistics and Overall Recycling."

37. Yeo, "What Rubbish."

38. Cheam, Chua and Yong, "Recycling in Singapore: 10 Years On—Chute, We Got It Wrong."

39. Several town councils in Singapore have provided designated bags at the doors of residents to fill with recyclable items, which will then be picked up. Ironically the bags that have been provided by the town council are plastic bags.

40. Ibid.

41. Ibid.

42. Cutter, *Race, Class and Environmental Justice.*

43. See Bullard, "Race and Environmental Justice in the United States"; Bullard, "Blacks and Latinos on the Frontline for Environmental Justice: Strengthening Alliances to Build Healthy and Sustainable Communities"; and "Race, Class and the EPA's Environmental Justice Strategic Plan."

44. See Briggs, Abellan, and Fecht, "Environmental Inequity in England: Small Area Associations between Socio-Economic Status and Environmental Pollution"; and Bulkeley and Walker, "Local Environment: The International Journal of Justice and Sustainability."

45. Stephen, Bullock, and Scott, "Environmental Justice: Rights and Means to a Healthy Environment for All."

46. Schlosberg, "Reconceiving Environmental Justice: Global Movements and Political Theories," 517–40

47. Ibrahim and Chung, "Quality of Life of Residents Living Near Industrial Estates in Singapore," 203–25.

48. Hobson, "Enacting Environmental Justice in Singapore: Performative Justice and the Green Volunteer Network."

49. Ibid., 673.

50. Teng and Escaler, "The Case for Urban Food Security: A Singapore Perspective," 4.

51. Ong, "Singapore's Inequality Battle."

52. Kaur and Switow, "Panel on The Economics of Living: Discussing Poverty in Singapore."

53. The Soup Kitchen Project.

54. Ibid.

55. Ground-up Initiative.

56. Dobson, *Justice and the Environment: Conceptions of Environmental Sustainability and Theories of Distributive Justice*, 244–45.

57. For examples, see "Catholic Perspectives on Faith and the Environment: A Bibliography," The Yale Forum on Religion and Ecology and Khalid, "Islam and the Environment."

58. For a sample of faith-based initiatives, see Schwencke, "Globalized Eco-Islam," "Faiths and Environment—World Bank Support 2000–2005," "Faith in Action: Communities of Faith Bring Hope for the Planet," and the Islamic Foundation for Ecology and Environmental Sciences.

59. Project ME: Muslims and the Environment.

60. Faith in Ecology: Go Green and Say Amen to Jesus!

61. Kamarul, "Singapore's First Eco-Friendly Mosque to Open on Friday."

62. Aziz, "Poh Ern Shih: The High-Tech, Eco-Friendly Buddhist Temple."

Bibliography

Animal Concerns Research and Education Society (ACRES). "How It All Began." http://www.acres.org.sg/aboutacres/howitallbegan.html (Accessed June 19, 2012).

Animal Concerns Research and Education Society (ACRES). "Our Achievements." http://www.acres.org.sg/aboutacres/ourachievements.html (Accessed June 19, 2012).

Avadhani, P.N., Y.C. Wee, L.M. Chou, C. Briffett, R. Hale, H.C. Ho, K. Lim, K.K. Lim, and R. Subaraj. *Master Plan for the Conservation of Nature in Singapore.* Malayan Nature Society, Singapore Branch.

Aziz, Hatta. "Poh Ern Shih: The High-tech, Eco-friendly Buddhist Temple." CNN GO, January 2 2010. http://www.cnngo.com/singapore/none/poh-ern-shih-ecofriendly-temple-801480

Briggs, D., J. J. Abellan, and D. Fecht. "Environmental Inequity in England: Small Area Associations between Socio-Economic Status and Environmental Pollution. *Social Science & Medicine* 67, no. 10 (2008): 1612–1629.

Brook, B. W., N.S. Sodhi, and P.K.L. Ng. "Letters to Nature, Catastrophic Extinctions Follow Deforestation in Singapore." http://www.ebd.csic.es/carnivoros/personal/eloy/landuse/brook.pdf

Bulkeley, H., and G. Walker. "Environmental Justice: A New Agenda for the UK." *Local Environment: The International Journal of Justice and Sustainability* 10, no. 4 (2005): 329–32

Bullard, Robert D. "Blacks and Latinos on the Frontline for Environmental Justice: Strengthening Alliances to Build Healthy and Sustainable Communities." The National Black Latino Summit, Working Paper, http://www.jointcenter.org/hpi/sites/all/files/04-BLS_enviro%20justice.pdf (Accessed September 27, 2012).

Bullard, Robert D. "Race and Environmental Justice in the United States." *Yale Journal of International Law* 18 (1993): 319–35.

Cheam, J., G. Chua, and N. Yong. "Recycling in Singapore: 10 Years On—Chute, We Got It Wrong." *The Straits Times,* December 12, 2010, http://wildsingaporenews.blogspot.com.au/2010/12/recycling-in-singapore-10-years-on.html

Chua, G., and K.S. Goh. "Pushing Hard for Singapore Eateries to Think Green." *The Straits Times,* January 28, 2011, http://wildsingaporenews.blogspot.com.au/2011/01/pushing-hard-for-singapore-eateries-to.html (Accessed June 19, 2012).

"CNG Cars in Singapore: Moment in the Sun Over." *The Straits Times,* September 21, 2010, http://www.eco-business.com/news/cng-cars-singapore-moment-sun-over/ (Accessed June 19, 2012).

Cutter, S.L. "Race, Class and Environmental Justice." *Progress in Human Geography* 19, no. 111 (1995).

Dobson, A. *Justice and the Environment: Conceptions of Environmental Sustainability and Theories of Distributive Justice.* Oxford: Oxford University Press, 1998.

ECO-Singapore. "Report on Re-Live @ Heartland Choa Chu Kang." September 22 to October 6, 2010, http://www.scribd.com/doc/39150907/Re-Live-Heartlands-Choa-Chu-Kang-Report

Energy Market Authority of Singapore. "Electric Vehicles Test-Bedding." http://www.ema.gov.sg/ev/ (Last modified September 15, 2012. Accessed September 17, 2012).

Ermisch, J., and W. Huff. "Hypergrowth in an East Asian NIC: Public Policy and Capital Accumulation in Singapore." *World Development* 27, no. 1 (1999): 21–38.

Faith in Ecology: Go Green and Say Amen to Jesus!, Feb 3, 2012, Archdiocese of Singapore, http://www.catholic.sg/webevent_detail.php?EventID=2123

Green Drinks, http://www.greendrinks.org/ (Accessed June 19, 2012).

Green Drinks: The Need for Conservation of Singapore's Wildlife, January 27, 2011, The Artery @ Red Dot, Singapore, http://www.facebook.com/events/145890058797938/

Green Drinks, Singapore, http://sggreendrinks.wordpress.com/

Green Solutions. "The Green Corridor." http://www.thegreencorridor.org/about/ (Accessed June 19, 2012).

Ground-up Initiative, http://www.groundupinitiative.org/ (Accessed June 19, 2012).

Hobson, K. "Enacting Environmental Justice in Singapore: Performative Justice and the Green Volunteer Network." *Geoforum,* 2006, 671–81.

Hua, D., and Y.W. Loh. "Managing Storm Water in Urbanised Singapore for Flood Control." Presented at Flood and Storm Surge Control Training, Tokyo, Japan, January 25–27, 2011. http://www.asianhumannet.org/db/datas/1102/singapore.pdf (Accessed January 19, 2012)

Hugh T.W. Tan et al. *The Natural Heritage of Singapore,* Third Edition, p. 78. Singapore: Prentice Hall-Pearson Education South Asia Pte Ltd, 2010. http://www.dbs.nus.edu.sg/staff/details/hugh_tan/Ch4%20Waste-%20and%20Reclaimed%20Land%20p.%2078.pdf

Ibrahim, M.F., and S.W. Chung. "Quality of Life of Residents Living Near Industrial Estates in Singapore." *Social Indicators Research* 61, no. 2 (2003): 203–25.

Islamic Foundation for Ecology and Environmental Sciences, http://www.ifees.org.uk/ (Accessed June 19, 2012).

Kamarul, N. "Singapore's First Eco-friendly Mosque to Open on Friday." Channel NewsAsia, May 14, 2009.http://www.channelnewsasia.com/stories/singaporelocalnews/view/429210/1/.html (Accessed June 10, 2012).

Kaur, D., and M. Switow. Panel on "The Economics of Living: Discussing Poverty in Singapore," One Singapore, http://www.onesingapore.org/news/the-economics-of-living-discussing-poverty-in-singapore/

Larson, N., M. Story, and M. Nelson. "Neighbourhood Environments: Disparities in Access to Healthy Foods in the US." *American Journal of Preventive Medicine* 36, no. 1, (2006): 74–81.

Lee, H.L. "A Home with Hope and Heart." National Day Rally 2012, Prime Minister Lee Hsien Loong's Speech in English, August 26, 2012, http://www.channelnewsasia.com/annex/ND2012_PMLEE_ENGTEXT.pdf (Accessed August 20, 2012).

Lee, K.Y. *From Third World to First: The Singapore Story: 1965–2000,* pp. 188–211. Singapore: Times Media, 2000.

Lee, P.O. "Water Management Issues in Singapore." Paper presented at Water in Mainland Southeast Asia, November 29 to December 2, 2005, Siem Reap, Cambodia. http://www.khmerstudies.org/download-files/events/Water/Lee%20Nov%202005.pdf (Accessed April 20, 2012).

Martínez-Alier, J. *The Environmentalism of the Poor.* Cheltenham UK/ Northampton MA: Edward Elgar, 2002.

Ministry of National Development, Singapore. "Written Answer by Ministry of National Development on the Reviewing of Genetically Modified Food." May 14, 2012.http://app.mnd.gov.sg/Newsroom/NewsPage.aspx?ID=3554&category=Parliamentary%20Q%20&%20A&year=2012&RA1=&RA2=&RA3= (Accessed June 19, 2012).

The National Religious Partnership for the Environment. "Catholic Perspectives on Faith and the Environment: A Bibliography." http://www.nrpe.org/index.php?option=com_k2&view=item&layout=item&id=264&Itemid=1212 (Accessed June 19, 2012).

Nature Society (Singapore), 2011, The Green Corridor: A Proposal to Keep the Railway Lands as a Continuous Green Corridor, http://nss.org.sg/documents/TheGreenCorridor101103.pdf (Accessed June 10, 2012).

Nature Society (Singapore), http://www.nss.org.sg/about.aspx?id=3 (Accessed June 10, 2012).

Nature Society of Singapore (NSS), 1990, Master Plan for the Conservation of Nature in Singapore, http://www.nss.org.sg/about.aspx?id=2 (Accessed June 19, 2012).

Ong, E. "Singapore's Inequality Battle." *New Mandala,* December 23, 2011, http://asiapacific.anu.edu.au/newmandala/2011/12/23/singapores-inequality-battle/ (Accessed June 19, 2012).

The Online Citizen. "Call for Moratorium over Plans for Bukit Brown." March 21, 2012. http://theonlinecitizen.com/2012/03/call-for-moratorium-over-plans-for-bukit-brown/ (Accessed September 1, 2012).

Ooi, G.L., and K.H. Phua. "Urbanization and Slum Formation." *Journal of Urban Health* 84, Suppl 1 (May 2007): 27–34. Pickrell, J. "Singapore Extinctions Spell Doom for Asia?" *National Geographic News,* July 23, 2003, http://news.nationalgeographic.com/news/2003/07/0723_030723_singapore.html

Project ME: Muslims and the Environment, Facebook, http://www.facebook.com/Muslims.and.Environment (Accessed June 10, 2012).

Public Utilities Board, Singapore. "Managing Flash Floods—Overview." http://www.pub.gov.sg/MANAGINGFLASHFLOODS/Pages/default.aspx (Accessed June 19, 2012).

Pulau Ubin. "Chek Jawa on Pulau Ubin." http://www.pulauubin.com.sg/Chek-Jawa-On-Pulau-Ubin.html (Accessed June 19, 2012).

The Roots of Singapore's Tree Planting Season, *MNDLink,* http://www.mndlink.sg/2009/2009_Dec/NParks_article.htm

Schlosberg, D. "Reconceiving Environmental Justice: Global Movements and Political Theories." *Environmental Politics* 13, no. 3, (2004): 517–40.

Schwencke, A.M. "Globalized Eco-Islam: A Survey of Global Islamic Environmentalism." Leiden University, http://media.leidenuniv.nl/legacy/report-globalized-eco-islam-a-survey-schwencke-vs-24-february-2012-pdf.pdf (Accessed June 19, 2012).

Sierra Club. "Faith in Action: Communities of Faith Bring Hope for the Planet." http://www.sierraclub.org/ej/downloads/faithinactionreport2008.pdf (Accessed June 19, 2012).

Sim, F. "Carrefour Says No to Shark's Fin." *Yahoo News,* January 7, 2012. http://sg.news.yahoo.com/ntuc-s—no-shark-s-fin—policy-to-kick-in-by-end-april.html

Sodhi N.S., L.P. Koh, Barry W. Brook, and Beter K.L. Ng. "Southeast Asian Biodiversity: An Impending Disaster, Trends in Ecology and Evolution." http://faculty.jsd.claremont.edu/emorhardt/159/pdfs/2006/Sodhi.pdf

The Soup Kitchen Project. https://www.facebook.com/pages/The-Soup-Kitchen-Project-Singapore/300766934440 (Accessed June 10, 2012).

Statistics Singapore—Key Annual Indicators, http://www.singstat.gov.sg/stats/keyind.html#popnarea

Statistics Singapore—Time Series on GDP at 2005 Market Prices and Real Economic Growth, http://www.singstat.gov.sg/stats/themes/economy/hist/gdp1.html

Stephen, C., S. Bullock, and A. Scott. "Environmental Justice: Rights and Means to a Healthy Environment for All." ESRC Global Environmental Change Programme, Special Briefing No. 7, 2001, http://www.foe.co.uk/resource/reports/environmental_justice.pdf (Accessed June 19, 2012).

Teng, P., and M. Escaler. "The Case for Urban Food Security: A Singapore Perspective." *NTS Perspectives,* no. 4, p. 4. Singapore: RSIS Centre for NTS Studies, 2010.

Urban Habitat. "Race, Class and the EPA's Environmental Justice Strategic Plan." http://urbanhabitat.org/epa (Accessed September 27, 2012).

Urban Redevelopment Authority of Singapore. "The Rail Corridor: About the Project." http://www.ura.gov.sg/railcorridor/about.html (Accessed June 19, 2012).

National Environment Agency. "Waste Statistics and Overall Recycling." http://app2.nea.gov.sg/topics_wastestats.aspx (Accessed June 20, 2012).

World Bank. "Faiths and Environment—World Bank Support 2000–2005." 2006. http://siteresources.worldbank.org/GLOBALENVIRONMENTFACILITYGEFOPERATIONS/Resources/Publications-Presentations/EAPFaith&Environment.pdf (Accessed June 19, 2012).

Yale University. "The Yale Forum on Religion and Ecology." http://fore.research.yale.edu/ (Accessed June 19, 2012).

Yeo, S. J. "What Rubbish." *The Straits Times,* June 15, 2009. http://wildsingaporenews.blogspot.com.au/2009/06/non-recyclable-waste-in-recycling-bins.html (Accessed June 10, 2012).

PART III

Oceania

Chapter 13

The Australian Environment Movement: Between Conflict and Cooperation

Cassandra Star

Introduction

In this chapter I explore the political importance, and current state, of the contemporary environment movement in the Australian context. To provide a clear picture of the movement, its efficacy and its political influence, it is necessary to examine the key historical formative elements of the movement in Australia and how these laid the groundwork for ongoing strengths and weaknesses. The two preceding political eras—the periods of the Hawke and the Howard governments—are also explored to interrogate the nature of the relationship between the movement and the state during these eras, and to understand how this has shaped the tactics and the efficacy of the movement and its actions in the current era. Subsequently, the movement's internal dynamics of conflict and cooperation are examined in relation to external political and policy effectiveness. In addition, the relationship between the environment movement and other progressive social movements is examined to consider the role of potential alliances to increase public reach and legitimacy.

Overall, in this chapter I argue that the current political efficacy of the movement is compromised from a number of angles: the failure to capitalize on the successes of the Hawke era; the failure to learn from the challenges of the Howard era; and the need to move beyond the past focus on "wilderness" issues to forge systematic social, economic, and political reform. At present, the Australian movement remains stuck between a "frontier conservationist/preservationist" movement and a modern green movement.

The Current Shape of the Movement

There has been recent growth in membership of the environment movement in Australia, up to 7.2 percent in 2007, though it is well behind other Western nations.[1] The contemporary Australian environment movement can be thought of as constructed from several important types of groups. This diversity of the movement, with its competing goals and approaches, has been and remains a clear strength. I characterize these different types of groups as grassroots and community-based action groups, youth groups, professionalized groups or environmental movement organizations (EMOs), and green political parties.

While I class grassroots and community-based action groups together, they are subtly different. Grassroots groups are locally based and often form in reaction to a local issue or problem, but they can also spontaneously form from a group of concerned citizens in response to the interest and instigation of a single individual. Examples of these include a multitude of local climate groups, such as the Bayside Climate Change Action Group in Melbourne's St. Kilda and Bayside. Such groups provide a local focus, often creating and delivering information for the local community on environmental issues of interest. Community-based action groups are a small variation on this, with support and/or facilitation by a larger organization or group, such as Greenpeace Australia-Pacific's support of grassroots citizen action groups within its organization. These groups are essentially decentralized and operate on a larger issue, but within a local community.

A relatively recent development in the Australian environment movement is the formation of specific youth environmental groups. These groups have rapidly taken on an influential role in the movement, which has been accused by some of stagnation and a lack of room for succession for younger activists, and is still dominated by leaders of the 1970s and 1980s.[2] The creation of a space specifically for youth has broadened the profile of the movement and scope for influence. A key player in the youth environmental space is the Australian Youth Climate Coalition (AYCC) formed in 2006; other groups include the Australian Student Environment Network (ASEN). The focus of youth organizations is not only to engage and educate youth but also to provide a clear voice for youth on environmental issues. An example of this is AYCC's 2011 "Meet your member" campaign encouraging young people to meet their elected representatives, share their opinions on climate change, and deliver signatures on postcards of other young people supporting their views.

However, the focus of most literature on the Australian environment movement is squarely on professionalized groups or EMOs. This includes a range of organizations, such as Friends of the Earth (FoE), which has less than a handful of full-time staff, all the way through to the Australian Conservation Foundation, World Wildlife Fund, and Greenpeace, which employ a significant number of professional staff, including researchers with PhD qualifications. These groups are the main face of the movement and include public action and media work as part of their core business. Despite being the "professionalized" end of the movement, almost all EMOs in the Australian context are also reliant heavily on the work of volunteers to conduct their day-to-day business due to funding constraints.

Green political parties, while having arisen from the environment movement, by virtue of their participation in the formal political process are no longer considered part of the environment movement per se. Indeed, in Australia, the precursor to the Greens political party was established in 1972 in the wake of the Lake Pedder campaign, and has in recent years come to achieve significant mainstream political influence with 13 percent of the vote in the Australian Senate, a first for any Australian minor party, and their first House of Representatives seat in the hung parliament following the 2010 election. However, it is important to keep in mind that while the existence and mainstream influence of the Greens has removed the movement's political leverage of the green vote in the formal political process, the Greens do not speak for the movement, or vice versa.

History of the Australian Environment Movement

There is a long history of environmental concern and activism in the Australian context. Early nature preservation and conservation can be traced back to the 19th century, largely through the influence of wealthy or influential hunters or botanists,[3] which mirrors early developments in the United States. The birth of the contemporary Australian environment movement is much more recent. As in most Western countries, the development of modern environment movements is tied to the surge of other counterculture movements of the 1960s, including civil rights, feminism, and antiwar movements.[4] Thus, I mark the beginning of the contemporary Australian environment movement from this time.

Several significant or influential moments since the movement's inception highlight critical issues politically and demonstrate some ongoing challenges for the movement. I outline three cases in the following paragraphs to bring these issues to the fore.

The concerted, but ultimately futile, campaign to save Lake Pedder in the Australian state of Tasmania from hydroelectric development is a key moment in Australian environmental history. It is credited with being the earliest national environmental campaign,[5] as a precursor to the more successful campaign to save the Franklin River in the 1980s. Additionally, it spurred the beginnings of the Green Party in Australia, and the public career of its now former leader, Bob Brown. The fight to save the Franklin River displayed some key strengths of the tactics and strategy of the movement. The campaign utilized cooperation between local environmental groups and larger national groups to gain a public profile and support. The political strategy of playing the local state (Tasmanian) government against the national government was crucial in preventing the flooding of the Franklin. Lastly, the mobilization of public concern and potential electoral impact as leverage for government influence was a very effective strategy in this campaign, and is attributed as significant in the Hawke government's win in the 1983 federal election.[6] These successful tactics have remained in the movement's playbook, but as I shall explore later, the efficacy of these tactics is significantly reduced in the current era. The less successful elements of this case—including a lack of reform of the underlying development approach of Tasmania's hydro-industrialization, and the federal government's provision of monetary compensation to the Tasmanian government, which was used to build subsequent hydroelectric capacity at alternate wilderness locations—highlight the symbolic nature of the celebrated win, which has been undermined by a lack of conversion of public concern into lasting societal or policy change.

A similar lesson can be drawn from the environment movement's involvement in the ecologically sustainable development (ESD) process initiated by the Hawke federal government in the early 1990s. While I address later the relationship between the Hawke government and the environment movement, this particular example demonstrates a similar dilemma, despite the fact that the movement was invited "inside" the state, rather than working to leverage popular concern from the "outside." In the period after the 1992 Earth Summit, organized by the United Nations in Rio de Janeiro, the Australian government commenced a process (as did others) aimed at reform of policy and important economic sectors, to achieve ecological sustainability in line with the summit's official embrace of sustainable development. The process focused on bringing together key political and economic stakeholders, including business, unions, and the environment movement, in a consensus-based forum to deliver recommendations to move the country and economy toward ESD. The final outcomes produced by the included groups, especially for the time period, were very commendable. Less successful was the watering down of the recommendations and the lack of implementation by public sector agencies.[7] This example again highlights the "moments" of success the movement is capable of achieving, but the undermining of these by the failure to secure systematic reform and change, even while operating "inside" government. This process also highlighted and reinforced tensions and divisions within the movement between those who were "insiders" and "outsiders" to the process, and the government. These tensions continue.

The final case that holds clear and continuing lessons for the environment movement and demonstrates ongoing challenges is the fight against the Jabiluka uranium mine in the remote Northern Territory of Australia. The campaign against the mine, located against the boundaries of the Kakadu National Park World Heritage area, was the largest social movement campaign in Australia in the late 1990s and early 2000s. A key element of the campaign, like that of Lake Pedder, was the call by local activists, in this case the Mirarr indigenous owners of the site, to national groups and activists to support the campaign and build public support through a blockade at the site. What sets apart the Jabiluka campaign is the cooperation and alliance politics between the environment movement and indigenous activists in this case. While the campaign has ultimately been successful to date (the mine leaseholder, Rio Tinto, says the mine could reopen one day), it highlighted significant tensions and limitations in the environment movement's approach to working with other groups. Paasonen articulates the problems associated with such cooperation between movements of affluence (the environment movement in this case) and movements of crisis (the Mirarr people), where an imminent and direct threat drives movements of crisis, which obviously have more to lose.[8] Other commentators are less generous, and observe that just as in the United States, environment movement organizations in Australia are dominated by white, male, middle-class members and their preoccupations, and are thus blind to environmental justice concerns.[9] Indeed, Pickerill highlights that among ongoing issues and tensions between green and indigenous interests in Australia, power and ownership and economic issues are key, with a need to move beyond colonial paternalism required to nurture progress, despite clear common ground between the movements.[10]

Thus, these critical cases in the history of Australia's environment movement highlight key challenges for the movement in the current era and moving forward. The movement has engaged in difficult campaigns, producing high-profile wins that have, however, been significantly undermined by an inability to convert symbolic wins into systemic reform while operating both outside and inside the state. In addition, ongoing tensions exist during and between campaigns, both between groups inside the movement and with other significant social movements, including the indigenous movement. This suggests the need to look more closely at the tactics and political strategies of the contemporary movement, as well as how the movement relates to itself and other movements in the Australian political landscape.

Tactics, Political Strategies, and Impact of the Movement

The current tactics, political strategies, and impact of the environment movement in Australia are significantly shaped by the recent history of the movement and its relationships with the state during these periods. One of the reasons that the relationship between environmental nongovernmental organizations (ENGOs) and the government during the Howard era is of such interest is the contrast it represents with the relationship under other governments. In this section, I provide some context for considering the plight of ENGOs in the Howard era. Perhaps in most stark contrast to the Howard years is the preceding Hawke era.[11]

During the Hawke era, ENGOs enjoyed significant legitimacy and they enacted a pivotal role in public policy debates and political battles. While at least some of this insider status was clearly pragmatic—the Hawke government took office on green votes[12] and it was very cognizant of this—the engagement of ENGOs went beyond this. Hawke and several of his ministers undertook high-level engagement and consultation with key ENGOs[13] and there was notable innovation in environmental policy and environmental institutional arrangements during this time. There is evidence that ENGOs were able to enact significant influence on a number of critical environmental issues of the time. A number of key figures in the movement also had authority with the government. Phillip Toyne, at the time the executive director of the Australian Conservation Foundation (ACF; 1986–1992), went on to become deputy secretary in the Commonwealth Department of Environment from 1994 to 1997 in the Hawke government. During his time at the ACF he was able to contribute significantly to public debate, as well as to engagement with the public sector, on many environmental issues.[14] However, the place of ENGOs in policy debates during the Hawke era is representative of a different approach of the government and the state to interacting with civil society during this era.

For ENGOs, two cases exemplify the approach taken. The first case again recalls the government's ESD process. The process was designed to be an exercise in collaborative policy deliberation as a precursor to the government's response to international policy moves on sustainable development. This case highlighted the significance of ENGOs in two ways. Firstly, the legitimacy of their policy agenda was confirmed by engaging ESD as a policy issue, as well as the inclusion of ENGO representatives in the process. Despite the eventual weaknesses of the exercise,[15] it still highlights a more engaged approach to government-ENGO relations.[16] The second case is that of the Resource Assessment Commission (RAC). The role of the RAC during the Hawke era was to be an institutional mechanism to resolve the disruptive resource development conflicts of the time. The RAC was created as an independent institution, with an independent chairperson, charged with arbitrating resource-development issues where economic benefits clashed with environmental concerns.[17] By institutionalizing recognition that the economic benefits associated with resource exploitation meant challenging tradeoffs between two sets of important societal goals, it again provided a clear demonstration of ENGOs' legitimacy and of the place of their voice in policy. These are just two examples that demonstrate a quite different relationship between ENGOs and the government preceding the Howard era. However, regardless of this legitimacy, the failure of the movement to convert this into systematic reform is problematic.

The contrasting relationship between the Howard government and Australian ENGOs can be illuminated by examining two key cases—the funding of NGOs and the Natural Heritage Trust. An early indicator of the Howard government approach toward civil society was to restrict funding to groups with a "political" purpose. For ENGOs, this move impacted each group differently. For example, Greenpeace Australia-Pacific, which draws substantial revenue from its membership base, was probably least impacted of all Australian ENGOs.[18] Conversely, FoE Australia, a relatively small (but active) ENGO, was substantially affected, losing significant revenue and could no longer support one of its few full-time paid campaigners.[19] Thus, ENGOs and civil society overall were adversely affected and faced a changed arena with a new government that had substantially different views, and thus intentions, about the role of civil society *vis-à-vis* public policy engagement and political debate. The example of the Natural Heritage Trust, the Howard government's key environmental funding delivery mechanism, demonstrates this clearly. The sale of Telstra generated substantial funds for environmental projects during this time.[20] However, guidelines for eligibility to apply to the trust introduced by the Howard government precluded groups that had "political" purposes.[21] This restriction effectively precluded Australian ENGOs from proposing projects, despite the fact that they had a significant history of expertise in the priority areas funded. The exclusion of this and reduction of other opportunities for funding brought many ENGOs close to financial ruin.[22] It also brought into question the legitimacy, expertise, and capacity of the trust as a whole. Thousands of small community groups and regional organizations were funded to deliver projects across the country,[23] with good intentions but often little or no environmental knowledge, required expertise, or appropriate experience. The outcomes of the change to a strained relationship under the Howard government were challenging for ENGOs, but also for society as a whole, leading to a lost decade of environmental policy. How the movement has, or can, move forward from this point is considered next.

Cooperation and the Movement

Aside from the environment movement's tactics and political strategies, a critical influence on its importance and impact on green politics in the Australian context is the engagement between groups within the movement, as well as the engagement with other movements, on issues of common ground. The ability to act in concert and build productive campaign alliances, especially in times of hostile government, is important for success. Thus, I next examine dynamics within the movement between different groups, and between the movement and related progressive social movements.

During the lost decade of the Howard years, and also since the return of the Labor government to power in Australia in 2007, the importance of the environment movement's ability to present, if not a united front, then a strong cooperative one that represents significant public support, cannot be overstated. In an era where environmental concerns, especially represented by groups with "political" aims, have been actively excluded from policy dialogue, cast as the "enemy" of conservative governments and de-funded, the movement has been required to adopt a strong cooperative approach to maintain strength and relevance in public debate. In the United Kingdom, similar pressures during the reign of Margaret Thatcher and beyond led to a strong, enduring civil society, with an environment movement in recent years characterized by open collaboration on joint campaigns with other key movements.[24] However, similar approaches in Australia have been embryonic at best. This can be attributed to factors both internal to the movement and within the relationship between the state and the movement.

A lasting legacy of the Hawke government's neo-corporatist approach to environmental policymaking in Australia, despite its achievements, has been a pattern of engagement by the movement with subsequent governments characterized by groups that enjoyed insider status under the Hawke government attempting to reclaim this status, in the face of criticism and resentment from outsider groups. Obviously, under a hostile government, such engagement is ineffective. However, these groups faced a second difficulty with the Howard government's decisions around funding (or rather de-funding) long-standing ENGOs and Conservation Councils, creating a new schism in the movement between the "political" ENGOs and the subsequently funded "task-oriented" or "on the ground" ENGOs. These disconnects within the movement itself made it less likely for groups to come together in a coordinated or cooperative way, especially for those groups not political in nature; indeed, to do so would have threatened their newly acquired funding. Groups with a political agenda suffered a lack of funding, impacting their ability to undertake core activities, let alone new cooperative agendas. Significantly, peak and regional groups, historically the keepers of any coordinated or cooperative efforts, were substantially defunded from the mid-1990s.

In the ensuing years, unlike the outcome of the Thatcher years in the U.K. there was a weakening rather than a strengthening of the movement, and a widening of the schisms between "insider" and "outsider," "political" and "on the ground," and "respectable" and "radical." This effectively reinforced government exclusion of these groups from policy debates, with green discourse and symbols co-opted by government and business. Public concern for environmental issues, as reflected in polling, subsequently dropped up until 2004. From its election in 1996, the Howard government pursued a policy of systematically silencing significant political dissent and, in this case, was wildly successful.[25] Given the power the environment movement possessed under the Hawke government, it represented a clear threat to be neutralized. The scars of this concerted campaign still remain on the Australian environment movement, as evident in its current tactics and political strategies. Thus, the pendulum swings closer to internal conflict than it does to cooperation.

In addition, while adverse conditions in other countries have led to strong and robust connections between movements, this is not evident in the Australian experience. Relationships between the environment movement and unions, social justice and indigenous movements in the Australian context remain fragile, with much work to be done to move toward productive cooperative or coalition actions. While there are examples of isolated pockets of good work in these areas, including work in the Pacific by FoE on climate justice and by Greenpeace on renewable energy capacity, these have tended to represent actions with specific local communities, rather than ongoing cooperative partnerships between movements. Despite the early promise of cooperation, the relationships between the environment movement and unions in Australia remains fraught, through the union movement's Green Bans on construction in significant Sydney green locations, led by Jack Mundey in the 1970s. Part of the impasse has been, and remains, a focus of the environment movement on "green" wilderness issues that are frequently at odds with the focus of the union movement on jobs, often in extractive industries. This conflict is also reflected in the broader framing of environmental issues in the Australian public debate, notably the construction of an inverse relationship between jobs and the environment, which is perpetuated in political debates. In failing to evolve to embrace environmental issues beyond a wilderness or frontier understanding, the movement fails to challenge this framing and thus reduces its chances to form worthwhile alliances with the union movement on urban and industrial environmental issues, including environmental health and safety. Green wilderness issues are easy to conceptualize and solve in symbolic, piecemeal ways—a characteristic approach of the Hawke era—they can be "negotiated" and bargained. Brown or urban issues, however, present fundamental challenges to urban life and require systemic reform: a key area of weakness for the Australian environment movement.

Two other movements of key cooperative potential for the environment movement are the broad social justice movement and the indigenous movement. The environment movement's inability to engage in effective cooperation with these movements beyond short-term coalitions of convenience can also be traced to the key weakness highlighted above, stemming from its overwhelming focus on "green" rather than "brown" issues; environmental issues are predominately framed around the "environment" as "nature," or a place where people are *not.* Thus, despite issues of commonality and the environment movement's recognition of the political value and potential strategic importance of such collaborations, meaningful engagement with other movements where the focus is on people, and social justice within human societies, is fraught and fundamentally difficult. In many ways, such engagement, especially with social justice movements, would be of significant benefit to the green movement, due to the greater societal acceptance and legitimacy of aid and development groups compared to the perceived "radicalism" of the environment movement and its portrayal as such in the Australian media.

Thus, overall the Australian environment movement can be broadly characterized as one in which coalitions and collaborations are limited and short term at best. This is not due to a lack of desire by the movement or individual players to capitalize on the potential strategic value of such cooperation. It still carries the scars of the Howard government's deliberate campaign of exclusion, division and systematic extinguishing which, rather than strengthening the movement, left it broken, fragile and divided. Coupled with the environment movement's propensity to a "green" focus, the marginalization of "brown," human and social justice issues leaves little room for meaningful, ongoing cooperative relationships within the movement or with other progressive movements.

The Movement and Green Politics in Australia

Well established, the environment movement in Australia is currently operating with a more receptive Labor government that has stated its commitment to major environmental issues, including climate change. Despite this, and the potential and promise of strong public concern about environmental issues, the movement still faces major challenges in terms of its efficacy and ability to positively influence policy in the Australian context.

The first of these challenges is the rebuilding of public support, not just for environmental issues, but also for the movement itself and its mandate. This necessitates the movement maintaining some "outsider" status through rejection of a sole focus on engagement with the government and the public sector. It also requires the movement to challenge and reject its framing by government and media as "radicals" far removed from the reality of modern Australian working-class life. To actively break down this perception, the movement does indeed need to address its bias toward a white, male, middle-class leadership and membership focused on wilderness.

In addition, the movement needs to remake its image in the eyes of government, and its role in the wider societal milieu. While a return to the key role the movement played during the Hawke neo-corporatist era of environmental policy is unlikely due to its loss of leverage of the "green vote," an improvement upon the Howard government's view of the environment movement as "just another rent-seeking lobby group" is needed. The Rudd and Gillard governments have made progress on this ideological position, but the movement has some way to go before it is seen as an interest group motivated by the interests of the broader society, rather than sectional interests. Achieving this will hopefully provide the movement with greater access to policy dialogues and membership of policy networks.

Once entry into government dialogues and networks has been reestablished for those groups within the movement that seek it, there are further gains to be achieved. While insider status can bring significant wins, the lessons of the Hawke era need to be learnt. If the movement relies too heavily on insider engagement to reach its aims, it will again be exposed to developing weaknesses by becoming distanced from the public and its concerns, and will possibly experience a loss of rigor and resilience. The challenges faced during the Howard years demonstrate this. In addition, the insider status of the movement during the Hawke years and an analysis of its effectiveness during this time demonstrate a key weakness in its ability to convert key symbolic wins into long-term systematic, policy, and institutional reform. This weakness needs to be overcome if the movement is to achieve the true potential of insider status.

In order to achieve this, the movement needs to reflect upon, and embrace, the internal changes it is required to undergo to meaningfully engage in cooperation and collaboration, both internally and externally. While internal changes will be less fraught, they will require the movement to genuinely embrace the value of its diversity. Externally, the challenges are more significant. While the environment movement sees the strategic political value in engaging with other movements, moving beyond tokenistic engagement will be challenging. Identifying and nurturing common ground will require the movement to shed its frontier environmentalism in favor of a focus on modern "brown" urban environmentalism that centralizes human, social justice concerns as its driving concerns.

Engaging other social movements is a key step for the environment movement toward achieving a level of potential political influence. To truly achieve this, the task of rebuilding civil society in the Australian context is needed. Civil society has been ravaged and hollowed out by the Howard government's sustained attack on political dissent, its divisive approach to NGO funding, and the embrace of wider neoliberal approaches within the Australian context. The increasing importance placed on the contracting of services by the public sector has contributed to a decline of the political and civic engagement of NGOs, ENGOs included. In addition, the shift in emphasis from citizen to consumer in the public sector has changed public perceptions and expectations of involvement in civil society and public life.

To capitalize on the significant potential of the environment movement to have an impact on policy and politics in the Australian context, civil society in Australia must be re-imagined and reinvigorated. This remains a significant challenge for the environment movement, and one that cannot be undertaken alone, necessitating the formation of strong and enduring cooperative relationships with other progressive social movements.

Conclusion

The key message of this chapter, and the key learnings from the recent history of the environment movement in Australia, demonstrates that the political climate in which this movement and its political influence once prospered has significantly changed in ways that the movement neither anticipated, nor has substantially adapted to. To regain an influential place in the policy dialogue and political milieu of Australian civil society, the environment movement needs to find itself—or needs to find the modern incarnation of itself—to remain relevant in the contemporary era. To do so will require the movement to move beyond the tactics and strategies that were once successful in the neo-corporatist Hawke era and to heal the damage of the Howard era. This will require internal examination of the movement and collaborations that embrace its diversity, to overcome painful and harmful divisions. It will also necessitate reaching out to other progressive social movements to begin to nurture meaningful collaborations on the basis of common ground. To capitalize on such common ground will entail a significant repositioning of the movement from its traditional "green" wilderness focus toward a more human-centered focus. This will create an opportunity for the movement, in partnership with other progressive social movements, to heal and grow Australian civil society into a strong, productive, and transformative space for the Australian public.

Notes

1. Tranter, "Environmental Activists."
2. Ibid.
3. Dunlap, "Nature and the English Diaspora."
4. Hutton and Connors, "A History of Australian Environment Movement."
5. Ibid.
6. Davis, "Wilderness Conservation in Australia."
7. Christoff, "Whatever Happened to ESD?"
8. Paasonen, "Between Movements of Crisis."
9. Doyle, *Green Power.*
10. Pickerill, "Finding Common Ground."

11. While the relationship between the government and ENGOs cooled during the Keating years, I have not addressed that element in this chapter—but it is possible to speculate that this was partly due to the difference in Keating and Hawke's personal leadership styles (e.g., see Brett, "Graham Little's Theory"), and can also be partly attributed to the "recession we had to have" (Conley, "Globalisation and the Politics").

12. Kelly, *The End of Certainty;* Marsh, "Australia's Political Cartel?"; Norton, "Tasmanian Forests."

13. Downes, "Neo-Corporatism and Environmental Policy"; Haward, "Institutional Design"; McEachern, "Environmental Policy in Australia."

14. Toyne, *The Reluctant Nation.*

15. While the key stakeholders involved were able to come to meaningful policy recommendations at the end of the deliberative dialogue, the recommendations and final reports were ultimately watered down by public servants prior to their public release. This left many participants in the process angry and disillusioned. For a fuller account of the exercise see Christoff, "Whatever Happened to ESD?"

16. This is despite the split within the movement over whether or not to participate inside the process—while four groups initially participated, two groups withdrew their support for the process during the middle of the dialogue. Ibid.

17. Economou, "Australian Environmental Policy-Making."

18. Danny Kennedy, interview by the author, 2005, transcript.

19. Stephanie Long, interview by the author, 2006, transcript.

20. There has been a significant amount of scholarship about the Natural Heritage Trust and the impact on environmental policy in Australia. In certain ways it does represent a policy innovation in terms of the delivery of funds and environmental services at the local level; however, it also suffers from fragmentation, lack of overview and expertise (Crowley, "Effective Environmental Federalism?"; Head, "Participation or Co-Governance?"); but it also has a focus on environmental "clean-up," including tree planting, waterway rehabilitation and other physical works (Hajkowicz, "The Evolution of Australia's") at the expense of other important environmental priorities.

21. Hamilton and Maddison, *Silencing Dissent*; Peatling, "Green Army Warned"; Tomar, "Redefining NGOs."

22. For a full discussion of funding issues see Staples, "NGOs Out in the Cold."

23. For example see Hajkowicz, "The Evolution of Australia's"; and McAlpine et al, "Regional Planning."

24. See Star, "Dancing Together."

25. Manne, "The Howard Years."

Bibliography

Brett, Judith. "Graham Little's Theory of Political Leadership." *International Journal of Applied Psychoanalytic Studies* 6, no. 2 (2009): 103–10.

Christoff, Peter. "Whatever Happened to ESD?" *Cappuccino Papers* No. 1. Melbourne: Imagine the Future Inc, 1995.

Conley, Tom. "Globalisation and the Politics of Persuasion and Coercion." *Journal of Social Issues* 39, no. 2 (2004): 183–200.

Crowley, Kate. "Effective Environmental Federalism? Australia's Natural Heritage Trust." *Journal of Environmental Policy and Planning* 3 (2001): 255–72.

Davis, Bruce. "Wilderness Conservation in Australia: Eight Governments in Search of a Policy." *Natural Resources Journal* 29, no. 1 (1989):103–13.

Downes, David. "Neo-Corporatism and Environmental Policy." *Australian Journal of Political Science* 31, no. 2 (1996): 175–90.

Doyle, Timothy. *Green Power: The Environment Movement in Australia.* Sydney: UNSW Press, 2000.

Dunlap, Thomas. *Nature and the English Diaspora: Environment and History in the United States, Canada, Australia and New Zealand.* Cambridge: Cambridge University Press, 1999.

Economou, Nick. "Australian Environmental Policy-Making in Transition: The Rise and Fall of the Resource Assessment Commission." *Australian Journal of Public Administration* 55, no. 1 (1996): 12–22.

Hajkowicz, Stefan. "The Evolution of Australia's Natural Resource Management Programs: Towards Improved Targeting and Evaluation of Investments." *Land Use Policy* 26, no. 2 (2009): 471–78.

Hamilton, Clive, and Sarah Maddison, eds. *Silencing Dissent: How the Australian Government Is Controlling Public Opinion and Stifling Debate.* Sydney: Allen and Unwin, 2007.

Hannan, Ewin. "Business Rebuffs Howard over IR Ad Blitz." *The Australian,* April 5, 2007.

Haward, Marcus. "Institutional Design and Policy-Making 'Down Under': Developments in Australian and New Zealand Coastal Management." *Ocean and Coastal Management* 26, no. 2 (1995): 87–117.

Head, Brian. "Participation or Co-Governance? Challenges for Regional Natural Resource Management." In *Participation and Governance in Regional Development,* edited by Robyn Eversole and John Martin. Ashgate: Aldershot, 2005.

Hutton, Drew, and Libby Connors, eds. *A History of the Australian Environment Movement.* Cambridge: Cambridge University Press, 1999.

Kelly, Paul. *The End of Certainty: Power, Politics and Business in Australia.* Sydney: Allen and Unwin, 2008.

Manne, Robert. *The Howard Years.* Melbourne: Black Inc, 2004.

Marsh, Ian. "Australia's Political Cartel? The Major Parties and the Party System in an Era of Globalisation." In *Political Parties in Transition,* edited by Ian Marsh. New South Wales: The Federation Press, 2006.

McAlpine, Clive, Sonja Heyenga, Bruce Taylor, Ann Peterson, and Geoff McDonald. "Regional Planning in Queensland's Rangelands: Challenges and Prospects for Biodiversity Conservation." *Geographical Research* 45, no. 1 (2007): 27–42.

McEachern, Doug. "Environmental Policy in Australia 1981–1991: A Form of Corporatism?" *Australian Journal of Public Administration* 52, no. 2 (1993): 173–86.

Norton, Paul. "Tasmanian Forests, the Environment and Voting in the 2004 Australian Federal Election." *Australian Journal of Political Science* 41, no. 4 (2006): 597–614.

Paasonen, Karl-Erik. "Between Movements of Crisis and Movements of Affluence: An Analysis of the Campaign against the Jabiluka Uranium Mine, 1997–2000." PhD diss., School of Political Science and International Studies, University of Queensland, 2007.

Pickerill, Jenny. "Finding Common Ground? Spaces of Dialogue and the Negotiation of Indigenous Interests in Environmental Campaigns in Australia." *Geoforum* 40, no. 1 (2009): 66–79.

Peatling, Stephanie. "Green Army Warned on Politicking." *Sydney Morning Herald*, April 12, 2005.

Staples, J. "NGOs Out in the Cold: Howard Government Policy towards NGOs." Democratic Audit of Australia Discussion Paper No. 19/06, 2006.

Star, Cassandra. "Dancing Together: Environment, Development, Aid and Faith Organisations in Climate Politics in the United Kingdom." Paper presented at the Australasian Political Studies Association Annual Conference, Melbourne, Australia, September 24–26, 2007.

Tomar, Ravi. "Redefining NGOs: The Emerging Debate." Current Issues Brief 52003–4. Canberra: Department of the Parliamentary Library, 2004.

Toyne, Phillip. *The Reluctant Nation: Environment, Law and Politics in Australia.* Sydney: ABC Books, 1994.

Tranter, Bruce. "Environmental Activists and Non-Active Environmentalists in Australia." *Environmental Politics* 19, no. 3 (2010): 413–29.

Chapter 14

The Land Has Eyes and Ears

Mesake Koroi and Robert Wolfgramm

Introduction

I met Mesake Koroi in 1990. It was an early Friday evening and he was sitting in his home kitchen with a bottle of whisky. At that time, Mesake was working as public relations officer with the Fiji Development Bank. He was well known throughout the country as a print and radio journalist having worked at the *Fiji Times* newspaper and at the Fiji Broadcasting Commission during prior decades when Fiji was monopolized by one newspaper and one radio station. I was at Mesake's door because I was doing fieldwork for my doctoral research problem (*Kai Viti?*)[1] and had been directed to consult him as a potential resource, if not as a prospective research subject. He turned out to be both, but our opening conversation centered on an invitation to join his family on a forthcoming visit to his island home of Vanua Balavu—some 20 hours by boat east of urban Suva, Fiji's capital, to sample "real Fijian village life." Three years later, I did. What was fascinating about Mesake and his fellows from the village of Mavana was that they constituted a group who were plugged into the dynamics of modernizing and Westernizing Fiji, but who remained integrated, rooted, grounded (as metaphors) in the contradictory values and *weltanschauung* of rural agricultural cycles and village subsistence economies—of the kind Mesake describes in this chapter.[2]

No one lives "in harmony" with nature or with anyone else for that matter, and the tensions Mesake notes in his negotiation of his dual worlds, past and present, is something I can identify with. Though not being myself "*iTaukei* Fijian," like Mesake, I was born in Fiji and have some Fijian ancestors. My childhood was similarly spent in rural isolation prior to receiving a grammar school education in Fiji, and then an Australian one.

My movements between Australia and Fiji from 1963 (when I departed for an Australian education) to 2004 (when I retired back to Fiji) were brief encounters characterized by a romantic gloss. That is, I saw Australia as increasingly polluted by industrialization and cultural loss; Fiji as an increasingly attractive soft coral, green jungle, a haven for ritual reinforcement of lost identity.

Mesake too speaks in his terms of this kind of contrasting experience through his emphasis of the link between environment and indigenous (*iTaukei*) identity. That link was forged in a preliterate era, prior to contact with Westerners, when indigenous domination of the physical world led to settlements throughout the southwestern Pacific, 3,000 years ago.[3] By the late 18th century, contact between them and European sailors, adventurers, traders, and mission-minded Christians led to early-19th-century exploitation of Fiji's reef and forest resources: *beche-de-mer*, sandalwood, and pagan hearts being the chief targets.[4] By the mid-19th century, land was being cleared for cotton planting (which was tried and found wanting), followed by copra for oil, and sugarcane production.[5] By the middle of the 20th century, gold was being mined (and still is).[6] The dominant discourse through all of this was one of modernize, Westernize, "catch-up," exploit all our available resources. This gospel was so imbedded; I grew up in the 1950s and 1960s with the concept of "progress at any price" as an unquestioned orthodoxy. The environment was there to make Fiji rich. And that was good, full stop.

In keeping with this, following independence in 1970, Fiji began to entertain tourist visitors on a more industrial scale, with hotels and resorts taking their place on hitherto uninhabited bushy islets and beaches. The idea was to promote Fiji as a postcard destination that American soldiers who had been posted here during World War II had already sampled. In 1986, Pope John Paul II visited Fiji and nominated it as "the way the world should be"—a phrase that was captured by the tourism industry but which has haunted Fiji ever since (the fact of four subsequent coups in 20 years speaks for itself).[7] Parts of that postcard were four rainforest "nature reserves," which had been defined into existence from the late 1950s (I grew up in the midst of one of them, Nadarivatu—even while logging proceeded in the neighboring Koro'o hills). Other parts were the characteristic white sandy beaches, swaying palms, and sealife galore. But change was coming and the appearance in 1977 of Fiji's first management and conservation of fisheries legislation was a start—even if it was primarily aimed at protecting our economic zone rather than designed for ecological preservation or awareness.[8]

The 1980s brought great economic indicators for the nation, then economic collapse with Fiji's first two military coups. But tucked away in there was a new consciousness: sustainable development. I liked the phrase, having read by then Rachel Carson's *Silent Spring* (1962) and Jonathon Schell's *The Fate of the Earth* (1982). I was also being briefed by Aboriginal activist elder Gubboo Ted Thomas of the Yuin people of South New South Wales who had prostrated himself between bulldozers and loggers in his dreamtime country to "stop the desecration of mother earth" (as he put it to me).[9] The feminist witch Starhawk had also impressed me at the time as did the Catholic mystic, Matthew Fox.[10] Somewhere in this new-age mix of divergent spiritual traditions was a common ecological concern and "sustainable development" was to become part of a compromise that third world and fourth world peoples were to make to "advance," and "progress"—but gently, as it were. As Gubboo used to say: "I always pray over a tree or sapling before I chop it down." (I don't know how saplings felt about that, but it definitely made the 1980s more conscionable for the rest of us).

In Fiji, the 1980s also saw the first reports that by their very existence defined the environment as an ecopolitical issue. Using research data obtained from waterways and coastal strips not far from where I had grown up in the 1960s, Lal warned of uncontrolled pollutants contaminating our foreshores and putting our marine life at risk.[11] In 1988, we got our first declared "national park" in Sigatoka where its sand dunes were marked for protection.[12] By the early 1990s, this concern was taken up by Johnson Seeto in his investigation of pollution affecting Laucala Bay and Suva Harbour.[13] A wider report in 1992, *Environment Fiji: The National State of the Environment Report*, produced by Dick Watling and Stuart Chape, was the first of its kind. More "forest parks" that had come into existence through Fiji's National Trust authority during the 1980s were added to our heritage during the 1990s.[14] "Threatened," "rare," "endangered," and "vulnerable" entered our vocabulary to refer not so much to indigenous peoples but to our ecosystems and flora and fauna species. Urbanization, soil erosion, pesticide pollution, waste pollution, deforestation, and overfishing were all put on the table of public awareness and political discussion. Eventually the 1990s produced parliamentary legislation—the Sustainable Development Bill (1997, 1999)—to enable the state to create new laws and enable authorities to deal with these matters.[15]

The 1990s also saw the rise of consciousness of environmental issues and the integration and transposition of these hitherto private-spiritual issues into major political ones. A key player in this was the Pacific Concerns Resources Centre (PCRC). The PCRC was formed from the remnants of the

1970s Nuclear Free and Independent Pacific lobby that had been provoked into voice by French tests in its Polynesian territories, and Paris's responses to *kanaky* liberation movements. In 1993 it relocated from Hawaii to Suva. "Environment" and "sustainable development" were among the register of explicit "Pacific concerns" that reached us at that time.[16] I was fortunate to meet with officers and associates of PCRC during its establishment, including Stan Simpson, Prossey Mailau, Francis Sokonibogi, and later, Rex Rumakiek. Their enthusiasm was as high as their funding was low. Perhaps there is an inverse correlation there.

The timing back then coincided with my introduction into Mesake Koroi's world and the work of another who deserves mention—American-born and America-trained marine biologist Ed Anderson, who worked at the University of the South Pacific. Ed's ongoing testing of the quality of the water around Suva's harbor, rivers, and foreshores were best summed up in his words to me a few years back: "don't drink it." Not coincidentally perhaps, I first met Ed in Mesake's ancestral village at the end of 1994. He had been invited to Mavana by one of Mesake's cousins to do what I was there for: participate in and observe an ancient, secretive, fishing ritual they call the *uli drano.* Ed and I planned to write an academic paper about the ritual, but for a number of reasons, it ended up on the shelf. The ritual has ecological dimensions that Mesake does not discuss here, and it is not my place to raise them. But readers will see there is already plenty in Mesake's worldview that provides food for further thought. Indeed, if the environment matters, and if the health of humans is intertwined with the health of Fiji's environment, then there is much work to be done at all levels of action from macro to micro across all age groups, religions, genders, classes, ethnicities, and political ideologies, to improve and keep improving.

Robert Wolfgramm
May 2012

Fiji and Its People

The Fiji group consists of over 500 islands, 100 of which are inhabited. The two main islands are Viti Levu and Vanua Levu and there are two of moderate size, Taveuni and Kadavu. The total land area is 18,272 square kilometers.[17] The islands are scattered over a large area of the Pacific Ocean. They extend from longitude 177 degrees west to longitude 175 degrees east, some 603 kms, and between south latitudes 15 degrees and 20 degrees, 482 kms. Suva, the capital, is about 2,000 kms from the Equator, 3,153 kms northeast of Sydney, Australia, and 2,120 kms north of Auckland, New Zealand. The 180-degree Meridian passes through this tropical "paradise."

The islands were first sighted by Europeans as long ago as 1643 by the Dutch navigator Abel Tasman.[18] Captain Cook sighted Vatoa, one of the most southerly of the Fiji islands, in 1774, during his second visit to the Pacific.[19] But no attempt for a meaningful European settlement came until the early 1800s with news of the abundance of sandalwood trees in Fiji.[20] The sandalwood trade perhaps put Fiji on the international map as traders from Australia, India, and China successfully navigated the Pacific Ocean making Fiji their destination in search of that exotic sweet-smelling wood fetching hefty prices in the Oriental markets of China and India.

Fiji's total population according to the census of 2006 was around 846,085 people, of which 55 percent were Fijians and 37 percent were Indians; the rest consisted of people of different races, mostly Chinese, part-Europeans, and other Pacific Islanders.[21] There is a small European population of mostly professionals and businessmen.

Note that there is a sizable Indian population in Fiji. The Indians were first brought to Fiji by the colonial government under what was then known as the *girmit* system.[22] Under the system, thousands of Indians were recruited from mainland India to work in the sugarcane plantations with the option of making Fiji their home or returning to India at the end of their five-year contracts. Most of them of course decided to stay and now form a very significant component in the population mix of multiracial Fiji.

Similar recruits were also made from among Solomon Islanders and from Vanuatu.[23] Their descendents now form part of that racial mix. While there has been a considerable amount of intermarriage between indigenous Fijians and people from other Pacific Islands, there is hardly any intermarriage between Indians and Fijians and thus social interaction between these two main racial groupings remains noncommittal.

As far as the indigenous or *iTaukei* people of Fiji goes, Sir Alan Burns, in his book *Fiji,* summed it up when he wrote:

> If European knowledge of Fiji was extremely limited up to the eighteenth century, still less was known of its inhabitants and even today still less could be said with any confidence of the origins and early history of the Fijian people. It is perhaps that successive waves of immigrants of Melanesian stock were among the earliest to arrive in the islands. They probably came in their canoes from the westward in small groups, independent of and hostile to each other and established themselves in various parts of Fiji. They have left

> no written records and little even of reliable oral legend. The present Fijians are descended mainly from these dark skinned, fuzzy-haired people.[24]

There is, however, an important strain of Polynesian blood in the small eastern islands and on the windward coasts in the larger islands, having a considerable effect on the appearance and culture of the people in those areas. I myself am an *iTaukei* progeny of that Polynesian influence that dates back to migrating pottery-people, the Lapita (as scientists call them), in about 1200 CE.[25]

The Economy

What was once a leading economy in the South Pacific since independence from Britain in 1970, the Fijian economy is rather sluggish now due mainly to the unstable political climate that has plagued the country over the last 20 years.

Fiji is primarily an agricultural country with lush vegetation and very fertile land. Therefore since it was colonized in the early 1800s the economy had been styled on agricultural produce with sugar, copra, and banana as the main export produce.

Tourism earnings however have now formed the backbone of the economy taking over from the country's main exports—sugar, gold, timber, garments, and fish. Another major money earner for the country now is remittances from Fijians working overseas.

Since independence in 1970, there has been much emphasis on the manufacturing sector to boost the economy. The government offered various incentives and concessions through the establishment of tax-free zones to attract investors. Similar incentives were also extended to investors in the tourism business taking advantage of the country's beautiful environment that they could mint money out of.

It worked with Fiji now boasting one of the most sought-after tourist destinations in the South Pacific. The income generated from tourism now earns 16 percent of GNP and is responsible for about one-fifth of the total production and gross exchange earnings.[26]

If the government's annual budgetary provision, which is in excess of 1 billion Fijian dollars, is to be used as a yardstick in measuring the economic performance of the country, then Fiji is still far better off than most of the countries of the South Pacific despite its unstable political situation.[27]

The country also boasts modern internationally recognized standards in telecommunication, aviation, shipping, health, road infrastructure, industrial and manufacturing sector, and modern educational facilities, which include three universities.

"The Environment" as a Pacific Issue

But considered physically and geographically, Fiji is just one small Pacific Island nation among hundreds of others and the thousands of atolls spread around 30,000,000 square kilometers of the vast Pacific Ocean—a surface of salt water covering nearly 40 percent of the planet's surface.[28] We are but one language group among 2,000 others scattered across the Pacific and we represent around one-tenth (850,000) of the almost 8 million people who call the Pacific Islands their home.[29]

Our relative small size and isolation in the midpoint of the southwest Pacific gives us a different perspective of global issues like the environment. Although time begins with us because, as I say, we are on the 180-degree longitude where each day begins, we tend to see things from the tail end of scientific research and Western progress because we are at the "end" of their worldview rather than being at the forefront of it. And while we may not be at the cutting edge of science, we are definitely at the cutting edge of the global effects of science.

It has been that way since the latter part of the 20th century when the United States, France, and Great Britain began using our atolls and reefs for test blasting their way into the nuclear future. Bikini Atoll, Rongerik, and the *Rainbow Warrior* (ship) became household names as our Micronesian and Polynesian brothers and sisters to Fiji's north and east were the first to be relocated or guinea-pigged to make way for "progress." What effect that nuclearization of our ocean and atmosphere had on the genetic makeup of Pacific peoples is not as open and shut a case as the Americans and the French would have us believe. What long-term impact all that exploding had on fish below and the ozone above is still unknown.

By the end of the last century, we were told that the sea level of the Pacific Ocean was rising due to a phenomenon called "global warming"—later this phrase became "climate change." The media shocked us into a state of numbness by its insistence that soon we were all going to be drowning thanks again to the scientific progress led by Europeans and Americans. These peoples were churning out so much carbon into the atmosphere they were creating a hothouse, greenhouse effect that was going to raise the ocean above the high-tide mark by a meter or so in a very short time.

Then came the so-called scientific skeptics who called into question all the data accumulated by world weather, climate, and environmental authorities.[30] The warming that was said to be human-created—anthropogenic—was now said to be a natural long-term cycle that was unavoidable and that our Pacific ancestors had probably been through some of it before us. So we had better get used to it—resistance is futile seemed to be the message.

So how all this mix of information is to affect us is a debatable issue. Do we Pacific Islanders keep doing what we are doing and have done for thousands of years? Or do we change our consumption patterns too? How much concern and resources should governments therefore allocate to environmental improvements? Do they urge us to reduce our populations to make the future more sustainable? Should we start working on reducing our carbon footprint? (We didn't know we had one, until recently). Island-based pollution in the Pacific adding to the warming of the world? By how much? What about balancing the livelihood of our peoples against the sustainability of marine and river stocks? Or what should we do about invasive species of life that are creeping up on our nature ever since the period of precolonial contacts? The answer to all of these issues awaits final resolution not by Western scientific progress alone but by the response of Pacific peoples, and by Fijian people particularly, in my case.

Helping us work through these environmental issues for the past 30 years has been SPREP—the South Pacific Regional Environmental Programme—which was established after the 1982 Conference on the Human Environment in the South Pacific that was held in Rarotonga (Cook Islands).[31] As a sub-organization within the SPC—Secretariat of the Pacific Community—SPREP was given the work of ensuring that the Apia Convention on the Conservation of Nature in the South Pacific was observed. This convention was formalized in 1976 and a similar Convention for the Protection of the Natural Resources and Environment of the South Pacific was adopted in Noumea in 1986.[32]

Noting the growing importance of the environment as a concern for Pacific peoples, SPREP was reviewed in the years 1990–1993 and an action plan was drawn up to make its work materially relevant to us all.[33] This plan included coordination of action, monitoring environmental concerns, promoting species protection programs, ensuring ecologically sustainable use of resources, reducing marine pollution, and addressing the need to get the message out to governments, NGOs, and people on the ground, ordinary people of the region, that unless we acted to save our precious world, it would soon disappear.

In 2003, the action plan was further refined in Barbados and Johannesburg by relevant officials and organized into three main areas of concern:

natural resources management, pollution prevention, and climate impacts.[34] Under the first concern came issues like biodiversity loss; under the second issue came things like how to deal with hazardous waste; and under the final concern were issues like banning chlorofluorocarbons (CFCs).

Right now, we are in the middle of a 10-year adaptation of the original action plan that SPREP now refers to as "Island Ecosystems (Programme 1)," and "Pacific Futures (Programme 2)."[35] These programs are important bureaucratic ways of restating the same things that concerned them in 1982. We Fijians and Pacific Islanders may have never spelled out "action plans," "conventions" and "programs" in our approach to the environment, but I believe we have always had overlapping if not synonymous concerns with SPREP and these concerns surfaced not just since 1982 but since our ancient ancestors first set foot on their atolls thousands of years ago. Let me explain.

The Fijian View of Our Environment

There is a common belief among indigenous Fijians that the land has "eyes and ears." It is sacred. As such the land and its surrounding environments are a living spirit. It has the power and ability to strike back against those who do not treat the land and its environment with respect.

While the Fijian transformation from being a deceitful heathen to a God-fearing Christian since the advent of Christianity in 1835 is a remarkable story in itself, the belief in the land and its environment has never strayed.[36]

Fijians regard land as their heritage, a gift from God if you like. And as such, it should be treated with reverence, due caution, and utmost care. They believe that land has some kind of a magical power that may be quickly unleashed against those who want to personally benefit by manipulating the land-owning system without the authority of the owning clan members. It usually results in mysterious death. Call it superstitious or whatever—that's a fundamental belief of the *iTaukei* Fijian.

To demonstrate the passion with which Fijians treat their prized possession, land is owned communally—never individually. Although 92 percent of the total land area in Fiji is owned by Fijians as "native land," individual Fijians do not own it; in a sense we are landless, unless of course if we are fortunate enough to afford a freehold block of land.

All native lands are held in trust by the Native Land Trust Board[37] under specific legislations that do not allow any native lands under their control to be sold in the open market.

Under the Native Land Trust Act (1940) and the Native Lands Commission Act (1905), blocks of native lands are owned by indigenous *mataqali* (clans). What this means is that if a portion of the clan's land is to be leased or the forest on it logged, 60 percent of the members of the clan have to agree. It can be deduced therefore that under the land ownership system in Fiji one can appreciate the passion with which Fijians treat their land and environment.

From a development perspective, it can be argued that the land tenure system in Fiji (the majority of the land being native-owned) is indeed cumbersome. By that I mean that trying to lease native lands can be a very frustrating exercise considering that the developer would like to get on with his or her business as quickly as possible. Sometimes it takes years to get the legal documentation done. But from the environmental preservation point of view delays in the land lease process could be a blessing in disguise.

Look at it this way. The Pacific Rim has seen many natural disasters of late, the most recent one being the 8.6 magnitude earthquake that shook Japan in early March 2011 with the ensuing tsunami generated by it causing billions of dollars worth of damage to life and property. Coming further south there was the earthquake and the tsunami that swept islands in Indonesia in 2004. And coming even closer to Fiji there was the tsunami that ravaged Samoa in 2008 and the volcano eruptions in Tonga the same year. And of course, the earthquake that caused considerable damage in Christchurch, New Zealand, and the devastating floods in Australia early this year. These are disasters within striking distance of Fiji and yet we have managed to steer clear of these.

The questions that are being tossed around the kava "grog" bowls[38] in Fiji in response to all of this is how has the country managed to escape these natural disasters? Is it just lady luck smiling on us or is there something supernatural about it? Has Fiji's unique position, the fact that it is graced by the 180th Meridian, something to do with it?

The Villager's Perspective

Life in Fiji is still perhaps what one may describe as a mixture of modern (urban living) and semi-modern (rural) living. I was born and bred in the rural area until I was 10 years old. I left that rural scene because I needed an education that was available only in the urban centers—thus my moving to Suva, and I have lived there since.

But I still vividly remember growing up in the village among 300 other members of my own tribe, *Yavusa Qalitu*, of Mavana village on the island of

Vanuabalavu. Literally translated as the "long island," Vanuabalavu is on the northeastern part of the Fiji archipelago. Those fond memories I will cherish for the rest of my life as I believed it laid the foundation of what I am today. There are 14 villages on the island, Mavana being one of the bigger ones. Vanuabalavu Island has a population of approximately 3,000 people living there permanently.[39] However, there are more from the island that now live elsewhere in Fiji and in other parts of the world.

Vanuabalavu is among a cluster of small islands (100 in all—30 of which are inhabited) that occupies the eastern part of the Fiji archipelago.[40] These islands form what is now known as the Lau group of islands. Most of these islands are a combination of coral and volcanic texture in their formation. There is no record of volcanic eruptions in Fiji though, but numerous craters and volcanic plugs can be seen as evidence of intense volcanic activity in the past.

In Vanuabalavu itself, obviously what was once a volcanic crate is now filled with water forming an inland lake home to a species of huge mullet fish locally called *yawa,* which are usually harvested once a year. When the waters of this inland lake are disturbed the nauseating odor of decomposing vegetation transforming itself into various forms of gas fumes rises to the surface suffocating the fish and making easy catch for the villagers.

My own island, Vanuabalavu, is a fine example of this natural mix process. The northern tip of the island is pure volcanic, encompassing breathtaking sights with little rocky islands of different shapes and sizes jutting out of the tranquil deep blue sea, some rising to a height of 283 meters.[41] These clusters of molten volcanic rocks are clothed with wild vegetation now home to scores of different species of birds. These rocky pinnacles are in a lagoon forming the "bay of islands," now a tourist attraction and a haven for visiting yachtsmen. The middle section of the island is a purely coral formation and the southern tip of the island is a mixture of volcanic and coral rock formation again. Most of the islands in the group are similar in nature.

In view of global warming or climate change, if any part of Fiji is to be swamped by the rising sea level due to environmental neglect, the Lau group of islands will be among the first to go under. Be that as it may, to an ordinary Fijian like me, brought up in such a natural habitat, I lived a life that would blend in with our environment, not by choice, but because it has to be done that way. It may sound far-fetched, but I say that with all sincerity considering that that's the only way.

Some Historical Background

The first Fijians, as we have pointed out, are known to have colonized Fiji (according to *Lapita* pottery discoveries) around 3,000 years ago. They lived off the land and the sea in conjunction with the environment they enjoyed at the time.

History tells us though that my early ancestors "clubbed and ate their way to power" in their pursuit to control and extend their influence among their foes.[42] That was the norm, we are told, until the advent of Christianity. It was through this phase of spiritual evolution or uplifting that the warlords gave up their cannibalistic clubs for the Holy Bible.

Then the colonial powers moved in, and thus began the massive raping of the environment as they tried to turn this tropical paradise into a concrete jungle in the name of modern-day development. They cannibalized the rain forests, stripping it of prized species of the much sought-after sweet-smelling sandalwood trees and other valuable species of hardwood.

The rich delicacies of the surrounding reefs, like the *beche-de-mer* (sea cucumbers), also fell victim to the invading intruders. There was hardly any resistance, as the much-feared warring tribes of "the Feejees" (now Fiji) would soon be subdued by the early missionaries advocating the God of peace and the "love thy neighbor" philosophy.

In the meantime, the invaders' interests were basically to grab as much as they could from the land and sea while the going was good. Taking advantage of the locals' vulnerabilities the invaders even went to the extent of enticing the locals to work for them in return for token gifts of mirrors and smoking pipes, but mostly by stuffing them with "fire water" (whisky and rum).

The unsuspecting locals likened these invaders to Biblical manna falling from heaven.[43] The magic of the musket that kills by simply making a loud noise was simply too much for them. Seeing your own image in the mirror was in many cases flabbergasting for the natives. And the sensation of being drunk with fire water was something else. Little did they realize that these men were nothing but wolves dressed in sheep's clothing.

The invaders' only interest was to strip the land and sea of its prized natural resources—selling their loot in the Oriental markets in India and China for huge profits. Local interest was never in contention at all. And for that matter, I suppose the expression "preserving the environment" was not even thought of as science, and they were obviously oblivious to the harmful effects of excessive industrial pollution emitted into the atmosphere.

Traditional Factors

My own island Vanuabalavu was not spared in that "gold rush fever" by the invaders either. After stripping its barrier reefs of pricey sea slugs they moved to establish the island as the seat of Fiji's government. Fiji's first currencies were printed on the island. The colonizers tried to set up a sugar and cotton industry on the island. Luckily, this ambitious adventure was short-lived, as the struggle for political control among the colonizers intensified. I shudder to think what Vanuabalavu would look like now if that move had been successful. This is just a soft glimpse into Fiji's history that really started the degradation of the local environment.

Now, fast forward those thoughts to the village of Mavana on Vanuabalavu Island a century and a half later to when Pope John Paul II referred to Fiji as "the way the world should be."[44] He was referring, of course, to the environment in general terms where the political climate was conducive to multiracial living, economic and social stability were the order of the day, and there existed a clean environment with hardly any pollution.

By the time I grew up in the village, Fiji of course had already gone through many of these changes I have mentioned. The country was already a British colony with the seat of government shifted from Levuka to Suva. But life in the village had not been much affected. Yes, there was a school already in existence and there was some kind of economic activity through the village cooperative store. However, village life was still one slow drag with villagers living on a day-to-day basis. There was no hurry. Time was more or less at a standstill. People were doing their own thing without a care in the world. It certainly was an extension of the African ideology of *Akuna matata*, no worries at all. Life was just one big fun without a care in the world.

For instance, although my uncle slept the whole day, the family didn't lack the basic necessities of life. When it was time for a meal there was an abundance of food on the table and no one bothered to find out what he did the whole day or whether he had contributed toward the family life at all. It was a unique sort of life that I suppose is the envy of many now.

As I grew up in the village, one of the first things I learned was to blend my lifestyle with that of my natural environment. By that I meant that how I lived my life would always be in conformity with that of my environment. It came naturally to me. It appears that there is a kind of bond that binds my life with how I live among the animals, the birds, the sea, the forest, the river, and of course my fellow humans.

Another example of this was that when sleeping toward the wee hours of the morning and the cock crowed for the first time, it signaled to me that the tide is coming up and it's time to get the boat ready if we were to go out trawling in the dawn twilight. And that, of course, is dependent on the cycle of the moon, the direction of the wind, and the condition of the sea—the rougher the sea the better it is. Otherwise I would stay in bed until the other bird species started singing, signaling that it's time to get out of bed as the sun is almost up.

Although there were no lectures on how to live or work your life with these natural environmental phenomena, the learning process came by simply observing and following what the old folks did. The learning process was more practical rather than theoretical. You walk your talk rather than talk your walk.

As I now sit back and cast my mind back to my early days in the village, I can only marvel at the farsightedness of the old folks in how they blended their lifestyles to work harmoniously with their natural environment. I mean my grandfather didn't have an education as far as I know, but he had this ability to precisely predict what would happen in a particular month by simply observing the change in the characteristics of certain trees, the behavior of wild birds and animals, the weather patterns and the lunar cycles. All these behavioral patterns were interlinked with our environment on the island.

I am still mesmerized by the fact that when the *drala* trees along the shorelines flowered, the old folks would immediately venture out into the reefs to fish for octopus. Obviously the flowering of the *drala* tree coincided with the mating season of the sea creatures bringing them out in their hundreds onto the open reef. The fishermen (*gonedau*) though would take only what they required. Similarly, when certain species of hardwood in the forest blossomed, a certain species of fish in the sea would swim closer to shore in their hundreds—thus, the time to go out fishing. Other phenomenal activities in the forest would also coincide with the activities of different varieties of sea creatures on the reefs, providing our families with an easy source of protein.

Even planting of food in the gardens was not done just anyhow. It was planned to coincide with certain phenomenal movements that relate to the environment. The planting of root crops like different varieties of yams and *dalo* was also dependent on certain activities of the lunar circle, the high tide, and forest activities. The planting of different varieties of bananas was dependent on the lunar cycle and weather pattern as well.

My grandfather could also foretell disasters by simply observing how insects and birds built their nests. If stinging hornets built their hives closer to the ground there was sure to be a hurricane or cyclone. Likewise, if fruit trees like mangoes had an unusually heavy fruiting season, it was almost certain that there was going to be a natural disaster.

The weather and the lunar cycle also dictated where to fish and when to go out fishing. When the wind was blowing from a certain direction and at a certain position of the moon a certain variety of fish would be easy catch at a certain fishing passage. And at a certain wind direction when the new moon was about to appear, another species of fish would be easy catch. Sometimes, I wondered whether the fish had become lunatics in their behavior as we caught them by the hundreds. Even the type of bait used would determine the size of the catch.

In the evenings when the tide was out, one could see dozens of lights lighting the foreshore as family members went out onto the reefs hunting for fish, sea crabs, and other reef delicacies trapped in ponds of shallow water as the tide went dry. And boy oh boy, how I always looked forward to a good healthy breakfast of fresh fish cooked in thick coconut cream with a dash of lemon and chilies added, the next morning. Such a style of fishing did not happen all the time. It only happened once in a while as and when the tide and the moon cycle dictated.

Even the behavior of animals, domestic or otherwise, was linked to environmental phenomena. But you would need to know these signs so you could tell what was to come. For example, when a dog was on its back with all four legs pointed up to heaven with its tongue wagging out in desperation, it was sure to rain in the afternoon!

One thing I observed as my grandfather and I went out gardening was how very selective he was as to which tree to cut down and which tree to preserve. We did not cut trees indiscriminately. Likewise, burning of forests was strictly prohibited. However, the burning of a newly cleared area for a new plantation was a must—a view directly opposed, incidentally, to good farming husbandry I later learned in school.

But if we cleared an area for a yam plantation, you could rest assured that coconut trees and other varieties of fruit trees would be replanted once the harvest was done. It was quite noticeable that the size of our harvest was dependent on the amount of work that we put into the preparation of the yam garden. It was also dependent on which month of the year the yam was planted in, and the state of the tide and the moon cycle faced at the time of planting. The process of cleaning the garden once the yams started growing and how the yam vines were allowed to creep onto tree branches that were prepared for such purposes also played a pivotal role in the size of the harvest.

I remember my grandfather giving me strict orders never to allow women into the yam garden at any time for fear that their menstrual period could affect the yams. He claimed that yams were living organisms that reacted unfavorably to such natural systems including the cleaning of the garden during rainy weather. I don't know whether those claims are scientifically true or not, but one thing I know is that by observing all the taboos, the dos and don'ts imposed, my grandfather and I harvested giant-sized yams—which were usually the talk of the village.

The point is: everything we planted was done according to its season.

For example, Christmas is usually a time when members of my tribe return to the village for their holiday. Therefore for those of us who remain in the village, we would time whatever we planted to be harvested around that time. Here I am talking about pineapples, watermelons, bananas, sugarcane, and so on. Fruit trees that grow wild usually bear fruit around that time anyway. So when the rest of the Mavana tribe returned, the fun was usually not so much in the feasting but in our adventure through the forests picking up berries, mangoes, pawpaws, bananas, and other fruits.

In the island environment, we were but subsistence farmers. We would plant just for our daily needs and a little left over for village traditional obligations. It therefore meant that the mass destruction and the cannibalizing of the forests was nothing we could conceive. Our farming activities were on a smaller scale compared to the commercial farming activities that are driven by the hunger for more profit. Yes we did need money in the village situation but not as a means of survival. It was simply to buy the basic necessities like sugar, washing soap, and kerosene for lighting; otherwise one could survive by simply living off the land while making sure that the environment was well cared for.

When we were out sailing in the ocean visiting outer islands, my grandfather would use the stars and the mountains as a guide to steer us through reefs and passages to get ashore. The direction of the wind and the currents were also taken into consideration.

In our situation in the village we lived as close to nature as possible. "Toilet" was where you did it when the need arose, so to speak, with Mother Nature recycling the freckles just like she did the droppings of birds and animals. Yet people lived a healthy life.

And what a life. Here I am reminded of the Garden of Eden in the *Book of Genesis,* where God planted a garden and put Adam and Eve to live in it and eat of the fruit of all the trees except the fruit of the tree of life and the tree of knowledge of good and evil.[45] Imagine if Adam had not sinned by disobeying God's command. The type of lifestyle I encountered while living in my village at my early age would be quite similar to the type of lifestyle Adam and Eve would have enjoyed—only better I suppose.

The lifestyle I enjoyed in my village was seemingly laidback, without a care in the world as they say. There was plenty to eat, tons of laughter, and I enjoyed life to the full. I was doing my own thing. If I wanted to eat fish for dinner I just had to consult the weather and the position of the moon to know exactly where to fish and what species of fish I would be having. We had to live a lifestyle that synchronized with the requirements of our environment. Even domesticated fowls and ducks roamed the village freely while pigs and cattle were part and parcel of the equation. It was a life next to godliness.

My Present Situation

Now, 50 years down the line, the situation I encountered back then is certainly a far cry from what I am subjected to. Of course it would be naive to believe that the situation then could still apply now.

Again, like the world over, there are many factors that have contributed to the general degradation of our environment. These would be population growth, urbanization, industrialization, increased poverty levels, uncontrolled pollution and, most importantly, ignorance.

Fiji's evolution from when the country was ceded to Queen Victoria of Great Britain in 1874 to what it is today is quite exceptional in some respects. It has been a marathon race taking in its stride political, economic, and social developments that would be the envy of our Pacific neighbors.

Yet the past 106 years have seen Fiji evolve from uncharted waters feared by many explorers to becoming the hub of the Pacific—the crossroads—if you like, boasting two modern cities and towns enjoying a living standard comparable to those of other modern societies.

But perhaps therein lies our environmental problem. In our haste to catch up with the rest of the modern world, fascinated with their achievements in modernizing their standard of living, we as a country have failed to take due recognition of the side effects such developments have on the environment.

For example, urbanization is now regarded as a major problem facing Fiji at the moment.[46] More and more people have opted for a life in the towns and cities, mostly relying on the meager income of their relatives for support while they seek employment, something that has been hard to come by recently.

Another factor that has been forcing the steady increase in the urban drift is the search for higher education. Most, if not all, of the major learning institutions in Fiji are urban-established.

As such, the influx of students from rural areas in search of higher education automatically drifts them to the urban centers with high hopes of a better education.

But more often there is disappointment than joy. Most of these "higher learning" facilities are not equipped to absorb the high migration of students, thus creating further problems among their hard-pressed relatives trying to make ends meet in urban centers.

In 1966 only one-third of Fiji's population (33.4 percent) lived in urban areas; in 1986, this had grown to 38.7 percent, and in 1996, 45.4 percent.[47] 10 years later, more than half of Fiji's population lives in urban areas. It can be seen from these figures that the urban drift is perhaps the most telling factor that confronts the environment in the Pacific island states like Fiji. The facilities are not geared to cater to such a high migratory trend, thus giving rise to the mushrooming of squatter settlements that lack proper infrastructure.

It is common knowledge that in such a desperate situation, people will do anything to provide their basic needs. They will ignore all as long as they can have a roof over their heads and food in their stomach and money in their pockets to carry on.

It is also a known fact that such squatter settlements rear their own sanitation problems, high incidence of pollution, very unhealthy living standards with people exposed to communicable diseases, social and cultural disorientation, and breakdown in law and order. All these, small as they may seem, are contributing factors toward environmental seepage.

No doubt, Fiji's population will continue to increase. It is estimated to reach the 1 million mark by 2030.[48] Unless the government plans carefully to cater to the influx of the new generation to come we are only compounding our problem.

As the population continues to increase so does the need to provide more food. And to produce more food there is a need to clear more land of its tropical forests. And the cycle continues. As it is now the environment is being ravaged; at some point, perhaps even beyond recognition. Forests have been cleared exposing bare hilly soil planted with crops like ginger, *dalo,* and cassava for quick cash.

Even when these cash crops are harvested and sold in the markets, bare hills are allowed to remain fallow. Reforestation is the furthest thing on their minds. The hunger for money has forced people to ignore natural laws of sustainability, leaving the environment exposed, a direct threat to wildlife and other microorganisms that are direct ingredients for the sustainability of life itself.

In many ways, what is happening now is history repeating itself when the invaders raided these islands, stripping them of natural resources and cannibalizing the environment to the point of extinction. But what they did then was small scale compared to what is happening now. In the name of development, humankind has completely ignored the "cycle of life," which involves living in conformity with our environment.

The new order now is that money can buy anything. Without money you are nothing—a far cry from the environmentally sustainable life I used to enjoy in the village. Perhaps the saddest thing I am now experiencing in my new environment is humanity's hunger for wealth, to the extent that they have completely turned a blind eye to the requirement of sustaining the environment. The wealthy are getting wealthier simply by taking advantage of the laxity in policing the implementation of legislation that governs the protection of the environment. Under the guise of development, these unscrupulous businessmen lobby politicians, flashing dollars here and there, as long as they get government approval for their projects.

Those who are familiar with the topography of land in both Viti Levu and Vanua Levu (the two main islands in the Fiji group) will notice that the vegetation is equally divided into two. In Viti Levu the western side is mainly grassland devoid of trees (except the replanted pine forests). Many believe that the grassland in Fiji's west and the depletion of birdlife were caused by human impact, especially the use of fire for slash-and-burn agriculture (a process that started perhaps with the *Lapita* pottery people and is still in use today) and the destructive effects of introduced rats and pigs.

However, this view is contrary to that of physical geographer Patrick Nunn of the University of the South Pacific, who argues that the human impact on the environment is much overstated and that man is more of a victim than an agent of environmental change.[49] He points to nature-induced fires and in particular to the importance of the "little climatic optimum" (LCO) from 750 CE to 1250 CE, and the 1300 CE event. During the LCO, temperatures and sea levels rose, increasing aridity and leading to water-conserving strategies; and smaller human groupings gave way to larger groupings and settlements.

Nunn describes the 1300 CE event as a burst of environmental change and massive and eroding social disruption arising from a rapid drop in temperatures, a fall in sea level, and the exposure and near death of near–sea shore coral reefs, more storms, and falling water tables. The net result was a sharp decline in available reef foods and crop yields. He estimates that food resources readily available to coastal dwellers fell by around 80 percent. Across the Pacific warfare, cannibalism and headhunting became commonplace; long-distance voyaging ceased and inter-island trading was curtailed. A rather interesting argument I think.

Fiji's Legal Position

Whether one takes Nunn's view or my own, to combat the ongoing destruction of our environment, Fiji is a signatory to many protocols and agreements that deal with the environment. But the major problem in Fiji now is the lack of legislation that would ensure all developmental and business-oriented projects are environment-friendly. Although Fiji is among the signatories to the Kyoto Protocol that commenced in 1997, our country lacks the necessary legislations that would make sure that principles enshrined in the protocol are adhered to.

And even if there are legislations already in existence, unfortunately they do not go far enough to drive the Devil's fear into these business tycoons to respect the environment. All they are interested in is the return on their investments. Those of us who do not control the economic or political power can only watch from the sidelines with horror knowing that one day we will all become victims of our own doing in the degradation of the environment.

Not satisfied with logging the country's natural tropical forests and stripping the reefs of their sea slugs and fish population, mushrooms of fly-by-night businessmen have converged on Fiji taking advantage of the rather fragile and unstable political and economic climate the country is currently going through. With the promise of lending a helping hand to get the economy back on even keel these fly-by-night operators offer big money to locals so they can buy container loads of live reef corals leaving large portions of Fiji's coral reefs a virtual desert. A group of these businessmen was recently also buying different tiny species of "live colored coral fish" for overseas indoor aquariums. Some of them have taken out native leases, using heavy machines to bulldoze the land and selling the same for a profit leaving the land bare and untenable. They are even selling soapstone that is the land.

The irony of all this is that there is no legislation to cover such degradation of the environment. Trees that took 100 years to grow are being logged and sold off as timber again under the guise of development. Even the mangrove sumps have not been spared. Presently, anyone can take out a two-dollar license to harvest and log tons of mature green mangrove for fuel (firewood).

The logging of mangrove swamps is really a pity considering that many varieties of fish use the mangrove swamp as a breeding ground. Mangrove swamps are feeding grounds to millions of tiny sperm fish as part of their life cycle. Is it any wonder then that the cost of fish in the local market is skyrocketing? Life is certainly getting unmanageable these days all because we have turned our backs on maintaining and preserving the environment for the sake of a few dollars more.

The Political Factor

But perhaps the biggest environmental degradation Fiji is facing at the moment is the rather unhealthy and unstable political climate the country is facing. It is certainly unpredictable and can be very frustrating living in such a political climate. In short, Fiji has been living under some form of dictatorship since the military overthrew a democratically elected government on December 6, 2006. The country's parliamentary process was systematically and forcefully dissolved as a military government took control of government.[50]

It would be fair to say though that since 1987, Fiji's democratic government system has been wrought with instability due mainly to military coups. A country with a population of less than a million people yet supporting a history of countless coups and countercoups in 20 years is indeed inviting disaster. We are now paying the price socially, politically, economically, and environmentally for the ambition of a few power-hungry people under the pretext of a better tomorrow.

The burning question though is, does government have the political will to tackle the environmental issue head on as a matter of urgency? Well, that, I will leave it to your own imagination. But as for me, oh! how I long for those wonderful environment-friendly days back in Mavana village of those yesteryears.

Mesake Koroi
2012

Notes

1. Wolfgramm, "Kai Viti."

2. Incredibly, from this one remote village, among Fiji's many hundreds, were found leaders in business, politics, the military, the public service, and the media. Indeed, Mavana village has uniquely produced both a head of the hegemonic Methodist Church of Fiji (subscribed to by about 70 percent of the indigenous *iTaukei* Fijians), and a Prime Minister of Fiji. Both of them, by even rarer

coincidence, suffered the ignominy of being couped from office—Reverend Koroi in 1989 by clerical supporters of Colonel Rabuka's coups; and Laisenia Qarase in 2006 by Commodore Frank Bainimarama.

3. Geraghty, "Fijian," 389; Walsh, *Fiji*, 394, 395, 405; Thorne, *Man*.

4. Walsh, *Fiji*, 114, 59.

5. Allen, *Cyclopedia*; Ali, *Plantation*; France, *Charter*; Gillion, *Fiji Indians*; Grav elle, *Fiji's Times*, 155–60; Lal, *Girmityas*; Naidu, *Violence*; Scarr, *Fiji*, 78–82, 95–103; Thomson, *Fijians*; Walsh, *Fiji*, 110–13.

6. Emberson, *Labour and Gold*.

7. No reference to this phrase appears in the official Vatican record of the Pontiff's speech of November 21, 1986, but it is well quoted by both secular and Catholic media—see for example Sam Howe Verhovek, "Ethnic Crisis," and Father Patrick Colgan, "Week."

8. Watling and Chape, "Environment Fiji"; Chape, "Review."

9. From 1983 through 1990, I conducted annual weekly workshops for tertiary students with Aboriginal elder Gubbo Ted Thomas in his ancestral Yuin country centered at Wallaga Lake on the South NSW coast. One of those excursions was documented in a student-produced short film, *Umbara Dreaming* (1987)—a digitized copy of which is now held in my collection.

10. Starhawk, *Dreaming*; Fox, *Coming*.

11. Watling and Chape, "Environment Fiji."

12. Institute, "Priority," 10–17.

13. Watling and Chape, "Environment Fiji."

14. Convention, 13–17.

15. Institute, "Priority," 10–17.

16. Pacific Concerns, "Submission," 1.

17. Walsh, *Fiji*, 402.

18. Ibid., 405.

19. Ibid., 405.

20. Ibid., 405.

21. Ibid., 402; Government "Statistics."

22. Ali, *Plantation*; France, *Charter*; Gillion, *Fiji Indians*; Gravelle, *Fiji's Times*; Lal, *Girmityas*; Naidu, *Violence*; Scarr, *Fiji*; Thomson, *Fijians*; Walsh, *Fiji*.

23. Walsh, *Fiji*, 122.

24. Burns, *Fiji*, 25.

25. Walsh, *Fiji*, 394, 395, 405.

26. Ibid., 335.

27. Ibid., 403; Lal and Naidu, *1987*; Fraenkel and Lal, *2006*.

28. Chape, "Review," 6.

29. Ibid., 12.

30. Lord Monckton is the most prominent of these climate skeptics.

31. Chape, "Review," 1.

32. Ibid., 1–6.

33. Ibid., 3–6.

34. Ibid., 4.

35. Ibid., 5, 6.

36. Calvert, *Missionary Labours*; Mann, *Religion and Symbolism*; Thomson, *Fijians*; Tomlinson, *God's Image*; Vernon, *James Calvert*; Webb, *Observations*; Williams, *Fiji*.

37. Under the current political regime, all references to "Fijian" and "native" that refer to or mean indigenous, are now named *iTaukei* (meaning landowner). The motive is to nationalize the concept of "Fijian." That is, all citizens of the state are now "Fijian," while the original, colonially defined Fijians are now specifically *iTaukei* Fijians.

38. Kava, "grog" or *yaqona* in *iTaukei* Fijian, is a social, ceremonial sedative root drink of the nation. See Clunie, *Yalo*; Thomson, *Fijians*; Tomlinson *God's Image*; and Tora, *Yaqona*, for useful descriptions of Fijian kava rituals.

39. Estimates vary—see for example http://vanuabalavu.com.fj/theisland.html

40. Derrick, *Fiji*, 294.

41. Ibid.

42. Waterhouse, *The King*; Calvert, *Missionary*.

43. The New Fijian Translation Bible, *Exodus* 16.

44. Attributed to the Pontiff during his historic visit to the nation in 1986.

45. The New Fijian Translation Bible, *Genesis* 1–3.

46. Griffin and Monsell-Davis, *Fijians*; Walsh, *Fiji*.

47. http://www.statsfiji.gov.fj/Census2007/Release 1—Population Size.pd cf. Walsh, *Fiji*, 76, 77.

48. Ibid., 43.

49. Walsh, Fiji, 394, 395.

50. Lal, Chand, and Naidu, 1987: *Fiji Twenty Years On*; Fraenkel, Firth, and Lal, *The 2006 Military Takeover in Fiji*.

Bibliography

Ali, Ahmed. *Plantation to Politics*. Suva: University of the South Pacific, 1980.

Allen, Percy, ed. *Cyclopaedia of Fiji*. Suva: Fiji Museum/Oceania, 1907/1984.

Amrattal, J. *Women's Role in Fiji*. Suva, Fiji: South Pacific Social Sciences Association, 1975.

Brookfield, H.C. *Melanesia: A Geographical Interpretation of an Island World*. London: Methuen, 1971.

Burns, Sir Alan. *Fiji*. London: Corona Library, 1963.

Calvert, James. *Missionary Labours among the Cannibals*, Vol. 2 of *Fiji and the Fijians*, with Thomas Williams, edited by George Stringer Rowe, Third Edition. London: Hodder & Stoughton, 1870.

Carson, Rachel. *Silent Spring*. Boston: Houghton Mifflin, 1962.

Chape, Stuart. "Review of Environmental Issues in the Pacific Region and the Role of the Pacific Regional Environmental Programme." Paper from Workshop and Symposium on Collaboration for Sustainable Development of the Pacific

Islands: Towards Effective e-Learning Systems on Environment, Japan, February 27–28, 2006.

Clammer, John. "Colonialism and the Perception of Tradition in Fiji." In *Anthropology and the Colonial Encounter,* edited by T. Asad. London: Ithaca, 1973.

Clifford, James. *The Predicament of Culture.* Cambridge, MA: Harvard University Press, 1988.

Clunie, Fergus. *Yalo i Viti, Shades of Fiji: A Fiji Museum Catalogue.* Suva: Fiji Museum, 1986.

Colgan, Patrick. July 29, 2010. http://columban.org/3208/regions/fiji/fiji-updates/a-week-in-the-life/

Convention on Biological Diversity 1997 National Report to the Conference of the Parties by the Republic of Fiji. Suva, Fiji: Department of the Environment, 1997.

Crocombe, Ron. *Asia in the Pacific Islands—Replacing the West.* Suva: IPS Publications, 2007.

Dean, Eddie, and Stan Ritova. *Rabuka No Other Way.* Sydney: Doubleday, 1988.

Deane, (Rev.) Wallace (Principal, Teachers' Training College Davuilevu Fiji). *Fijian Society—or the Sociology and Psychology of the Fijians.* London: Macmillan & Co., 1921.

Derrick, R.A. *A History of Fiji.* Suva: Government Printing and Stationery Department, 1946.

Diole, Philip. *The Forgotten People of the Pacific.* London: Cassell, 1977.

Earle, Timothy. "The Evolution of Chiefdoms." In *Current Anthropology* 30, no. 1, February 1989.

Eliade, Mircea. *From Primitives to Zen.* London: Collins 1977.

Emberson-Bain, Atu. *Labour and Gold in Fiji.* Australia: Cambridge University Press, 1994.

Flannery, Tim. *The Future Eaters—An Ecological History of the Australasian Lands and People.* Sydney: Reed Books, 1994.

Fox, Matthew. *The Coming of the Cosmic Christ.* Melbourne: Collins Dove, 1988.

Fraenkel, Jon, Stewart Firth, and Brij Lal, eds. *The 2006 Military Takeover in Fiji, A Coup to End All Coups?* Canberra: ANU E-Press, 2009.

France, Peter. *The Charter of the Land.* Melbourne: Oxford University Press, 1969.

France, Peter. "The Kaunitoni Migration Notes on the Genesis of a Fijian Tradition." In *Essays from the Journal of Pacific History,* edited by Barrie McDonald, [no listed publisher], 1979.

Garrett, John. *To Live among the Stars: Christian Origins in Oceania.* Suva: WCC Publications, 1982.

Geraghty, Paul. "Fijian Communalects." In *Fiji: An Encyclopaedic Atlas,* edited by Crosbie Walsh. Suva: University of the South Pacific, 2006.

Geraghty, Paul. "The Ivolavosa and the Codification of Fijian." In *Directions: Journal of Educational Studies* 27, no. 1.

Gillion, K.L. *The Fiji Indians.* Canberra: ANU Press, 1977.

Government of Fiji, Bureau of Statistics, Suva. http://www.statsfiji.gov.fj

Gravelle, Kim. *Fiji's Times, A History of Fiji—In Three Parts.* Suva: Fiji Times and Herald, 1981.

Gray, Elizabeth Dodson. *Green Paradise Lost.* Massachusetts: Roundtable Press, 1981.

Griffin, Chris, and Mike Monsell-Davis, eds. *Fijians in Town.* Suva: Institute of Pacific Studies, 1986.

Hocart, A. M. *The Northern States of Fiji.* London: Royal Anthropological Institute, 1952.

Institute of Marine Resources. *Priority Environmental Concerns in Fiji.* Suva: SPREP, IWP-Pacific Technical Report No. 6, 2004.

Kasper, Wolfgang, Jeff Bennett, and Richard Blandy. *Fiji Opportunity from Adversity.* St. Leonards, N.S.W.: The Centre for Independent Studies, 1988.

Lal, Brij V. *Girmityas: The Origins of Fiji's Indians.* Canberra: Australian National University Press, 1983.

Lal, Brij V., Ganesh Chand, and Vijay Naidu, eds. *1987 – Fiji Twenty Years On.* Lautoka: Fiji Institute of Applied Studies, 2008.

Legge, J. D. *Britain in Fiji 1850–1880.* London: Macmillan, 1958.

MacNaught, T. J. *The Fijian Colonial Experience.* Canberra: ANU Press, 1982.

McLuhan, T. C. *Touch the Earth.* London: Abacus, 1973.

Mageo, Jeannette Marie, and Alan Howard, eds. *Spirits in Culture, History and Mind.* London: Routledge, 1996.

Mann, Cecil W. "Religion and Symbolism in Fiji." In *The Journal of General Psychology* 23 (1940): 169–84.

Martinez-Alier, Joan. "Environmental Justice as a Force for Sustainability." In *Global Futures—Shaping Globalisation,* edited by Jan Nederveen Pieterse. London: Zed Books, 2000.

Mitchell, Andrew. *A Fragile Paradise—Nature and Man in the Pacific.* London: Fontana, 1989.

Moody, Roger. *The Gulliver File, Mines People and Land: A Global Battleground.* London: Minewatch, 1992.

Morgan, Sally. *My Place.* Australia: Fremantle Arts Press Centre, 1988.

Morrell, W. P. *Britain in the Pacific Islands.* London: Oxford University Press, 1960.

Naidu, Vijay. *The Violence of Indenture in Fiji.* Suva: University of the South Pacific, 1980.

Nation, John. *Customs of Respect.* Canberra: ANU Development Studies Centre, 1978.

Nawadra, Tevita. *Fijian Cultural Notes.* Suva: University of the South Pacific, undated.

Nayacakalou, R. R. *Tradition and Change in the Fijian Village.* Fiji: South Pacific Social Sciences Association, 1978.

New Fijian Translation Bible, Nai Vola Tabu. Suva, Fiji: Ah Koy Christian Trust, 2011.

Pacific Concerns Resource Centre. *Submission to the Session of the Working Group on the Universal Periodic Review on Fiji.* Geneva: February 2010.

Pacific Rituals—Living or Dying? [no authors listed]. Fiji: Institute of Pacific Studies in association with Pacific Theological College, 1986.

Ravuvu, Asesela D. *The Fijian Ethos.* Suva: Institute of Pacific Studies, University of the South Pacific 1987.

Ravuvu, Asesela. *Vaka I Taukei: The Fijian Way of Life.* Suva: Institute of Pacific Studies of the University of the South Pacific, 1983.

Readings on Continuity and Change, Volume one: Pacific Island Studies. [compilation—no authors listed]. Frankston State College, undated.

Robertson, Robbie. "The 2006 Election: The End of Communalism." In *Fiji: an Encyclopaedic Atlas,* edited by Crosbie Walsh. Suva: University of the South Pacific, 2006.

Robertson, R.T., and Akosita Tamanisau, eds. *Fiji: Shattered Coups.* NSW: Pluto Press, 1988.

Roth, G.K. *Fijian Way of Life.* Melbourne: Oxford University Press, 1953/1973.

Routledge, David. *Matanitu: The Struggle for Power in Early Fiji.* Suva: Institute of Pacific Studies, 1985.

Sahlins, Marshall. *Tribesmen.* New Jersey: Prentice Hall, 1968.

"Saving Our Small World." Videotape, *Monday Conference on Population, Ecology and Resources.* Adelaide: ABC, 1974.

Scarr, Deryck. *Fiji: A Short History.* Sydney: George Allen & Unwin, 1984.

Schell, Jonathan. *The Fate of the Earth.* New York: Knopf, 1982.

Sercombe, Bob, and Anthony Albanese. *Our Drowning Neighbours.* The Australian Labor Party's Policy Discussion Paper on "Climate Change in the Pacific"—Part 2, accompanying "Toward a Pacific Community"—Part 1.

Sokonibogi, Francis. *Fiji Indigenous People and Mining.* Paper prepared by Fiji Indigenous Ownership Rights Association, Suva, 2007.

South Pacific Sea Level and Climate Monitoring Project—data reported ABC online, www.abc.net.au/news, 2006.

Sprout, H. and M. *Toward a Politics of Planet Earth.* New York: van Norstrand, 1971.

Starhawk. *Dreaming the Dark: Magic, Sex and Politics.* Boston: Beacon Press, 1982.

Thomas, Nicholas. "Sanitation and Seeing: The Creation of State Power in Early Colonial Fiji." In *Comparative Studies in Society and History,* Vol. 32, No. 1, January 1990.

Thomson, Basil. *The Fijians: A Study of the Decay of Custom.* London: Dawsons, 1908/1968.

Thorne, Alan. *Man on the Rim.* Videotape, TV documentary series on Pacific peoples.

Tomlinson, M. *In God's Image–the Metaculture of Fijian Christianity.* Berkeley: University of California Press, 2009.

TORA, Veramu. "Yaqona in Fiji." In *Pacific Rituals Living or Dying?* Suva: Institute of Pacific Studies of the University of the South Pacific in association with the Pacific Theological College, 1986.

Umbara Dreaming. Video-documentary. Melbourne: Chisholm Student Minorities Society, 1987.

Verhovek, Same Howe. "Ethnic Crisis in Fiji Threatens a South Seas Eden." *New York Times,* June 7, 2000.

Vernon, R. *James Calvert: From Dark to Dawn in Fiji.* London: S.W. Partridge & Co., undated.

Walsh, Crosbie. *Fiji: An Encyclopaedic Atlas.* Suva: University of the South Pacific, 2006.

Waterhouse, Joseph. *The King and People of Fiji (Containing a life of Thakombau; with Notices of The Fijians, Their Manners, Customs, and Superstitions Previous to the Great Reformation of 1854).* London: Paternoster Row, 1866.

Watling, Dick, and Stuart Chape (compilers). *Environment Fiji: The National State of the Environment Report.* Suva, Fiji: Government of Fiji & IUCN, 1992.

Webb, (Rev.) Arthur J. *Observations on the Hill Tribes of Navitilevu, Fiji.* [unplaced], Australia & New Zealand Association for the Advancement of Science, 1890.

Williams, (Rev.) Thomas. *Fiji and the Fijians* (2 vols.). Suva: Fiji Museum, 1858/1982.

Wolfgramm, Robert. "Australia and the Pacific: Polity, Territoriality, Spirituality." Presentation to Pacific Island Students Association, Victoria University, St Albans Campus, September 14, 2004.

Wolfgramm, Robert. "Australia and the Pacific: Saviour or Neo-colonial sheriff?" Presentation to Australia and the Pacific Symposium, Bob Hawke Centre, University of South Australia, Adelaide, November 15, 2003.

Wolfgramm, Robert. "Kai Viti: On Being Fijian without Being 'a Fijian.'" PhD diss., La Trobe University, 1994.

Wolfgramm, Robert. "On Customary Law." Presentation to Fiji Indigenous Owners Rights Association, Suva, Fiji, October 2005.

Wolfgramm, Robert. "Religious Function and Meaning of Yaqona/Kava." Presentation to the International Kava Conference, Pacific Island Forum Secretariat, Suva, Fiji, November 30 to December 2, 2004.

Young, John. "The Response of Lau to Foreign Contact." In *The Journal of Pacific History,* Volume XVII, No. 1, January 1982.

Chapter 15

Environmentalism in New Zealand

Ton Bührs

Introduction

Often depicted as a "clean and green" country, New Zealand has a strong international environmental reputation. While this image is very contestable, and the country's environmental conditions and the environmental performance of its governments leave much to be desired, it can be plausibly argued that environmentalism has been a significant force in New Zealand from the 1960s. The rise of environmental concerns and awareness, expressed through a variety of groups and ways, has been prominent and, in some respects, at the cutting edge internationally. New Zealand gave birth to one of the first green parties in the world, and a strong antinuclear movement has pushed the country to its nuclear-free status. Nature conservation, also, has received strong and widespread support. However, despite its strengths, environmentalism in New Zealand has only been able to scratch the surface of the environmental challenges that the country faces, in part because of the way(s) that these challenges are predominantly conceived, and because of the dominant political-economic structures that continue to add to the environmental pressures.

This chapter will, first, sketch the context within which environmentalism in New Zealand has emerged and evolved. Then, it will describe the variety of streams of environmentalism, and the main differences between these in outlook and concerns. Third, it discusses the changes and continuity in environmental issues and activism throughout the last 50 years, divided into four periods. Finally, it concludes with an assessment of the outcomes of 50 years of environmentalism in New Zealand and the ongoing challenges.

New Zealand: A Brief Overview

New Zealand, consisting of two main islands and several smaller islands, together covering some 267,000 square kilometers, stretches roughly between the 34 degree latitude and the 47 degree latitude in the Southwest Pacific. It has a population of 4.3 million (2011), a temperate climate, and an economy largely dependent on natural resources, including agriculture (dairy products, meat, and wool), forest products, minerals, and tourism. Two-thirds of the population is ethnic European (mostly descendants of immigrants from the United Kingdom), 14 percent indigenous Maori, 9.2 percent Asian, and 6.9 percent from the Pacific Islands.

As a former British colony, New Zealand's political system is of the Westminster style, with a parliament traditionally dominated by either of the two main political parties (National and Labour). Although mitigated by a three-year electoral cycle and, in the 1990s, the introduction of a proportional electoral system, political power tends to be highly concentrated, to the extent that the country has been characterized as an "elective dictatorship."[1] Governments are commonly able to introduce major policy and institutional changes, however, often to see these reversed again by a successive government, making policy continuity rather than policy gridlock (as in the United States), a main political challenge.

In the mid-1980s, the ruling (Labour) government initiated a comprehensive reform of environmental institutions that, by and large, still prevails. The reform, driven by neoliberal ideas and principles, led to the creation of an organizational and legislative framework that assigns most responsibility for environmental decision making to territorial (district and regional) councils and an Environment Court.[2] Although the sustainable management of resources provides the overarching principle of the main environmental law, the Resource Management Act 1991, in practice, given the absence of clear parameters or guidance on what is "sustainable," the system has not prevented a continuation of environmental degradation in many areas and aspects of the environment. Arguably, the neoliberal reforms shifted the main thrust behind "development" from the state to the market, but it did not imply or bring about a "greening" of government policies and practices in areas from which most environmental pressures and problems originate, including economic, agriculture, energy, urban development, and transport policies.

The emergence and evolution of environmentalism in New Zealand should be seen in this broader geographical, political-economic, and sociocultural context. Many of the problems that gave rise to environmental concerns, and that often still continue to do so, stem from the structure of the economy, and from the political-economic "imperatives" that held and still hold governments, or more broadly the state, in their grip. Many of the environmental issues that sparked off New Zealand's environmental movement, and that still tend to dominate the environmental agenda, have their origin in state-led or state-sponsored development, notably of agriculture, energy, transport, and mining, which has led, and continues to lead, to the destruction of nature (forests, wetlands, wildlife) and the degradation and decline of resources (soil erosion, biodiversity, water).

Environmentalism in New Zealand: Six Streams

Environmentalism is a very big church. It comes in many shapes and colors. Although often referred to as a "movement," it encompasses a broad array of very diverse groups and individuals, inspired by different issues, beliefs and causes. Arguably, the only thing "environmentalists" have in common is that they have (not necessarily share) a concern, or concerns, about (an aspect of) "the environment," itself a potentially all-encompassing concept. But they differ widely in their interpretations of "the environment" and environmental problems, the solutions they advocate, and the actions and strategies to advance these.

Environmentalism (and environmentalists) has been classified in different categories and by different labels, including "reformist" and "revolutionary,"[3] "deep" and "shallow,"[4] "radical" and "moderate,"[5] "technocentric" and "ecocentric,"[6] while some prefer to distinguish between "environmentalists" and "deep ecologists" as fundamentally different categories.[7] While this distinction between environmental streams on ideological grounds can be insightful and useful, there are also other differences between groups or categories of environmental advocates that are important to understand their role, actions, influence, and effectiveness in environmental matters, such as their sociocultural background, their socioeconomic position in society, and their political orientation and inclination (for instance, with regard to types of activism, adversarial or cooperative strategies, working from outside or from within "the system"). As the characteristics of environmental advocates and groups can overlap in a variety of ways, classifications can become elaborate and complex if their diversity is to be covered fully. This chapter takes a pragmatic approach by distinguishing six streams of environmentalism that can be seen to have played a significant, and often prominent, role in the advancement of environmental concerns in New Zealand:

1. Maori
2. Grassroots environmentalism
3. Environmental organizations
4. Professional/expert-based environmentalism
5. The political (Parliamentary) wing
6. Institutional environmental advocates

While this classification, represented in Table 15.1, is not systematic (based on a single criterion or set of criteria), these streams will be described with regard to their environmental ideologies (which are diverse also within these streams), their relevance (in terms of the kind of issues that they tend to focus on; their programs), and the main strategies used in the pursuit of their objectives. Although this description does not claim to be fully comprehensive of New Zealand environmentalism, the aim is to draw attention to the more influential forms, some of which tend to receive less attention than they deserve.

Maori

The Maori, the indigenous people of New Zealand, are considered here because their worldview, as that of many other indigenous peoples, is often considered to be more closely aligned with nature than that of people in modern societies. The traditional Maori worldview can be characterized as holistic in that it sees living humans as inherently connected with their ancestors as well as the rest of nature, the land, water, sea and sky, and everything contained therein. *Mauri* or life force resides in all parts of nature, and hence all things natural, from rocks to trees, birds and fish, must be respected, also when used by humans to meet their needs. From a Maori perspective, humans don't and can't own nature, but have a responsibility of guardianship (*kaitiakitanga*), and obligation to manage resources in ways that ensure their indefinite existence. Resources and resource decision making are common and collective affairs, not the subject of individual ownership or rights.[8]

Maori views and interests, although not universally and equally respected among the wider population, have received significant recognition within New Zealand's political and environmental institutions, from the

Table 15.1. Branches of the Environmental Movement in New Zealand

Groups/Branches	Orientation/ Main Issues	Main Strategies	Most Influential Period(s)
Maori	Traditional (holistic) views, but not homogeneous; pragmatic; land ownership and traditional rights	Legal and direct action; electoral and parliamentary action	Pre-colonial period; revival from mid-1970s and growing ever since
Grassroots environmentalism	Reactive to development projects/proposals; "greening" of local communities and practices	Direct and practical action and initiatives	1960s and 1970s; 2000s
Environmental organizations	Largely pragmatic, focused on issues, notably nature conservation and antinuclear	Protest and direct action; media and lobbying	Early 1970s to mid-1980s
Professional/ expert-based environmentalism	"Wise use"/utilitarian, but ideologically diverse; concerns about environmental degradation	Professional practice	1960s; from 1990s with the introduction of the Resource Management Act ("expert witnesses")
Institutional environmental advocates	Pragmatic; focus on strengthening environmental capacity	Political persuasion based on formal roles	Early 1970s; mid-1980s
The political wing—Values Party; Green Party	Questioning of dominant values; Green Party principles: ecology; equity; nonviolence; democracy; broad range of issues	Electoral and parliamentary action; media	1972–1975; from late 1990s, following the introduction of proportional representation in 1996

Treaty of Waitangi in 1848, the creation of the Waitangi Tribunal in 1975 (dealing with Maori claims of past infringements of Maori rights under the Treaty), the designation of Maori seats in Parliament, and the integration of clauses in legislation (including a key body of environmental legislation, the Resource Management Act), requiring authorities to give recognition to the principles of the Treaty and to consult with Maori on environmental issues and decisions. However, to what extent the institutionalization of Maori views and interests has influenced environment decisions, policies, and practices is open to debate. In particular, government decision making regarding the seabed and the foreshore, which was seen by many Maori as trampling upon their traditional rights, provoked a strong backlash.

However, this depiction of "the" Maori worldview and interests should not be taken to imply unity or harmony among Maori with regard to environmental issues and politics in general. Politically, the Maori are as divided as the *pakeha* (white people) in their view of the present world, including on matters to do with social, economic, and environmental issues and policies, as reflected in the stances taken by and within the Maori Party (that supports a right-wing government led by the National Party), its recent more left-wing offshoot of the Mana Party, and by Maori in the Labour Party, the Green Party, the National Party, and other political parties. This diversity also is apparent with regard to many of the main environmental issues that have been prominent on the political agenda, as discussed in the second part of this chapter, raising the question of whether and to what extent it is (still) valid to regard the traditional Maori worldview as a common and unifying belief system underlying Maori environmental activism.

Grassroots Environmentalism

Here, grassroots environmentalism is interpreted as a broad and diverse category of forms of environmental advocacy and activism initiated by individuals or groups, often at the local level. Arguably, grassroots activism provided the cradle of the modern environmental movement (nature conservation organizations go much further back in history, but had, and often still have, a narrower focus) and was linked with the social movements that sprung up in the 1960s, addressing the many perceived shortcomings of Western societies and challenging authorities (like the *Bürgerinitiativen* in Germany), preceding the creation of national- and international-level environmental organizations in the 1970s. On the other hand, grassroots action was also triggered by the various forms of pollution and other environmental problems that were becoming increasingly apparent during the 1960s, raising health and other concerns. As such, this stream covers a wide array of reactive and ad hoc forms of environmental activisms (including local groups taking action against local pollution issues), as well as the more politically oriented groups that advocate fundamental changes of (consumer) society and institutions.

In New Zealand like in many other countries, grassroots environmentalism played a significant role in drawing public attention to a range of environmental issues during the 1960s. However, by comparison to more highly industrialized countries, grassroots environmentalists focused less on pollution issues, which were widely regarded as less serious than in other countries, if they existed at all. In New Zealand, nature conservation issues loomed much larger on the public agenda, including the protection of native forests from logging, the protection of lakes and rivers from hydropower schemes, and the protection of nature from the effects of mining. Many of these projects and proposals, although geographically local, provoked reaction and action on a far wider, often national, scale, but can still be considered forms of grassroots environmentalism as they were organized as spontaneous ad hoc campaigns in which hundreds or even thousands participated, often with success.[9]

Although grassroots environmentalism tends to be predominantly reactive, triggered by particular projects or proposals, some forms are more proactive and future-oriented, and engage in practical activities aimed at improving the environment and the management of resources and promoting sustainability at the local community level. Examples are the organic growing and permaculture movements, ecological restoration initiatives, and local social economy (alternative currency) and "greening cities" movements (coming under a range of labels, including the Transition Town movement). Such social-environmental "reconstruction" groups have gained in support and significance in New Zealand as well as many other countries.[10]

Environmental Organizations

The late 1960s and early 1970s saw the creation of many environmental organizations to provide a basis for sustained action on environmental causes, at the local, national, and international levels. Many of the organizations that were established then are still in existence, and some have become household names, such as Friends of the Earth and Greenpeace, also in New Zealand. Internationally, in the course of the past 40 years, the role and support for environmental nongovernmental organizations (ENGOs), has increased rather than decreased, as reflected in membership numbers and their often concerted activities aimed at addressing numerous environmental issues.[11]

In New Zealand, the most prominent environmental organizations are the Royal Forest and Bird Protection Society (RFSB), Ecologic, the

Environmental Defence Society, ECO (Environment and Conservation Organisation, an umbrella organization of some 60 groups), Greenpeace NZ, and Friends of the Earth NZ. The Royal Forest and Bird Protection Society, established in 1923, is the oldest New Zealand environmental organization, and still one of the biggest, focused foremost on the protection of nature. Ecologic also has its roots in concern about New Zealand's nature, especially indigenous forests. It was established in the early 1970s (first as the Beech Action Forest Council, changing its name to the Native Forest Action Council in 1975, then becoming the Maruia Society in 1988, and the Ecologic Foundation in 1999) as a more radical, direct action group than the RFSB, but has broadened its agenda over time. The Environmental Defence Society (EDS) was also set up in the early 1970s, modeled on its American namesake, to fight environmental issues in the courts. ECO was also set up in the early 1970s to promote cooperation and coordination of actions and campaigns between the bigger as well as smaller (local) environmental groups. Greenpeace NZ and Friends of the Earth NZ were also both established in the early 1970s as branches of these international organizations, the latter with a greater degree of autonomy in the selection of campaign issues than the former.[12]

As the focus of environmental organizations shifts from issue to issue, largely in response to initiatives and proposals from governments and the business sector, and on the basis of perceived priorities and strategic considerations, their programs appear to be largely reactive and pragmatic rather than inspired by visions of a "good (green) society." Even when promoting sustainable solutions for environmental problems, these tend to be, deliberately, of a kind that is acceptable to people from across the political spectrum (such as in the form of green technologies or better environmental management). Environmental organizations seldom put forward ideas or proposals for radical political-economic change, and hence may come across as "nonpolitical."

Nonetheless, this impression of political neutrality is misleading and environmental groups vary in their (often implicit) political-ideological and environmental-ideological orientation. Some, like Greenpeace, may appear more radical than others largely because of the forms of (direct) action that they take, but are highly pragmatic ideologically, giving scant attention to social, sociocultural, and equity considerations, or to fundamental, systemic political and economic issues (distribution of wealth and power, capitalism). Friends of the Earth International arguably is less radical in its actions (which are largely decentralized), but has a standpoint that encompasses the promotion of social justice, equity and human rights as well as environmental protection, and that opposes neoliberalism, and hence appears to be more "political" (on the political left) in orientation. In New Zealand, the EDS and the Ecologic Foundation have shed much of their confrontational tactics and have become much more cooperative and constructive in their dealings with government over time. Politically, the dominant players in these organizations are closer to the right than the left, and are sometimes referred to as "Blue-Greens" because of their ideological affinity with members of the "center-right" National Party. As an umbrella organization, ECO is even less ideologically pronounced and more explicitly focused on (pro) environmental issues than the other organizations. However, the frequent bickering within and between environmental organizations can be seen as a sign of a significant diversity of political-ideological views among the members, posing a continuous challenge in the development of strategies and campaigns that rely on widespread public support for their success.

Professional/Expert-Based Environmentalism

Professional experts, in areas such as forestry, land and water management, engineering, architecture, and planning, often were the first to express concerns about the environmental implications of the work that they were involved in, long before the rise of environmental groups. In the 19th century, forestry and land (erosion control) experts, concerned about the decline of resources as a result of expanding, indiscriminate and often wasteful use, were instrumental in the emergence of the "wise use" approach to the management of natural resources, mainly for utilitarian and economic reasons.[13] Some experts, inspired by early conservationists like George Perkins Marsh, also recognized the intrinsic value of nature and advocated or supported the protection of areas as reserves or national parks. In New Zealand, such concerns started to get some traction in the 1880s, and became an integral element of the philosophy of state agencies involved in the management of natural resources in the early part of the 20th century.[14]

In the late 1950s and 1960s, scientists played an important role in moving nature conservation issues further onto the government agenda, especially where the state was involved in development projects that were having detrimental environmental effects (such as the damming of rivers for electricity generation). "Conscience stricken" experts involved in "development," also in the urban environment (engineers, architects and planners), played a significant role in moving such issues onto the agenda of governments.[15]

In general, not just in New Zealand, scientists and professionals have played and still play a key role in the recognition and definition of environmental problems, and in the environmental policy process.[16]

However, scientists and experts are as diverse a group as any, and can be found on all sides of environmental argument and conflict. The role of science in the environmental policy process, although central, is subject to different interpretations and much debate, as reflected in the literature.[17] While science is often looked at, by governments and the public, for "objective" knowledge and guidance in environmental issues, scientific practice is heavily influenced by constitutive and contextual values and interests that orients it toward "technocratic" interpretations and solutions rather than toward questioning the underlying factors of a political-economic and sociocultural nature.[18]

The Political (Parliamentary) Wing

New Zealand was the birthplace of one of the world's first green parties, the Values Party, established in 1972. Despite the distrust many environmentalists held (and often still hold) of the established political institutions, especially at the national level, and a preference for working toward change at the local or community level, a growing number looked at parliamentary representation as a platform for advocating and promoting policy and institutional change. However, in New Zealand, gaining parliamentary representation proved to be difficult during the first few decades of the existence of the Values Party, in large part because of the nature of New Zealand's electoral ("First-Past-the-Post") system, which does not favor small parties. It was only in 1996, after the electoral system changed to a Mixed Member Proportional (MMP) system, that a reformed Green Party gained seats in Parliament, initially as a member of a coalition of parties (the Alliance). From 1999, the greens have contested elections successfully on their own force, gaining between 5 and 7 percent of the votes, jumping to 10.5 percent in the elections of 2012.

Although widely perceived as an "environmentalist" party, Values and other green parties have always advocated a broad agenda, comprising peace issues (based on the principle of nonviolence), enhanced democracy (greater public participation in decision making), social justice, as well as ecological sustainability.[19] However, notwithstanding these shared principles, ideological differences and disagreements on strategy have always existed within green parties, in particular between fundamentalists and realists (referred to as *Fundis* and *Realos* in Germany), and/or "left (or red) greens" and "right (or 'blue') greens," although the more pragmatic orientation has tended to dominate, especially in more recent times.

To what extent green parties have been effective in bringing about policy and institutional change aimed at achieving their goals, including environmental protection, is a question subject to debate, and dependent on the country involved. As we will see in the second part of this chapter, in New Zealand, the influence of the greens in policy debates and policy development has, in some areas, been significant and growing.

Institutional Environmental Advocates

Although government officials or agencies are normally not considered to be part of the environmental movement, but rather seen as target groups for the movement, there is a good argument for including some of them in the environmental camp. Government officials have a variable degree of autonomy in their thinking and doing, regardless of the mandates and lines of accountability under which they are operating, as illustrated in the *Yes Minister* television series, but also in the public policy literature.[20] This applies also to government officials who consider themselves environmental advocates "within the system," whatever department or agency they are working for, and may apply also to agencies as a whole, especially those with environmental mandates. Institutional environmental advocates can also be found in quasi nongovernmental organizations (Quangos), such as advisory bodies on environmental matters or in the form of National Sustainable Development Councils (NSDCs).[21]

In New Zealand, government officials, especially in environmental agencies, have played a significant role in advancing and protecting environmental values and interests from the early 1970s, such as in the Environmental Council, the Nature Conservation Council, and the Commission for the Environment, albeit with variable degrees of success and despite being heavily constrained by political-institutional and resource limits.[22] Environmental advocacy "within the system" was considerably strengthened as a result of extensive reforms of the public service in the 1980s, creating a Ministry for the Environment, a Department of Conservation (with an explicit advocacy role for nature conservation), and a Parliamentary Commissioner for the Environment as an independent "systems guardian" and watchdog for environmental interests.[23] This is not meant to say that New Zealand's political and policy system has been greened from within, and that environmental concerns and interests are systematically integrated (not just "considered") in the policy development process, let alone given priority. New Zealand has still a long way to go before any such claim can substantiated. But inasmuch as the green movement's "march through the institutions" is concerned, there has been significant development and change in New Zealand.

This brief overview of the various streams of environmental advocates indicates diversity in ideological orientation, of the relative importance attached to issues and objectives, of actions and strategies used in the pursuit of goals and objectives, and in social support basis. The diversity between and within the various streams of the environmental movement has been a subject of considerable debate, notably on the point of whether this is a strength or a weakness.[24] How the diversity within the environmental movement has played itself out in the New Zealand context will be discussed in the next section.

The Evolution of Environmentalism and Environmental Action

Although there have been some dominant themes in New Zealand's environmental agenda throughout the last 50 years, largely related to the structural/systemic characteristics described earlier, there also have been significant developments in the specific environmental issues, conflicts and battles over time, and changes in the relative importance of the various streams of the environmental movement in addressing these issues. To a large extent, these developments are related to changes in the political-economic context, notably with regard to the role of the state, but they are also influenced by broader developments (globalization, technological/media, and sociocultural change).

The discussion here will be structured chronologically, based on four eras:

1. Early 1960s until the mid-1980s: the era of "state-led development" and the rise of confrontational environmentalism;
2. Mid-1980s until 2000: the era of "radical neoliberalism" and the devolution of the role of the state; "pulling the teeth of environmentalism" by "new right" strategies;
3. 2000 to 2008: the era of "mitigated neoliberalism" in times of relative prosperity; the search for new ways of accommodating environmental demands while maintaining economic growth and development;
4. From 2008 until the present: "obstinate neoliberalism" in the face of global financial-economic crisis; the return of environmental confrontation.

During each era, different environmental issues tended to dominate the political agenda, and different parts (streams) of the environmental movement tended to play a more important (or even dominant) role.

The Early 1960s until the Mid-1980s: The Era of State-Led Development

From its creation, the New Zealand state has played a dominant role in the country's economic development. The state has been responsible for building infrastructural facilities (roads, railways, ports, hydrodams for generating electricity, and irrigation of farmland), but has also been involved in, or promoted, development in a range of other ways, such as the development of agriculture, forestry, mining, energy, and tourism, either directly through state-owned enterprises, or indirectly, by the use of financial incentives, research, advisory bodies, and conducive legal and regulatory frameworks. Many of the environmental issues that arose during the 1960s and 1970s find their origin in state-initiated or state-sponsored development projects, such as the proposal to raise the level of Lake Manapouri (to generate more electricity), which is often regarded as the starting point of mass-based environmentalism in New Zealand,[25] the damming of wild and scenic rivers (Save the Rivers Campaign), the logging of indigenous forests (Whirinaki and many other forests), mining (for gold and other minerals), and "Think Big" projects (including a proposal for an aluminium smelter at Aramoana, and the building of a liquid fuels plant using gas from the Maui gas field).[26] There was also growing concern about nuclear testing at Murorua by the French, and apprehension about the possibility of the introduction of nuclear power in New Zealand, even though a decision on this matter was deferred by the government.[27]

While concern about the environmental impacts of these projects was growing during this era, New Zealand governments remained firmly wedded to their active and often direct role in the development of New Zealand. The state's commitment to development was strengthened by the (looming) entry of the United Kingdom to the European Union, causing New Zealand to lose the United Kingdom as the main market for much of its exports, and the energy crises of 1973 and 1978, which exposed the country's vulnerability in terms of energy security. Growing environmental awareness on the one hand and intensified state-led development on the other created a stage for confrontation and numerous battles, stimulating the radicalization of environmental groups (for instance, the creation of the Native Forest Action Council, as the New Zealand Forest and Bird Protection Society was seen as too meek in its actions) and contributing to the already adversarial nature of New Zealand politics (which was arguably even more pronounced during the reign of Robert Muldoon, an authoritarian and polarizing prime minister between 1975 and 1984). During this period, much environmental action took the form of street protests, direct action and confrontation, complemented with legal action (notably by the Environmental Defence Society), lobbying (notably of the opposition party for much of this time, Labour, to incite a commitment to reversing decisions and policies if they were to regain power), and environmental education and mobilization at the local level, notably through environment centers.

To what extent the efforts of the environmental movement during this time were effective is debatable. Although some of the proposed projects (such as the proposed aluminium smelter at Aramoana, and the proposal for a nuclear power station) did not go ahead, this had as much if not more to do with economic reasons (overseas investors pulling out, costs and no perceived need for nuclear power in the near future) than with the scale of protests and the actions of environmental groups. Arguably, environmental action was most, but still only somewhat, successful in the area of nature conservation, such as the protection of wild and scenic rivers (by legislation adopted in 1982) and the protection afforded to some indigenous forests (Okarito, Pureora). However, it took a change of government, in 1984, for more significant environmental gains to be won.

Mid-1980s until 2000: Radical Neoliberalism "Pulling the Teeth of Environmentalism"

In 1984, the fourth Labour government came to power, heralding the start of a radical process of policy and institutional change, also in the environmental area. The main institutional changes included the creation of a Ministry for the Environment (previously, the main agency was a Commission for the Environment), a Department of Conservation, the Office of the Parliamentary Commission for the Environment (an independent watchdog agency), the creation of Regional Councils (a new layer of governance, with significant environmental responsibilities), and the introduction of the Resource Management Act, amalgamating a raft of previously fragmented environmental legislation. Altogether, the scale of environmental reforms was unprecedented and significantly strengthened the "environmental voice" within the machinery of government and decision making processes.[28]

The change of government also caused a change in relations between the environmental movement and the government. Environmental advocates, from environmental groups and academic/expert circles, were heavily involved in discussions about environmental reforms and successful in having some of their views and desires adopted, such as with regard to the creation of the Department of Conservation as an advocate for conservation. In these matters, the fourth Labour government took a far more open and participatory approach, and appeared much more sympathetic toward environmental causes, than its predecessor. As a result, the environmental movement abandoned, and arguably had less need for, a confrontational approach.

The change in political climate was, among other things, reflected in the West Coast Forest Accord, concluded in 1986 after consultation between the government and all stakeholders, including environmental representatives, which laid out a gradual process toward the sustainable management of indigenous forests, implying that most of these forests would gain protection and be added to national parks. Most other indigenous forests on public land received similar protection, while native forests on private land were brought under control with the adoption, in 1993, of the Forest Amendment Act, imposing sustainable management as a basic principle. Although these changes did not resolve all concerns regarding the protection of nature in New Zealand, they did, by and large, bring an end to one of the most prominent and contentious environmental issues, the logging of native forests.

The fourth Labour government also moved on another issue that many New Zealanders apparently felt strongly about: the nuclear issue. Although previous Labour governments also had expressed their dismay about nuclear testing in the Pacific, the fourth Labour government surprised many by going much further and by passing legislation to make New Zealand "nuclear free," banning even nuclear-powered (not only nuclear-armed) ships from visiting the country, and thus provoking a row and the disruption of military cooperation with the United States. However, the move enjoyed widespread public support, and its nuclear-free status, together with the bold environmental reforms referred to earlier in this chapter, added to the country's reputation as an environmental leader and a "clean and green" country in which New Zealanders found a new source of pride or even identity.

However, there was another side to the fourth Labour government's policies that was not commensurate with these environmental credentials, and that was to some extent masked by these prominent moves. Despite appearances, the government actually did little to resolve or even address the growing environmental challenges facing the country. In fact, it can be argued, it created an environmental policy vacuum in which the (market-led) drivers of environmental pressure could flourish. It did this by adopting neoliberal ideology and policies aimed at reducing the role of the state and by the devolution of responsibilities for the management of resources to the market and, to a lesser extent, the regional and local government.

Led by a troika of government ministers and a handful of Treasury officials, the fourth Labour government embarked on a reform of the public sector and service that arguably was the envy of Conservative leaders like Margaret Thatcher in the United Kingdom and Ronald Reagan in the United States. Given the extent to which power is concentrated in the New Zealand political system, the government was able to adopt and implement one of the most comprehensive and coherent neoliberal programs in the world, without being forced or pressured into this by the IMF or the World Bank and their "structural adjustment" agenda, as in many low- and middle-income countries. Where previously the New Zealand state took a leading role in the country's development, it now sought to transfer much of the responsibility for that to the market, within an "enabling" institutional and policy framework and with a minimum of regulation. Financial deregulation, and the "corporatization" and subsequent privatization of many state-owned enterprises were two of the main planks of these reforms. Two others were a shrinking of the public service and the adoption of the Resource Management Act.

Lack of space does not allow elaboration on these policies and their implementation. But it is important to refer to these as they significantly changed the context in which development progressed and associated environmental issues were (to be) addressed, with implications for the environmental movement. With the central government no longer playing a leading role in development, and the responsibility for assessing the environmental effects of proposals assigned to regional and local government, with little if any constraints imposed by national policies or standards, environmental advocates were compelled to disperse their attention and efforts from the national to the local level, and to shift their efforts from lobbying ministers and central government officials to using and building "expert knowledge" to challenge proposals in the Environment Court, where many issues were now decided. Thus, environmental conflicts became, to a large extent, decentralized, fragmented (issue/project focused rather than policy focused), and depoliticized (focused on physical, technical, and scientific aspects).

While environmental advocates may have won concessions in many of these (local) battles, these did not halt the trend of environmental deterioration, as development continued apace driven by market forces and facilitated (rather than steered) by the government through an enabling institutional and policy framework. At the national level, policies (energy, transport, agriculture, urban development) remained oriented toward boosting economic growth and development, with little if any integration of environmental concerns and interests. Let alone that there was a serious attempt toward the development of strategic (comprehensive, integrated, long-term) environmental policy to address the underlying drivers of environment pressures and problems. Hence, apart from the native forests issue, where a major policy change was achieved, the environmental movement continued to operate in a reactive, defensive, and fragmented mode with regard to the environmental challenge as a whole.

Both Labour (in power between 1984 and 1990) and National (governing between 1990 and 1999) were neoliberal aficionados. While the fourth Labour government had created the institutional and policy framework for the implementation of neoliberal principles and ideas, notably in the public sector, the public service and the economic realm, its National successor took the neoliberal agenda further in the areas of social policy, labor relations, health, and education. In the course of the 1990s, public discontent triggered by the pain (and little if any gain) caused by these policies, growing distrust of politicians, and political disillusion about the political system in general, led to a demand for a change of the electoral system. This resulted, after two referenda, in the abandonment of the First-Past-the-Post (FPP) system and the introduction of the Mixed Member Proportional (MMP) proportional system. This change, which took effect from the elections of 1996, was significant in terms of the conduction of politics, and the representation of the environmental movement in Parliament.

2000 to 2008: The Era of "Mitigated Neoliberalism"

The introduction of MMP heralded the end of one-party government and the beginning of the need for the two main parties to cooperate with other parties, as under the new system it is very difficult for one party to gain an absolute majority in Parliament. While proportional representation has been a long-standing feature of the electoral system in many countries, it is not of Westminster political systems, where FPP and single-party government has been the norm. Although it is too early to say whether the change toward MPP in New Zealand has brought about a change in political culture (from adversarial toward more cooperative, let alone corporatist policy styles), the change has been significant in terms of forcing governments to make concessions on some (policy) issues to smaller parties on which they rely for their ability to govern (based on "confidence and supply" agreements). MMP, as noted earlier, also made it possible for the Green Party, commonly perceived to be the parliamentary wing of the environmental movement, to gain representation within Parliament.

Although the fifth Labour government that came to power at the end of 1999 kept in place much of the neoliberal political-economic institutional and policy framework created by its predecessors, especially in the economic realm, it did mitigate, or at least not take further, the application of these principles in a range of policy areas, including social policy, labor relations, health, and education. Influenced by the "'Third Way" thinking that also inspired the Blair (Labour) government in the U.K. the Labour government of Prime Minister Helen Clark, in power until the end of 2008, tried to harmonize the perceived "economic imperatives" associated with the next stage in the globalization of capitalism with the "local" (national) impacts thereof, and with domestic needs and demands. As environmental concerns, notably about climate change, started to strengthen globally, and the adverse environmental impacts of continued development in New Zealand become more apparent, the fifth Labour government was virtually forced to take the environment more seriously.

Although the Green Party never entered into a formal government coalition with Labour, they concluded agreements with Labour that, over time, provided increased opportunity to the Greens to influence policy. Between 1999 and 2002, the agreement was informal, with the Greens giving support on "confidence and supply" in exchange for promises about consultation and minor policy concessions. Between 2002 and 2005, the two parties signed a cooperation agreement, again trading support on "confidence and supply" for some policy gains for the Greens. During the third term of the fifth Labour government, the agreement concluded between the parties became more comprehensive, giving the Greens government spokesperson roles on energy efficiency and a "Buy Kiwi-Made" campaign, and greater involvement in the development of a range of policies.[29] However, the growing influence of the Greens during this period should not be overestimated; the gains, although significant for some issues (like energy efficiency), did not change the basic growth-oriented tenets of the main policies, such as with regard to energy, transport, and agriculture. Nonetheless, the role of the Green Party as an arm of the environmental movement became increasingly important, as their representation in Parliament provided a significant platform for putting environmental issues on the political agenda, for gaining information that is more difficult to obtain from outside Parliament, and for enhancing the public profile of issues that were of great or growing concern to the environmental movement, such as genetic modification (on which the Greens got the government to appoint a Royal Commission), climate change, water pollution, and mining.

During this decade, with the rapid expansion of dairy farming and irrigation, freshwater pollution became an issue of foremost environmental concern for many New Zealanders,[30] and the subject of a ("dirty dairying") campaign by environmental groups. However, although the Labour government had promoted the idea that New Zealand's future was in the development of a "knowledge economy," the economic reality was that with the rising world prices for dairy products, dairy farming expanded rapidly to become the biggest export sector.[31] In accordance with the government's Third Way philosophy, the government addressed the issue by appealing to the cooperation of farmers rather than by taking recourse to stricter regulation of the dairy industry. However, the voluntary Dairy Accord adopted by the farming industry has had limited effect on the water quality of rivers, streams and lakes, which continue to be in poor condition in many parts of the country, especially those where dairying has expanded.[32]

Climate change also rose high on the political agenda during this period, albeit more because of the growing concerns about this issue internationally, and the potentially costly consequences of not meeting New Zealand's obligations under the Kyoto Protocol, which the government ratified in 2002, rather than because of domestic pressure or demands. Policy on the issue was developed largely within government circles, with two main options being considered: a carbon tax and an emissions trading scheme. The choice between these two options was not so much influenced by environmental groups or advocates (who were not unified in their views on the relative merits of both options), but by the degree of opposition encountered: farmers fiercely opposed the introduction of a carbon tax (proposed first), and although they were opposed to any measures that would imply a cost to farmers, there appeared to be less opposition from within business circles to the introduction of an emissions trading scheme (which potentially offered economic benefits, and from which agriculture would be exempted for several years).

During this period, the relations between the government and the environmental movement continued to be conducted in a cooperative or consultative rather than an adversarial way, fostered by the apparently environment-friendly rhetoric and stance of the government, and the now dominant ideology and practice of "network governance," which emphasizes the need for, and the superiority of, collaboration between stakeholders (including government officials, business representatives, and environmental advocates) in addressing environmental and other issues.

However, as Davies argues, in practice, the extent to which such consultative or deliberative exercises allow genuine democracy and input from the public is very limited, while states remain in control.[33] They are thus better seen as efforts to legitimize neoliberal policies and practices (assumedly involving decision making between equal partners), preventing environmental, and other social movements, from adopting more radical strategies that pose a threat to the dominant interests. The approach allows governments, especially in times of relative prosperity, to grant concessions to environmentalists while proclaiming a commitment to integrating or balancing economic, social, and environmental interests. However, given the limitations of network governance, and the growing gap between the imperatives inherent to the economic system, especially during times of economic crisis, and the measures needed to address ongoing environmental deterioration, it is unlikely that the support for, and viability of, this approach can be sustained, something that has become more apparent since the onset of the most recent financial-economic crisis.

From 2008 until the Present: "Obstinate Neoliberalism" during a Global Financial-Economic Crisis; the Return of Environmental Confrontation

The election of a new National government in 2008 and the concurrent onset of the global financial-economic crises marked not only a hardening of government commitment to economic imperatives and priorities but also a turning point in environmental policy. While the previous government at least held up the pretence that environmental interests were considered important and allowed environmental advocates to gain modest concessions, the new government blatantly relegated environmental concerns to a low order of importance, while reaffirming its commitment to promoting economic growth and neoliberal principles and policies. Not surprisingly, this puts considerable strain on the cooperative relationships that had evolved between governments and the environmental movement, and is leading to a revival of confrontation and environmental mobilization.

Soon after its election the National government of Prime Minister John Key launched a proposal for opening up parts of national parks that were up to then excluded from minerals prospecting. The rationale offered was that, in its pursuit of economic growth, New Zealand could not afford to ignore or continue to lock up the potential value of mining in these areas, estimated in the order of $54 billion.[34] These proposals caused a furor not only among environmental groups but also among New Zealand citizens at large. After 38,000 submissions, mostly opposed, a petition signed by nearly 50,000 citizens and a protest march by 15,000 people in Auckland (a level of protest on an environmental issue not seen in decades), the government backed off, although much of the national parks remain open to mining. More generally, the government continues to look at mineral exploitation (including of oil) as one of the main means for boosting New Zealand's economic growth, on land as well as offshore. Several initiatives and proposals, including mining for lignite in Southland, fracking for oil on the Canterbury Plains, and deep-sea oil exploration off the east coast of the North Island, have stirred up environmental protest.

Another blatant form of government intervention to support economic development was the dissolution of the Canterbury Regional Council and its replacement by appointed Commissioners on the grounds that the organization was dysfunctional and had failed to meet its responsibilities, notably with regard to water management. However, the move seems to have been inspired foremost by the government's desire to boost the further development of the dairy industry, requiring more water to be made available for irrigation than the Regional Council might have been willing to grant, as it considered that in some areas of Canterbury groundwater had already been over-allocated, based on an assessment of sustainable flows. The blatant pro-farming/irrigation stance of the government also undermined and almost torpedoed various forms of network governance on water management at the local and the national level, the outcomes of which now seem even less likely to be implemented if the government sees these as obstructive to the further development of the industry. The agriculture minister, David Carter, commented that the sacking of the Environment Canterbury councilors should be heeded by all regional councils "as a 'signal' to work more constructively with farmers" and, at the same time, announced that the government would make almost $1 million available in support of irrigation.[35] The power and privileged position of the farming lobby was also reflected in the accommodation of farmers' interests under the Emissions Trading Scheme.

The manifest pro-development stance of the National government, and the reaffirmation of its commitment to neoliberal principles and policies (tax cuts especially benefitting the rich, selling off more government-owned enterprises, concluding more free-trade agreements; downsizing of the public service and rationalizing social welfare policies), obviously makes it very difficult if not impossible for the environmental movement to continue a cooperative engagement with the government. Apart from re-igniting environmental protests on a range of projects, it leaves little if any scope for cooperative strategies aimed at avoiding further environmental degradation, let alone a strengthening of environmental policies and institutions. Environmental advocates are again, after some 15 years of strategic cooperation with governments, forced into a reactive and adversarial approach in their efforts to protect the environment.

In this light, it is perhaps not surprising that in the elections of 2011, the Green Party saw a considerable lift in support, obtaining 11 percent of the vote and 14 seats in Parliament compared to 6.7 percent and 9 seats in 2008. Although the Greens also have little chance of finding common ground with the National government, and hence of being actively involved in the development of policy, their stronger position in Parliament provides at least a fairly robust platform for challenging government policies, for advocating alternatives, and for mobilizing citizens on issues of environmental concern. Therefore, although environmental advocates are forced again to adopt reactive and adversarial strategies, the strong Parliamentary platform obtained by the Greens is unlikely to herald a return to the 1970s and early 1980s, when "direct action" was virtually the only strategy available to environmental advocates in their efforts to stop environmentally damaging projects and policies.

At the same time, the financial-economic crisis has sparked off a new, global form of "direct action" facilitated by the social media and aimed at addressing a broad range of political-economic, social, and environmental issues. The Occupy Movement, which started with the Occupy Wall Street movement in New York, puts particular emphasis on the greed and growing inequality that are perceived to be inherent to capitalism, issues that have attracted growing attention and public indignation at a time when the very rich and powerful keep on awarding themselves big bonuses and pay increases while a rising number of people struggle to make ends meet in times of economic hardship and austere government policies. Finding inspiration in the overthrow of authoritarian governments in Tunisia and Egypt by a mass protest movement, facilitated by social media, the Occupy movement arguably represents a revival of a belief in the "power of the people" in rich/Western countries, including New Zealand. However, by comparison to the movements in the United States and the southern European countries, the Occupy movement in New Zealand has attracted only relatively little support in the four main cities.

Conclusion

From its beginnings some 50 years ago, the environmental movement in New Zealand seems to have made a full circle. Starting off with mass protests against government-led proposals and developments that were deemed to be environmentally destructive ("state-sponsored vandalism"), followed by a period during which environmental activism was largely defused by neoliberal policies and strategies, more recently, it is again forced into more direct forms of activism as governments reassert more blatantly their commitment to the "imperatives" of an increasingly global economic system. However, there are also some significant differences between the environmental movement of the 1960s and that of the 2010s, notably with regard to the relative importance of the various branches of the movement, the main issues they focus on, their support base, and their potential to address environmental issues on the basis of a broader and stronger political and ideological platform.

Until the mid-1980s, direct activism and campaigns orchestrated by environmental organizations were the most important (if not the only) means by which environmental gains could be, and were, achieved, given the predominantly anti-environmental orientation within government circles, including government departments. Environmental battles were won or lost on the basis of the extent to which the environmental movement was able to mobilize mass protest. Although experts played a role in these battles, they became more important as a result of the reforms of the 1980s, especially under the new legal framework provided by the Resource Management Act 1991, which largely decentralized and depoliticized decisions on "development and environment" issues, forcing environmental advocates to rely increasingly on "expert witnesses" and "objective" science/knowledge to win cases. At the same time, at the national level, the reforms strengthened the role of institutional environmental advocates, notably in the Department of Conservation, the Parliamentary Commissioner for the Environment, and the Ministry for the Environment, although the role of the latter as an environmental advocate was, and remains, ambiguous. Although New Zealand was the first country where a national-level green party was created (under the name of the Values Party), it was only since the introduction of a proportional electoral system that it was able to gain representation in Parliament. Since then, the Green Party has steadily expanded its support base to become the third biggest party in Parliament. At the same time, although empirical evidence on that front is scarce, it appears that at the local/grassroots level, constructive forms of environmentalism, in the form of individuals and groups involved in organic growing, permaculture, the "Envirotown" (a variation on the "Transition Town") movement, and a large variety of environmental, ecological and/or sustainability initiatives (from zero-waste to sustainable housing/energy and ecological restoration projects) has grown in importance.

Overall, it can be concluded that the environmental movement in New Zealand is presently in a much stronger political position than it was in the 1960s and 1970s. Environmental interests and concerns have been institutionalized to the extent that they can no longer be ignored, while the actions of governments, at best often still not going beyond paying lip-service to those concerns, can be, and are, contested both from "below" and from within the political system. The rise in support for the Green Party, as well as the many forms of "constructive environmental activism" that have emerged across the country, mark a growth in the support base of the environmental movement.

Although, over the years, "new" environmental issues like climate change, genetic modification, and water pollution have become foci of concern and action for environmental advocates, nature protection has continued to be very important, despite the gains made in this area. This is illustrated, in particular, by the massive public reaction against government proposals to extend the areas in which mining is allowed within national parks, and also in strong regional protests against the "despoliation" by (proposed) wind turbines of landscapes considered to be of particular scenic beauty. In many cases, environmental activism still takes a reactive form, reacting *against* development projects or proposals, whether put forward by private companies, state-owned enterprises, or governments. Submissions, petitions, court action, lobbying, and protest demonstrations continue to be the much-used tools in efforts aimed at influencing decisions.

As noted earlier, in New Zealand as in many other countries, there has been a move by governments toward a more cooperative style of decision making, involving partnerships and discussion between stakeholders, including businesses and environmental groups, referred to as network governance. In New Zealand, this became the preferred approach of governments as well as many environmental advocates from the mid-1980s, coinciding with the rise of neoliberalism as the hegemonic political-economic ideology. However, in terms of substantive environmental outcomes, the approach has had limited, albeit not insignificant, results, notably in the area of the protection of native forests. It has not led to the adoption of policies and institutions that *reduce* or *prevent* further environmental deterioration from drivers like growing consumption, energy use, transport, and urban development. And with a government that continues to be imprisoned (and to imprison itself) by the imperatives of the prevailing economic system, even when the latest financial-economic crisis demonstrates once more its unsustainability (economically, socially, and environmentally), there is little prospect of such more fundamental policy changes being adopted.

Hence, although now in a significantly stronger position than ever before, the environmental movement in New Zealand, and elsewhere, continues to face the challenge of bringing about the kind of changes that address, effectively, the environmental problematique as a whole, rather than addressing issues on a case-by-case and ad hoc basis. If anything, this is the role and task of green parties, as they develop and advocate broader programs of political-economic and sociocultural change into which environmental imperatives have been integrated. In New Zealand, the Green Party (and its predecessor, the Values Party), while being perceived by many as being mainly or even only concerned about the environment, has always had a broader vision of the environmental challenge, as well as a program for change. However, such a program remains a work in progress, especially in terms of the development of environmentally sustainable economic institutions and policies, and even more so with regard to their introduction in the face of formidable vested interests and powers. Whether New Zealand's Green Party, with the support of the other branches of the environmental movement as well as the public at large, will be able to push through such an agenda for change remains anyone's guess.

Notes

1. Mulgan, *The Elective Dictatorship in New Zealand*; Palmer, *Unbridled Power.*
2. Bührs and Bartlett, *Environmental Policy in New Zealand,* 192
3. Devall and Sessions, *Deep Ecology.*
4. Naess, "The Shallow and the Deep."
5. Rootes, "Environmental Movements: From the Local to the Global."
6. O'Riordan, *Environmentalism.*
7. Sutton, *Explaining Environmentalism.*
8. Marsden, Palmer, and Goodall, *Resource Management Law Reform.*
9. Wilson, *From Manapouri to Aramoana.*
10. Seyfang and Smith, "Grassroots Innovations for Sustainable Development."
11. Dalton, "The Greening of the Globe?"; Keck and Sikkink, "Environmental Advocacy Networks"; Evans, "Counterhegemonic Globalization: Transnational Social Movements in the Contemporary Global Political Economy."
12. Wilson, *From Manapouri to Aramoana. The Battle for New Zealand's Environment,* 171–85.
13. Worster, *Nature's Economy.*
14. Roche, "The State as Conservationist."
15. Bührs, "Working within Limits," 59–65.
16. Ascher, Steelman, and Healy, *Knowledge and Environmental Policy*; Fischer, *Citizens, Experts*; Ingram, Milward, and Laird "Scientist and Agenda Setting";

Parliamentary Commissioner for the Environment, *Illuminated or Blinded by Science?*

17. Ascher, Steelman, and Healy, *Knowledge and Environmental Policy*; Fischer "Democracy and Expertise."

18. Longino, "Beyond Bad Science."

19. Dann, "From Earth's Last Islands."

20. Nordlinger, *On the Autonomy of the Democratic State*; Evans, Rueschemeyer, and Skocpol, *Bringing the State Back In.*

21. d'Evie et al., "NCSD Report."

22. Bührs, "Working within Limits."

23. Bührs, "Barking Up Which Trees?"

24. Norton, *Toward Unity among Environmentalists.*

25. Wilson, *From Manapouri to Aramoana,* 10–15.

26. Ibid., 69–96.

27. Ibid., 40–51.

28. Bührs and Bartlett, *Environmental Policy in New Zealand*; Memon, *Keeping New Zealand Green.*

29. Bale and Wilson, "The Greens," 401.

30. Hughey, Kerr, and Cullen, *Public Perceptions of New Zealand's Environment.*

31. Treasury, "New Zealand Economic and Financial Overview 2010."

32. Barnett and Pauling, "New Zealand's Free-market Reforms"; Deans, and Hackwell, "Dairying and Clean Streams Accord"; Ministry for the Environment, *Trends in River Water Quality.* TerraNature, "Water Quality."

33. Davies, *Challenging Governance Theory.*

34. Kay, "Mining in Conservation Land."

35. Macfie, "Taking the Waters"; Rodgers, "Critical Time for Water"; Williams, "Warning to Councils by Carter."

Bibliography

Ascher, William, Toddi A. Steelman, and Robert G. Healy. *Knowledge and Environmental Policy: Re-imagining the Boundaries of Science and Politics.* Cambridge, MA: MIT Press, 2010.

Bale, Tim, and John Wilson. "The Greens." In *New Zealand Government and Politics,* edited by R. Miller, 283–92. Melbourne: Oxford University Press, 2006.

Barnett, Jon, and Jonathan Pauling. "The Environmental Effects of New Zealand's Free-Market Reforms." *Environment, Development and Sustainability* 7, no. 2 (2005): 271–89.

Bührs, Ton. "Barking up which Trees? The Role of New Zealand's Environmental Watchdog." *Political Science* 48, no. 1 (1996): 1–28.

Bührs, Ton. "Working within Limits. Environmental Policy Development and the Role of Central Agencies: The Case of the NZ Commission for the Environment." PhD diss., University of Auckland, 1991.

Bührs, Ton, and Robert V. Bartlett. *Environmental Policy in New Zealand. The Politics of Clean and Green?* Auckland: Oxford University Press, 1993.

d'Evie, Fayen, Mary MacDonald, Francisco Mata, and Rocio Rodriguez, eds. *NCSD Report 1999–2000. National Experiences of Integrative, Multistakeholder Processes for Sustainable Development.* San Jose, Costa Rica: Earth Council, 2000.

Dalton, Russell. "The Greening of the Globe? Cross-National Levels of Environmental Group Membership." *Environmental Politics* 14, no. 4 (2005): 441–60.

Dann, Christine. "From Earth's Last Islands. The Global Origins of Green Politics." PhD diss., Lincoln University, New Zealand, 1999.

Davies, Jonathan S. *Challenging Governance Theory: From Networks to Hegemony.* Bristol: Policy Press, 2011.

Devall, Bill, and George Sessions. *Deep Ecology.* Salt Lake City, Utah: G.M. Smith, 1985.

Evans, Peter. "Counterhegemonic Globalization: Transnational Social Movements in the Contemporary Global Political Economy." In *The Globalization and Development Reader: Perspectives on Development and Global Change,* edited by J. T. Roberts and A. B. Hite, 420–42. Malden, MA: Blackwell Publishers, 2005.

Evans, Peter B., Dietrich Rueschemeyer, and Theda Skocpol. *Bringing the State Back In.* Cambridge: Cambridge University Press, 1985.

Hughey, Kenneth F. D., Geoffrey N. Kerr, and Ross Cullen. *Public Perceptions of New Zealand's Environment: 2004.* Christchurch: EOS Ecology, 2004.

Ingram, H., Milward, H.B., and Laird, W. "Scientists and Agenda Setting: Advocacy and Global Warming." In Risk and Society: the Interaction of Science, Technology and Public Policy, edited by Martin Waterstone, 33-53: Dordrecht: Kluwer Academic Publishers, 1992.

Kay, Martin. "Mining in Conservation Land—Proposal." *Stuff,* March 22, 2010. http://www.stuff.co.nz/business/3488434/Mining-in-conservation-land-proposal (Accessed November 26, 2010).

Keck, Margaret E., and Kathryn Sikkink. "Environmental Advocacy Networks." In *The Globalization and Development Reader: Perspectives on Development and Global Change,* edited by J. T. Roberts and A. B. Hite, 370–93. Malden, MA: Blackwell Publishing, 2007.

Longino, Helen. "Beyond 'Bad Science': Skeptical Reflections on the Value-Freedom of Scientific Inquiry." *Science, Technology & Human Values* 8, no. 1 (1983): 7–17.

Macfie, Rebecca. "Taking the Waters." *Listener,* April 17, 2010, 14–19.

Marsden, Maori, David Palmer, and Anake Goodall. *Resource Management Law Reform: Part A, The Natural World and Natural Resources, Maori Value Systems & Perspectives.* Wellington: Ministry for the Environment, 1989.

Memon, P. A. *Keeping New Zealand Green: Recent Environmental Reforms.* Dunedin, New Zealand: University of Otago Press, 1993.

Ministry for the Environment. *Nationwide and Regional State and Trends in River Water Quality 1996–2002. Executive Summary.* Wellington: Ministry for the Environment, 2005.

Mulgan, Richard. "The Elective Dictatorship in New Zealand." In *New Zealand Politics in Perspective,* edited by H. Gold, 513–31. Auckland: Longman Paul, 1992.

Naess, Arne. "The Shallow and the Deep, Long-Range Ecology Movement. A Summary." *Inquiry* 16, no. 95 (1972): 95–100.

Nordlinger, Eric. *On the Autonomy of the Democratic State.* Cambridge: Harvard University Press, 1981.

Norton, Bryan G. *Toward Unity among Environmentalists.* New York: Oxford University Press, 1991.

O'Riordan, Timothy. *Environmentalism.* London: Pion, 1976.

Palmer, Geoffrey. *Unbridled Power. An Interpretation of New Zealand's Constitution and Government.* Second edition. Auckland: Oxford University Press, 1987.

Roche, Michael. "The State as Conservationist 1920–1960—'Wise Use' of Forests, Land, and Water." In *Environmental Histories of New Zealand,* edited by E. Pawson and T. Brooking, 183–99. Melbourne, Vic.: Oxford University Press, 2002.

Rodgers, Murray. "Critical Time for Water." *The Press,* April 29, 2010.

Rootes, Christopher. "Environmental Movements: From the Local to the Global." *Environmental Politics* 8, no. 1 (1999): 1–12.

Seyfang, Gill, and Adrian Smith. "Grassroots Innovations for Sustainable Development: Towards a New Research and Policy Agenda." *Environmental Politics* 16, no. 4 (2007): 584–603.

Sutton, Philip W. *Explaining Environmentalism: In Search of a New Social Movement.* Aldershot, Burlington, Vt.: Ashgate, 2000.

TerraNature. "Water quality. How Clean and Green Is New Zealand. Freshwater Ecosystems at Risk." http://www.terranature.org/riversStreams.htm (Accessed September 22, 2005).

Treasury, New Zealand. *New Zealand Economic and Financial Overview 2010.* Wellington: New Zealand Treasury, 2010.

Williams, David. "Warning to Councils by Carter." *The Press,* April 29, 2010.

Wilson, Roger. *From Manapouri to Aramoana. The Battle for New Zealand's Environment.* Auckland: Earthworks Press, 1982.

Worster, Donald. *Nature's Economy: A History of Ecological Ideas.* Cambridge: Cambridge University Press, 1994.

Conclusion: Context Is All[1]: The Many Environmentalisms

Adela Alfonsi, Sherilyn MacGregor, and Timothy Doyle

Together, the contributions in this book demonstrate that environmentalism is a kaleidoscope: a continually changing pattern of shapes and shades. The 24 case studies presented here highlight the plurality of environmentalisms, the diversity of frames, and the many different approaches, actors, resource bases, and networks that make them tick. Each case study explores the content-specific responses of greens to the challenges presented by the hegemonic neoliberal development paradigm, along with their different track records and potential for success in the future. While there may be common themes and resonances across the chapters, the environmentalisms presented here have their own unique stories. These vary based on their geographical and geopolitical contexts, their own sociopolitical and cultural particularities, their experience of colonialism (both sides), the nature, type, and speed of their economic development, and the nature of their current relationship with the global economy and with modern cyber and social technologies. The overarching message of this collection is that environmental movements are shaped by their own social, political, cultural, historical, and institutional contexts. In this concluding chapter, we draw out what we see as the central themes and the points of convergence and divergence that have emerged in the course of this project.

Place Matters

The first theme is that it is impossible to think about environmentalisms in a vacuum. National and regional contexts remain important for organizing and understanding the many shades of environmentalism around the world, but so too are the more fundamental contexts of affluence and poverty that largely determine the ideas and the repertoires of green movements. We make so much at the beginning of Volume one about a world divided between the haves and the have-nots because this dichotomy—though far from perfect—emerges time and time again in the chapters that make up this collection of stories. Granted, the division is raw and overly simplistic. The construction of the global South, it can be argued, is too broad-brush. One could point out that there are profound differences, for example, between the "fourth world societies" of some Australian and Nigerian indigenous peoples who have been uprooted from their lands and communities and the experiences of peoples inhabiting the rapidly industrializing countries of China and South Korea. But we think this basic if imperfect line (drawn in the sand) between geopolitical and geo-cultural realities is a valuable heuristic tool for studying environmental politics in the face of a neoliberal orthodoxy that says we are all equal before the market that knows only Earth as corporation, and citizen as consumer. Environmentalists too, for that matter, often forget to celebrate diversity; there is a tendency to accept stereotypes and categories that come from the affluent world of the global North. To want to regard all the Earth's people as "one" is understandable—and sometimes laudable—but when this ostensible oneness—this "Universal We"—rides roughshod over regional and other differences, then this position outlasts its usefulness. Indeed, the cry of uniform experience is often a way of disciplining *other* experiences to conform to, and be disciplined by, an Anglospheric, often white and male, minority. What has become blatantly apparent for us in the process of compiling this work is that place, class, caste, gender, time and space really matter, not just for how environmentalism is understood and practiced, but for all aspects of political life. Where you are born and where you live will usually determine what work you are engaged in, what you eat, how healthy you will be, how long you will live for, and how you will die.

Simply put, there still exist profound differences in wealth on this planet. 80 to 85 percent of the world's resources are still consumed by 10 to 15 percent of the people. In fact, if anything, the global gap between the rich and the poor is accelerating in this current version of economic globalization. Obviously, those green movements emerging in the wealthy North will focus on different themes, ideas and issues, and do green politics very differently. Whereas in the North, green movements, traditionally at least, have been more interested in *lifestyle* issues, in the South, activists engage in *livelihood* issues. But, in more recent times, the social and environmental justice concerns of green movements in the global South—for example, concerns over who gets access to scarce resources, and how are these divided—are of more interest to Northern environmental movements. In social movements, power does not just flow one way. Although there can be no doubt that the North—even in movement terms—is still extremely powerful in agenda-setting and control of movement funds—there is mounting evidence that Southern concerns are exerting more influence on agendas in the North. As green movements from the affluent world begin to address the concerns of people in the global South, there is a growing realization that many environmental problems are inevitably shared. And yet there is also ample evidence of forms of *green imperialism*: the construction of postindustrial, affluent world issues like climate change as *the most important* green issue of our time. In the context of the global North, we have suggested that the acceptance of this dominant agenda may be leading to the depoliticization of environmentalism. Not so in the global South, where although often being pressured by Northern movements to join in the climate narrative, the struggle for environmental justice and sustainable livelihoods—the need for fresh and clean water; the need for, and ownership of food; access to energy and clean air—remain deeply political.

Alongside recognition of the global asymmetries of power and resources between North and South, it is also important to acknowledge the other fundamental division in the world: that between men and women. The socially constructed gender roles that shape daily life in most places on the planet and that have women shouldering a disproportionate share of life-sustaining labor also play a significant part in shaping the politics of environmental movements. However, the relative "place" of women and men is a factor that often gets overlooked in academic work in our field. The "Universal We" obscures over more than just place of residence. Admittedly, most of the chapters in the collection do not pay much attention to gender dynamics within the environmentalisms in question. However, two chapters—by Janet Muthuki on Kenya and by Woojoo Chang on South Korea—have made a deliberate choice to foreground the role of women and women's gender-based concerns and socially determined duties (these are very much related) in their stories. In both chapters, we are reminded of what feminist environmental scholars have been arguing for decades: women's unpaid caring labor (which is a free subsidy to capital at the same time as making politics in the public sphere possible) brings them into closer contact with environmental processes, and often gives them greater knowledge about environmental problems, than men.[2] As a result, women have taken on leadership roles in green movements in the North and the South. From the Green Belt Movement to food co-ops to campaigns against pesticides, they have brought their own approach to politics (often informed by maternal values or ethics of care) and been empowered in the process. And yet, men still tend to dominate the sphere of environmentalism and gender relations are seldom employed as an analytical tool in academic inquiries into its many dimensions (unless by ecofeminist scholars working on the margins). This is problematic because, as Muthuki writes, "gender relations are crucial to understanding [environmental issues] because gender shapes the opportunities and constraints that women and men face in securing viable livelihoods and strong communities across cultural, political, economic and ecological settings."

With or against the State?

Another common theme underlying the environmental movements represented here has been the struggle with the neoliberal development paradigm and the environmental consequences of privileging economic growth over humanitarian and environmental concerns. In some cases it would appear that, while not becoming irrelevant, environmental movements have been defanged, marginalized, or even co-opted by the alliance of capitalist business interests and their political supporters in governments. A key concern for some of our authors is whether environmental movements have lost their power to inspire and mobilize. While there is a diversity of origins, strategies, actors, and goals in environmental movements across the world, the picture that emerges in the English-speaking liberal democracies is one where the mainstream environmental movements are more disposed to working cooperatively with governments in the attainment of environmental goals, within the context of market-based deregulated contexts if need be, even if this is not the preferential option of the environmental movements concerned. Several of our authors suggest that possibly as a result, the political clout of the environmental movement has been eroded since the 1990s. For example, Laurie Adkin argues that Canadian environmentalism (notwithstanding a more recent shift toward more radical and confrontational approaches) has generally adopted a reformist "insider" approach, working collaboratively to influence government policy, and supporting market-driven approaches to resolving environmental issues such as reducing greenhouse gas emissions. However, given the current Conservative government's apparent lack of willingness to engage even with market-based approaches, the reader is left wondering whether the environmental movement in Canada will be able to meet the very pressing environmental challenges ahead of it.

The question of whether environmental movements have been defanged and need to be reoriented is taken up explicitly by Robert Brulle, who argues that environmental movements in the United States have become increasingly marginalized since the 1990s, despite the fact that they have a long and rich history, a strong organizational base, and strong public support. Thus, adjusting to the neoliberal consensus has led environmental movements in the United States to moderate their discourse and become highly professionalized and elitist, rather than having a transformative and grassroots approach. Specifically, Brulle identifies mainstream environmental discursive frames as being overwhelmingly reformist, and lacking the capacity to critique the discourse of neoliberalism or to provide a progressive, transformative alternative vision that can mobilize and inspire. The rise of what he calls "professional environmental advocacy" or "protest businesses" has led to a hierarchical, competitive, managerial form of social movement that prioritizes predictability and "reasonableness" over inspiration and grassroots participation, effectively condemning environmental movements to irrelevancy.

This rather bleak perspective has its echoes elsewhere, even though the institutional setting or context might be different. In the case of New Zealand, Ton Bührs demonstrates how, as a result of successive governments' commitment to neoliberal ideology, leading to the rolling back of the state, there is no central site for debating environmental policy anymore. Instead, environmentalism takes the form of ad hoc, reactive movements played out in local and regional ambits, which is where authority has been devolved. Bührs does not see New Zealand environmental movements as being in a state of stagnation and paralysis, but he does speak of a mood of disorientation; and given the urgency of environmental issues, it is also clear that measures taken to date have not been adequate. In other cases, governmental co-option of environmental movements is of a very different nature. In the case of Venezuela, for example, María Pilar García-Guadilla unambiguously states that "there is not a social movement with transformative potential." Her contribution reveals the very different context applicable to environmental movements in authoritarian or newly democratic states. Here, under the slogan of "Socialism of the 21st Century," the Chávez government presents itself as a valid alternative to capitalism. Its priority is the fight against poverty, for which purpose it is implementing a revolutionary social distributive policy, which falls into the contradiction of having to reconcile redistributive policies whose objectives are solidarity and social justice with a predatory resource extractivist developmental model. Venezuela's anti-imperialist discourse has also led the government to use oil as a weapon in a broader agenda to establish its own regional hegemony, thus providing added ideological legitimacy for the adoption of extractivist policies. In this way, development plans and policies, which purport to be able to carry out "ecological mining," have produced serious conflicts, abuses of human rights, principally among indigenous peoples, as well as serious environmental impacts. Hence the social justice rhetoric comes into fatal contradiction with government economic policy.

The discourse of social justice and sovereignty adopted by the "New Left" government of Hugo Chávez has served to divide and co-opt environmental movements in Venezuela. On the one hand, through its social redistribution policies, the government has set up its own ideologically allied networks and mechanisms for social movements to receive and distribute resources and funding. Co-option takes place quite readily by means of its populist policies and ideological rhetoric. If that fails, there is the option of criminalization of radical socio-environmental movements. Among activists themselves, there is division between those who believe environmental reform will come by means of achieving social revolutionary goals, and those who critique that the logic of the extractivist economy is inherently contradictory to the principles of social justice, and is possibly leading to violations of Venezuela's Constitution. While in Venezuela social movements and activism are harnessed to an explicitly ideological end, in Russia the opposite dynamic seems to be the case. While equally committed to resource extractivist developmental policies, the Russian government has, according to Oleg Yanitsky, increasingly showed its authoritarian face to democratic critiques. Under the Socialist regime, Yanitsky shows a Russian environmental movement in its origins heavily rooted in its intellectual milieu, and therefore of a scientific and professionalized culture, but which still managed to achieve public resonance and mass appeal, motivated by principles of self-actualization and civic responsibility. Under the increasingly hostile conditions of the 1990s, public participation and debate on environmental issues has been increasingly marginalized, leading Yanitsky to describe Russia in this market economy phase as a "democratic façade." The current state of environmentalism in Russia seems to be one of alienation and suspicion. Yanitsky describes the movements as supervised, in competition with each other for resources, and increasingly bureaucratized. Currently, the most effective work seems to involve heavily internationalized ENGOs, working with the resources of transnational ENGOs such as the WWF, usually in the context of managing the sustainable extraction and exploitation of natural resources. On the other hand, the "localists" (who defend local interests and environments) are more constrained and marginalized because of the power of business interests and governmental deafness to their claims.

In Singapore, on the other hand, we have a classic example of environmental movements shaping their campaigns to fit the state's economic development strategies. Although there are currents of opposition, and signs of a growing interest in environmental justice issues, the main strands of environmentalism in Singapore have been supported by the very urbanization policies that cause severe stress on the fragile ecosystem of the small city-state. As Sofiah Jamil explains, this close relationship to the state has led to weak levels of public support for environmentalism, with green groups having trouble making the case for how concern for the environment is relevant to poor and marginalized communities.

In all these cases, governments are fundamentally wedded to the causes of "modernization" and economic growth. But relations between environmental movements and states, and the degree of politicization of their relationship, are quite different depending on the context. In the mature liberal democracies of the United States, Canada, or New Zealand, mainstream ENGOs are working principally within a framework of strategic—albeit not always comfortable—cooperation with governments. The fact that they are involved in scientific and technical work displaces the debate away from an ideological one, leading to the depoliticization of their work. In Venezuela, environmental organizations and activists are actively marshaled to give support to a specific revolutionary project, and a certain form of politicization is precisely the point. In Russia, environmental movements and concerned citizens are actively discouraged from participation in environmental debates. In Singapore the debates are in many ways determined by the state's developmental agenda.

Cultural Frames

While complex processes of accommodation, negotiation, and resistance with or against the state have been the focus of some of our authors, others have been more concerned with the frames underpinning environmental movements. John Agbonifo argues that the study of Western social movements is ill-placed to explain "third world" environmentalism, because it is not able to engage with the metaphysical aspects of some environmentalisms. He makes a case for "varieties of environmentalisms" in order to guide against the common tendency to "substitute a part for the whole." His chapter reminds us that secular thinking will always struggle to understand certain strands of African environmentalism if it cannot accommodate the existence of metaphysical worldviews. Agbonifo acknowledges that Nigeria's environmental groups are not homogenous. Nevertheless, he describes the Ogoni struggle against the combined forces of Shell and the Nigerian state as not framed by Western materialist and utilitarian perspectives, but rather having a "cultural repertoire as frame." In this case, we are asked to consider a case study in which the local strategies of resistance and opposition are informed by a metaphysical worldview in which the lines between sacred and profane, natural and supernatural are blurred, and in which all things, animate and inanimate, exist as a related, interdependent whole. "Struggles over land and environmental resources," he affirms, "are simultaneously struggles over cultural meanings." The Ogoni collective action frame, therefore, is presented as a spiritually derived action frame. It challenges the "political opportunity" approach as being too limited because this approach elides the degree to which religious beliefs and convictions can embolden community resolve in a context where the political situation, objectively speaking, is not favorable to mobilization. The spiritual derivation also defines who the movement leaders might be, and how their collective demands are expressed (in this case, in the context of "the rightful demands of a landlord"). A similar principle is observable in South Korea, where, as Woojoo Chang explains, the religiously based philosophy of "Saengmyung" (life) is deployed as one of the discourses underpinning the environmental movements described by Chang. Like in Nigeria, this is a holistic discourse based on native philosophical concepts that are traditional, well understood, and familiar. They describe the interrelatedness of all things, animate and inanimate, and invest the whole with a divine force, so that all things are equally morally valuable and therefore must equally be respected for spiritual and metaphysical reasons. Another interesting perspective of how a holistic framework informs views on the environment can be seen in Mesake Koroi's deeply poignant and personal viewpoint from Fiji.

The case of Israel represents a different variant on the theme of culturally specific framings of environmentalisms. The paradox that Alon Tal presents us with is how it is that Israel, which has all the cultural, political, and institutional elements that should lead to environmentally excellent outcomes, has instead been failing to meet the challenges on a number of fronts. Tal points to Israel's extraordinary population density as one of the keys to deciphering this conundrum. The twin issues of the immigration of Jewish people to Israel and state policies supporting and fostering high birthrates are both issues of enormous political and religious sensitivity. Here, we are dealing with the "populate or perish" mentality, as well as the trauma of historical memory arising from the experience of the Holocaust. Another related issue is that of Israel's unique political culture, where its fractious system of small parties and coalition governments, and the overarching concern about security, has important implications for how environmental issues find their place in political debate.

Cultural issues are not specifically referred to by Sylvie Ollitrault in her chapter on France; and yet it is hard to avoid the conclusion that the famously French resistance to the undermining of national culture by foreign, principally Anglo-American influence, is at the root of at least some of the environmentalisms that she describes. While she identifies a complex interplay of global and local factors that underplay the French environmental movements, one of the strikingly domestic ones is the importance of a tradition of civil disobedience, an anarchist suspicion of the state, and a long history of "localized struggles." This becomes most strikingly apparent in the rural activism that defends local farmers and traditional forms of agriculture. The importance of a romanticized sense of local culture, heritage, and local cultural identity is also discernible in the background to some other environmental movements, such as in the cases of Liam Leonard's chapter on Ireland (i.e., the romanticized rural motif) and Cassandra Star's chapter on Australian environmentalisms (i.e., the "preservation of wilderness" approach).

These struggles for preservation or protection of often idealized local cultures is far removed from the narrative of cultural disruption that characterizes the colonized world. Janet Muthuki, for example, demonstrates how traditional African social structures, which permitted a gendered but balanced method of sustainably managing natural resources, was destroyed by colonialism, capitalism, and the current modernist development paradigm. In her assessment, colonialism has not only entrenched but also exacerbated patriarchy. Kenyan women are the primary managers of natural resources such as firewood, but they are no longer able to be effective managers because of the disruption caused to traditional systems. They are also the most vulnerable and disproportionately affected by the degradation of these resources. The Green Belt Movement (GBM) therefore puts the gender issue at the forefront, because gender is at the center of the nature of the environmental problem being faced in Kenya, as well as being the key to its solution. The holistic approach to grassroots activism that is fostered by the GBM's leader, the late Wangari Maathai, trains and empowers women, provides leadership and advocacy skills, and ultimately strengthens democratic reform and good governance. Maathai was not arguing that men should be marginalized—quite the opposite—but she did identify Kenyan women as being the nexus between defending human rights, building democracy, and sustainable development.

The issue of the intimate relationship between human rights and environmentalism is also pertinent to Mohamed Salih's discussion of how green issues are framed in Iraq. Here we are presented with an uncomfortable paradox. On the one hand, the case of the Marsh Arabs is a case of a people literally fighting for their survival. As Salih clearly points out, there is no way of separating demands for land, human rights, and self-determination from environmental rights in this case. Yet while Marsh Arabs took up arms supporting the coalition in the Iraq War of 2003, the major international environmental movements were stridently antiwar on environmental grounds, mostly based on the politics of climate change and its relationship with oil. This points to the danger of hegemonic environmental discourses from the minority world that are out of touch with the existential issues of the majority world.

Ashok Swain's exploration of forest management in India provides another example, similar to Agbonifo's, of the misapplication of Western frames to developing countries. Here, the international community establishes norms for conservation and preservation without understanding in what way these norms will be applied, or what the actual impacts on indigenous communities will be. In this case, the Indian government's ineffective adoption of Western practices of establishing protected area reserves has led to the perverse consequences of mass displacements, destruction of local cultures, and economic catastrophe for local communities. The affirmation that "global land conservation efforts are creating a growing class of invisible refugees in the third world" is disturbing indeed. In this case, what is striking is that for the indigenous peoples of Orissa, mobilization in defense of their interests has been galvanized not due to the work of traditional NGOs, and especially not because of relationships between international ENGOs and local ones, but because of the unexpected effects of external actors upon local populations. In this case, the presence of Maoists and Christian and Hindu preachers, both of whom, working in the pursuit of their own agendas, have unwittingly contributed to the formation of a new form of identity politics and different sorts of social bonds and networks. The Maoists have also provided the tribal populations of inland Orissa with a galvanizing, revolutionary frame. In this way, the appearance of unexpected actors interacting within a particular cultural, social, political, and institutional context has created some surprising results.

In a very different context, the role of international ENGOs also appears in Claudia Sepúlveda and Pablo Villarroel's chapter on Chile. These authors argue that as a result of the entrenchment of neoliberal economic developmentalism under the Pinochet regime, subsequently continued by democratic governments, Chilean environmental discourse has been channeled into expert, professional, and technocratic approaches, and robbed of its political content. Given the glaring failures of environmental policies, the current mood is of great interest in environmental issues, but the central problem is recovering a democratic space for public participation in debate and policy-making. Ironically, though, the possibility of forging this space for public participation has been hindered in Chile by the presence of large foreign NGOs, with their own agendas and approaches "disconnected from native perspectives," which have displaced local conservationist NGOs.

Repertoires

Many contributors to this collection have pointed to the necessity for a reinvigoration, emergence, or re-emergence of civil society, increased democratization and grassroots participation in environmental movements, and for an increased focus on justice and progressive, socially transformative action. Robert Brulle's call for expert knowledge to be placed at the service of and to be accountable to broader publics is one that would be shared by all authors in this book. But again, specific context is very important. Countries like the United States and the United Kingdom, with their long and rich history of social mobilization and with a great capacity to play an insider role in the policy process, are in a different position to places such as Turkey and Brazil, two countries transitioning from dictatorship to democracy in the 1980s. In this very different political and institutional setting, the use of legal repertoires to promote environmental causes has played a key part in securing environmental and human rights protections. In the case of Turkey, it is striking how newly emerged citizens' groups have been able to use legal processes, both within Turkish administrative courts and in the international legal arena. This has been possible, according to Gürcan Koçan and Ahmet Öncü, due to Turkey's international commitments according to the bilateral and multilateral accords that it has signed. In Brazil, campaigners trying to resist the construction of the Belo Monte campaign also applied institutional and legal approaches, principally on the grounds of Brazil's constitutional obligations to defend indigenous rights. In its early days, when a new, democratic constitution and legal framework was being developed, this strategy provided some striking successes, as described by

Benito Cao. Sarath Wimalabadara Kotagama's work on Sri Lanka also shows good examples of successful recourse to legal and institutional repertoires.

In Latin America, the capacity of large agribusiness, mining interests and the hydroelectric industry to countermobilize, however, has been very considerable. In light of the dispossession, public health disasters, and environmental devastation caused by failures of the state in Latin America, Mauricio Berger highlights the emergence of a more radical, confrontational, creative, community-based form of political action, the environmental rights networks. This is a different kind of political institution, made necessary because the state itself, co-opted by international corporative interests, is the key perpetrator of serious injustices and gross violations of human and environmental rights. The Brazilian, Mexican, and Argentine environmental justice networks provide a political space independent of the state, which, captive to capitalist forces, has lost its legitimacy and cannot represent its citizens. The networks are self-organized, inclusive, nonhierarchical with an open, autonomous nature and no formal structure. Clearly, this kind of approach has its inherent challenges. Still, Berger describes an organic process of "networking from below" that "unfolds the functions that correspond to the state but without the state." As a result, affected communities are able to make problems visible and organize defense, informational and educational activities, such as community monitoring and surveillance, conducting "popular" environmental impact assessments, compiling data and local knowledge, and most importantly, providing their own forms of political representation.

This is a repertoire that is very different from what is observable in some postindustrial European countries like Germany and France, where people may have internalized the idea of bringing about environmental change by being a "green" consumer, and exercise resistance and protest in more individualistic ways. As Jochen Roose observes, for example, environmentalism has become entrenched in the everyday behavior of Germans, who seem highly content to make personal sacrifices and to pay higher taxes to mitigate environmental harms. "Nearly 90 percent of the population," he says "thinks that ordinary people can help protect the climate (as the currently most pressing environmental problem) through their day-to-day behavior." Here, as in France to some extent, environmentalism operates at the level of cultural identity in a way that renders political movements and campaigns almost invisible. Institutionalization and individualization are classic qualities of environmentalism in many parts of the global North.

In other parts, collaboration and networking remain vital. Different sorts of possibilities are provided by international collaboration across international and local networks, and point to the possibility of an emerging *glocal* civil society. In his chapter, Christopher Rootes alludes to the fact that greater international collaboration may be the future for at least some of the British ENGOs. Such collaborations can take many shapes and forms. In China, the campaign against Apple examined by Da Mao represents an excellent example of how a fortuitous feedback loop can emerge between social and environmental organizations from different countries, in this case, China and the United States. The mobilization against the Belo Monte dam complex in Brazil, examined by Cao, provides another interesting illustration of the potential for the creation of complex glocal networks. His chapter shows how indigenous peoples and environmental activists (local, national, and international) as well as celebrities can be brought together around specific issues, forming a glocal network, and deploying a diversity of repertoires of action. Yet, the multiplicity of identities remains largely intact, and both the national (Brazilian) and local (Amazonian) context remains central to the struggle against the dam complex.

Conclusion

These examples of glocal networks suggest that even as environmental and social movements become increasingly networked internationally, the significance of place and local context cannot be underestimated. Indeed, the importance of focusing on contextual differences and the necessity of avoiding a unitary, universalizing narrative framework are two of the main insights from this edited volume—insights that we should do well to remember as environmental movements continue to network into the 21st century.

"Old" borders and boundaries that were meant to fade away into a new global order are not outdated at all, but are often the sites where the ravages of corporate and state-led globalization are challenged most vividly. In most parts of the planet, capitalism has not yet won but is currently being challenged by people's movements, many of which label themselves as *environmental.* Whereas in the global North, by contrast, environmental movements rarely use the "c" word, sometimes treating capitalism as an ally or partner in the move to "green" the processes of advanced industrialization from within market frameworks. But this is just one difference. As this collection demonstrates so clearly, there are many differences to be explored and analyzed, and it is our hope that scholars and activists in the field of environmental politics and social movement research will have gained new frameworks for thinking and a renewed sense of inspiration because of our efforts.

It has been the aim of these volumes to disrupt dominant conceptions of environmentalisms and to decenter the field by gathering together scholars from parts of the world that do not normally get included in such conversations. The fascinating chapters in our collection have been contributed by scholars who live, work, and sometimes engage in green political activism, in the countries about which they write. They offer vivid snapshots of what is happening on the ground and how local events are shaped by global forces and flows. We are particularly fortunate to have been able to include insights of majority world scholars whose case studies (of countries like Nigeria, Kenya, Fiji, and Iraq) are invaluable in shedding light on the blind spots that result from a global division of labor in academic knowledge production. The translation of words and ideas into common languages (not only English but also minority world academic discipline–speak) in order to be understood by a wide audience has been a rewarding yet challenging endeavor. It is not something that happens automatically in this business, which is part of the reason our field is so dominated by people like us. Inspired by the pioneering work of Arlene Tickner and Ole Waever, who have argued for a "worlding" of international relations "beyond the West,"[3] we have sought to challenge the exclusivity and asymmetry that shape the production of knowledge in environmental movement scholarship. We hope that, consistent with the "worlding" process, this collection serves to give voice to the many environmentalisms and ultimately contributes to "greater self-reflexivity"[4] among all of us who aspire to a truly global green perspective.

Notes

1. Atwood, *The Handmaid's Tale,* 144.
2. MacGregor, *Beyond Mothering Earth.*
3. Waever and Tickner, "Geocultural Epistemologies."
4. Tickner and Waever, "Conclusion: Worlding Where the West Once Was."

Bibliography

Atwood, M. *The Handmaid's Tale.* Toronto: McClelland and Stewart, 1985.

MacGregor, S. *Beyond Mothering Earth: Ecological Citizenship and the Politics of Care.* Vancouver: University of British Columbia Press, 2006.

Tickner, A. B., and O. Waever. "Conclusion: Worlding Where the West Once Was." In *International Relations Scholarship around the World: Worlding beyond the*

West, edited by A. B. Tickner and O. Waever, 328–41. London and New York: Routledge, 2009.

Waever, O., and A. B. Tickner. "Geocultural Epistemologies." In *International Relations Scholarship around the World: Worlding beyond the West,* edited by A. B. Tickner and O. Waever, 1–31. London and New York: Routledge, 2009.

Index

Page numbers in *italics* indicate tables.

"Activities and Issues of the Movement," Germany, 32
Adkin, Laurie, 362
Advocacy science/scientists, in Russia, 133–34
African environmentalism, 365–66
Agbonifo, John, 365
Agri-Food and Veterinary Authority, Singapore, 264
Agronomist Institute, France, 98
Air pollution: Sri Lanka, 190; *Waldsterben* (death of all forests in Germany) and, 30, *See also* Pollution
Alfonsi, Adela, 359–75
Al-Mawaddah mosque, Singapore, 275
Ambalavanar, V., 185
Anarchist view, of green activists, 4, 10
Anderson, Ed, 306
Ang, Wilson, 266
Animal Concerns Research and Education Society (ACRES), Singapore, 262–66
Animal rights ethics, 144–45
Animal rights movement: China, 218–21; France, 113; Turkey, 149, 153–55
An Taisce (the Irish Heritage Trust), 71
Anthropocentrism, 12, 142–45, *142*, 156–57, 310
Anti-developmentalism discourse in South Korea, 243–46; compared to *Saengmyung* (life), 252, *253*
Anti-Dong River Dam Campaign, South Korea, 241
Antiglobalization: France, 106–7; Russia, 128–29
Antinuclear energy movements/protests: France, 95, 97, 109–11; Germany, 27–31, *29*, 33–34, 37; historical background, 110–11; Ireland, 82, 84, 86, 87; New Zealand, 344; South Korea, 240; Turkey, 152–53; United Kingdom, 8
Anti-Saemangeum Reclamation Project Campaign, South Korea, 241, 242
Apia Convention on the Conservation of Nature in the South Pacific, 310
Apotheker, Arnaud, 109
Appiko forest movement, India, 171
Apple, anti-Apple supply chain campaign in China, 14–15, 224–26, 371
Architectural heritage, 71–72

Argentina, 370
Ariés, Paul, 112
Asia: overview, 12–15, *See also* China; India; Odisha (Orissa), India's protest against protected areas; Singapore; South Korea; Sri Lanka
Associations pour le maintien d'une agriculture paysanne (AMAP), France, 113
Attaturk, Mustafa Kemal, 11
Attitude research, 24–25
Australia, 5, 16, 287–302, 360; Australian Conservation Foundation (ACF), 289, 292; Australian Student Environment Network (ASEN), 288; Australian Youth Climate Coalition (AYCC), 288–89; Bayside Climate Change Action Group, 288; community-based groups, 288; cooperation and the movement, 294–97; current shape of the movement, 288–89; ecologically sustainable development (ESD), 290–91; environmental movement organizations (EMOs), 288–89; environmental nongovernmental organizations (ENGOs), 292–94; failures of the movement, 287–88; "frontier conservationist/preservationist" vs. modern green movement, 288, 299; grassroots groups, 288; Greenpeace, 288, 289, 293; green politics, 297–99; Greens Party, 289, 290; Hawke, Robert era, 287, 290, 293, 295, 298; historical background, 289–92; Howard, John era, 287, 292–94, 297, 298, 299; indigenous movement, 296–97; Jabiluka uranium mine campaign, 291; Lake Pedder campaign, 289–90; membership in environmental movement, 288; Natural Heritage Trust, 293; overview, 16, 287–88; Resource Assessment Commission (RAC), 293; social justice movement, 296–97; Star, Cassandra, 16, 287–99, 367; tactics, political strategies, and impact of the movement, 292–94; World Wildlife Fund in, 289; youth groups, 288
Australian Conservation Foundation (ACF), 289, 292
Australian Student Environment Network (ASEN), 288
Australian Youth Climate Coalition, 288–89
Authoritarian regimes, 7; China, 13–15; overthrow movements of by social media, 352; Russia, 10–11, 364; Singapore, 15–16; transitioning to democracy, 369, *See also* headings under specific countries

"Backyard Furnace" campaign, China, 210
Bayside Climate Change Action Group, Australia, 288
Beijing News, 227
Belo Monte campaign, Brazil, 369
Bentham, J., *142*, 143
Berger, Mauricio, 370
Berlin Wall, fall of, 28, 30, 32
Biological diversity, loss of in Sri Lanka, 190–91
Bioscientism ("Nature knows best"), Russia, 125
Birdlife International network, 105
Blair, Tony, 57
"Blue-Greens," New Zealand, 339, 340
Bonding circles, 168
Borders, types of, 6

Boreal (old-growth) forests, Russia, 135
Boreyko, Vladimir, 129
Borneo, 260
Bové, José, 106–9
Brazil, 369–70, 371
Brazilian Amazon, 371
Bring-Your-Own Bag/Receptacles, Singapore, 271
Britain. *See* United Kingdom (UK)
British Malaya, 260
British Trust for Conservation Volunteers, 48
Brown, Bob, 290
Brulle, Robert, 363, 369
Bührs, Ton, 17, 331–55, 363
Bukit Brown Cemetery (Singapore), 269
Bundesnetzwerk Bürgerinitiativen Umweltschutz BBU (Federal Network of Environmental Citizen Initiatives), Germany, 26
Bund für Umwelt und Naturschutz Deutschland BUND (Federation for Environment and Nature Protection), Germany, 26
Burns, Alan, *Fiji*, 307–8
Burren Action Group and the Mullaghmore tourism dispute, Ireland, 72–73, 82, 85
"Buy Kiwi-Made" campaign, New Zealand, 348

Callicott, J.B., *142*, 145
Campaign against Climate Change (CCC), UK, 61
Campaign to Protect Rural England, UK, 58–59
Camp for Climate Action, UK, 62
Canada, 5, 362, 365
Cao, Benito, 369–70, 371
Cap 21, France, 104
Capitalism, 3–4; global environmentalism and, 371
Captain Cook, 307
Carbon footprint, 310, *See also* Climate change
Carbon tax, New Zealand, 349
Carnsore Point nuclear protests, Ireland, 82, 84, 86, 87
Carson, Rachel, 112; *Silent Spring*, 49, 305
Categories for environmentalisms. *See* Environmentalisms
Celtic Tiger boom, Ireland, 75, 79, 87, 90
Chamber of Forest Engineers, Turkey, 147
Chamber of Mechanical Engineers, Turkey, 152
Chang, Woojoo, 15, 235–54, 361, 366
Chávez government, Venezuela, 363–64
Chemical fertilizers, China, 209
Chernobyl, Russia, nuclear accident, 27, 28, 37
Children's diseases, South Korea, 242
Chile, 368–69
Chilika Lake protest, Odisha, India, 173–74
China, 7, 13–15, 207–34, 360; anti-Apple supply chain campaign, 14–15, 224–26, 371; anti-dam campaigns, 221–24; anti-incinerator movement, 217–18; "Backyard Furnace Campaign," 210; *Beijing News*, 227; "China Water Pollution Map" (IPE), 224; Chongquing Green Volunteer Association (CGV), 215; citizen group formation, 218; Communist Party takeover (1949), 209; deforestation, 209, 220; Environmental Protection Administration, 211; environmental protection becomes government responsibility, 208–10;

environmental protection bureaus (EPBs), 214; *Exploring Yunnan Snub-Nose Monkey* (Xi), 220; Feng Yongfeng, 213; "Four Pests Campaign," 210; Friends of Nature (FON), 216, 217, 219, 220, 221, 225, 227; Global Village of Beijing (GVB), 217, 227; grassroots mass movements in, 217–18; Green Beagle, 218, 225; Green Camp, 215; Green Choice Alliance, 225; "Green Choice" campaign, 224–25; Green Earth Volunteers (GEV), 217, 223, 227; Guanting Reservoir DDT poisoning, 208–9; "Humans Do Not Have to Have Reverence for Nature," 227; individual actions, 214–15; Institute of Environment and Development (IED), 216; Institute of Public and Environmental Affairs (IPE), 218, 224; International Crane Foundation (ICF), 215, 219; Leading Group for Environmental Protection, 211; Liang Congjie, 227; Liao Xiaoyi, 227; Li Lailai, 216; Liu Detian, 216; Ma Jun, 224–25; Mao Da, 14, 207–29; ministry of environmental protection (MEP), 211; National Conference on Environmental Protection, 210, 211; Natural Resource Defence Council, 225–26; Nujiang River hydropower project, 222; overview and historical background, 13–15, 207–10, 228–29; paraxylene (PX) chemical plants protests, 217, 218; private-interest organizations, 215–16, 218; public-interest organizations, 215–18; Qu Geping, 211–12, 215; red-crown crane, 212, 219; reemergence of civil society/environmental nongovernmental organizations (ENGOs), 213–29; "Reform and Opening" policy, 212, 214; reverence for nature/environmental ethics public debate, 226–28; rivers and dam construction movements, 221–24; Sanmenxia dam on the Yellow River, 221; Saunders Gull Conservation Society of Panjin City (SGCSPJ), 216, 219; *Should the Three Gorges Dam Be Constructed: A Collection of Scholars Debate* (Dai), 222; snub-nose monkey, 219–20; State Environmental Protection Administration (SEPA), 211–12, 214; Tang Xiyang, 212–13, 215; Three Gorges dam on the Yangtze, 221–22; "Three Parallel Rivers of Yunnan Protected Areas" (World Natural Heritage), 222; Tibetan Antelope protection, 217; Wang Yongchen, 223, 227; Wen Jiabao, 222; wildlife conservation movement, 218–21; Wild Yak Troupe, 220–21; Wu Dengming, 212, 215; Xiaonanhai Hydropower Station (Yangtze River), 223; Xi Zhinong, 220; Xu Xiujuan, 212; Yancheng Nature Reserve (Jiangsu Province), 212; Yang Xin, 220–21; Zhalong Crane Nature Reserve (Heilongjiang Province), 212; Zhou Enlai, 208, 210

China Central Radio Station, 223

China Central Television, 220

"China Water Pollution Map" (IPE), 224

Chipko (Hug the Trees) movement in Uttar Pradesh, India, 170–71
Chirac, Jacques, 95
Chlorofluorocarbons (CFCs), 311
Chongquing Green Volunteer Association (CGV), China, 215
Choong, Olivia, 267
Christianity: in Fiji, 311, 314; in Odisha, India, 178, 368
Chun Do Hwan dictatorship, South Korea, 236
"Cities of Europe" project, 134
Citizen's Coalition for Economic Justice, South Korea, 239
Civil disobedience, France, 107–8, 111–12
Civil initiatives, Russia, 126, 130, 133
Civil society, 369; Australia, 298–99; destruction of in WWII, 26; global, 2; "independent, third, or voluntary" sector, 239; Ireland, 83; reemergence of in China/environmental nongovernmental organizations (ENGOs), 213–29; Russia, 118, 121–22; Singapore, 269; South Korea, 239–40, 244–46; Turkey, 139, 157
Clark, Helen, 348
Class analysis, 173–74
"Clean and green," New Zealand, 331, 345
Climate Action Network, France, 111, 113
Climate change, 3, 361; Fiji, 309–10, 313; France, 102; New Zealand, 348, 349, 354; North-South dichotomy, 5; post-politics of, 6; United Kingdom, 8, 60–63
Climate Justice Now, France, 113
Coal power plant program, Sri Lanka, 199–200
Cohn-Bendit, Dany, 101
Commission for the Environment, New Zealand, 341, 344
Commoner, Barry, 112
Commoning, Turkey, 147–50
Commons, Turkey, 147–50
Commons Preservation Society, UK, 47
Communist Party: France, 99; takeover in China (1949), 209; USSR, 10, 117–20
Communitarians (moderate red-green eco-patriots in Russia), 132
Conceptual borders, 6
Conféderation paysanne (Peasant Farmers' Union), France, 107–8, 109, 114
Conference on the Human Environment in the South Pacific, 310
Consequentialism, 141, *142*, 143, 156
Conservationists, Russia, 125–28, 134–35
Conservation of Nature in Singapore, 264
Continuums, 2
Convention for the Protection of the Natural Resources and Environment of the South Pacific, 310
Cooperatives: *Hansalim* cooperatives in South Korea, 248–52, 361; International Co-operative Alliance (ICA), 249; origin of, 249; utopian socialism and, 249
Costs, of environmentally friendly products, 35–36
Council for Environmental Conservation (CoEnCo), UK, 48, 57
Council for the Preservation of Rural England (CPRE), UK, 47–48, 58, 63
Counterculture hippy movement, France, 97
Critical Infrastructural Bill, Ireland, 73
Cultural borders, 6

Cultural-cognitive institutions, 24, 37
Cultural frames, 365–69
Cun, France, 107
Cyanide, 150

Dai Qing, *Should the Three Gorges Dam Be Constructed: A Collection of Scholars Debate*, 222
Dairy Accord, New Zealand, 349
Dairy industry, pollution in New Zealand and, 349, 351
Dams: anti-dam construction in China, 221–24; Anti-Dong River Dam Campaign in South Korea, 241; Anti-Saemangeum Reclamation Project Campaign in South Korea, 241; Guanting Reservoir DDT poisoning in China, 208–9; *Manifesto of Solidarity against the Reclamation Project in Saemanguem* (South Korea), 235–36
Da Zi Ran (Nature), 213
DDT poisoning: of the Guanting Reservoir, China, 209; of rice production, China, 209
Death of Nature, The (Merchant), 248
Deep ecologists (new pastoralists), Russia, 129
Deep ecology (Naess), 113, 145, 248, 333
Defanging of environmental movements, 362–63, *See also* Neoliberalism
Deforestation. *See* Forests/deforestation
Degrowth, 104–5, 112
Democratic churning, India, 169
Democratization in the South, protest movements and, 167–68
Deontologism, 141, *142*
Deontologist holist ecocentrism, *142*, 145, 156
Department of Conservation, New Zealand, 341, 344–45
Developmentalism: anti-developmentalism discourse in South Korea, 243–46; defined, 245; Muthurajawela ecological profile and development plan in Sri Lanka, 197–98; New Zealand, 339–40, 343–44, 350–52; political sub-group in Russia, 132
Dewey, J., *142*, 143–44
Die Grénen (West German Green Party), 7
Differentiation, environmental in Russia, 125–27
Displaced indigenous peoples, India, 166–67, 172–73, 176, 179, 368
Diversity, of green identities, 1–6
Donegal uranium mining issue, Ireland, 84
Doorae organic cooperatives, South Korea, 241
Doosan Electro-materials company, South Korea, 243
Doyle, Timothy, 1–18, 359–72
Druzhina movement (students' Nature Protection Movement, est. late 1960s), Russia, 117–18, 119–20
Dryzek, John, *Politics of the Earth: Environmental Discourses*, 243
Dublin, Ireland, 71–72
Duflot, Cécile, 101, 102
Dumont, René, 98–99

Earth First!, 53–54, 60, 89
Eastern Europe, 7
Eastern German Democratic Republic (GDR), 7–8; environmentalism in, 28
ECO (Environment and Conservation Organisation), New Zealand, 338–39
Ecocentrism, *142*, 143, 248

Eco-efficiency, Singapore, 266–68
Ecologically sustainable development, Australia, 290–91
Ecologic Foundation, New Zealand, 337–39
Ecologic protests, Germany, *29*, 30–31
Ecologie bleue, France, 104
Ecologie et survie Party, France, 98
Economic globalization, 360–61
Economy, Fiji, 308–9
Eco-patriots, Russia, 132
"Ecopolis" (eco-city), 134
Eco-politicians, Russia, 131
Ecopopulism in Ireland, 74–75, *76–77*, 78–79; ecopopulism defined, 74; outcomes, *80–81*, 82–83
ECO-Singapore, 262, 266–67, 268
Eco-technocrats, Russia, 127, 132–33
Ehrlich, Paul, 112
Elections, green performances in France, *103–4*
Emissions trading scheme, New Zealand, 349
EMOs. *See* Environmental movement organizations
Endangered Species (Import and Export) Act, Singapore, 264
Energy, in Turkey, 152–53
England, 46, *See also* United Kingdom (UK)
ENGOs. *See* Environmental nongovernmental organizations
Environmental Council, New Zealand, 341
Environmental Defence Society (EDS), New Zealand, 338, 344
Environmental ethical positions, taxonomy of, 140–45, *142*
Environmental impact assessments (EIAs), 153, 223
Environmentalisms, 359–72; classifications/categories of, 333; cultural frames, 365–69; defanging of movements, 362–63; defined, 139, 245; "fourth world," 360; gender relations, 361–62; geographical and geopolitical context, 359–62; green imperialism, 361; Left-Right spectrum, 2–4; metaphysical aspects, 365–66; North-South dichotomy, 2–6; with or against the state, 362–65; post-materialism, post-industrialism, and post-colonialism categorizations, 4–6; repertoires, 369–71; study of, 1; "third world," 365–66; "Universal We," 361; Western/Northern framing of environmentalism, 5; women and, 240, 241–42, 244, 251–52, 361–62, 367–68; "worlding" of international relations, 372
Environmental justice: defined, 271–72; in Singapore, 16, 271–76
Environmentally friendly behavior, Germany, 35–37
Environmentally friendly products, cost of, 35–36
Environmental Martyr (Xu Xiujuan), China, 212
Environmental Movement in Ireland (Leonard), 82
Environmental movement organizations (EMOs), 288–89
Environmental movements, private and public-interest types of, 215–18
Environmental nongovernmental organizations (ENGOs), 8; China, 214–29; in mature liberal democracies, 365; New Zealand, 337–38; United Kingdom, 47–48, *See also* headings under specific countries
Environmental Protection Administration, China, 211

Environmental Protection Agency, Ireland, 78, 83, 87
Environmental protection bureaus (EPBs), China, 214
Environmental racism, 271–72, *See also* Environmental justice
Environment Court, New Zealand, 332
Environment Fiji: The National State of the Environment Report (Watling and Chape), 305
Environment Protection Society, Sri Lanka, 185
"Envirotown" (variation on the "Transition Town") movement, New Zealand, 353
Eppawala phosphate mining issue, Sri Lanka, 198–99
Etatism: Russia, 129, 132; Turkey, 146
Ethical egoism, 143
Ethics: reverence for nature debate, China, 226–28; taxonomy of environmental ethical positions, 140–45, *142*
EU. *See* European Union
Europe: overview, 7–12, *See also* France; Germany; Ireland; Russia; Turkey; United Kingdom (UK)
European Court of Human Rights, 150
European Pressurized Reactor, 110–11
European project, 100
European Union (EU), 9, 100
Europe Ecologie—Les Verts (EELV), France, 96, 101–2, 111
Eurozone crisis, 9, 90
Everyday behavior, environmentalism in Germany and, 25, 35–37
Exploring Yunnan Snub-Nose Monkey (Xi), 220
Expressway development and the environment, Sri Lanka, 200–201
External actors/factors, 168–69; in Odisha, India protests, 165, 176–80
Extrinsic values, 141

Fate of the Earth, The (Schell), 305
Faucheurs volontaires, France, 107
Federation Nature et Environment (FNE), France, 105–6
Federation of Malaysia, 260
Feminism, 113
Feng Yongfeng, 213
Fiji, 16–17, 366, 372; Apia Convention on the Conservation of Nature in the South Pacific, 310; carbon footprint, 310; Christianity and, 311, 314; climate change, 309–10, 313; Conference on the Human Environment in the South Pacific, 310; contemporary situation, 319–22; Convention for the Protection of the Natural Resources and Environment of the South Pacific, 310; degradation factors, 319; economy, 308–9; environment as a Pacific issue, 309–11; *Environment Fiji: The National State of the Environment Report* (Watling and Chape), 305; exploitation of resources, 304; farming, 316–18; *Fiji* (Burns), 307–8; Fijian view of the environment, 311–12; Fiji Broadcasting Commission, 303; Fiji Development Bank, 303; *Fiji Times*, 303; fishing, 316–18; fishing ritual (the *uli drano*), 306; *girmit* system, 307; Gubboo Ted Thomas, 305; introduction by Robert Wolfgramm, 303–6; "Island Ecosystems," 311; *iTaukei* Fijian identity, 303–4, 307–8, 311; Koroi, Mesake, 16, 303–23;

Lapita pottery-people, 308, 313, 321; Lau group of islands, 313; legal position, 322–23; "little climatic optimum" (LCO) and the 1300 CE event, 321–22; *mataqali* indigenous clans, 312; military coups, 305; National Trust authority, 305; Native Lands Commission Act (1905), 312; Native Land Trust Act (1940), 311–12; overview, 16–17; Pacific Concerns Resources Centre (PCRC), 305–6; "Pacific Futures," 311; Pacific Islander population and geography, 309; the people, geography, and historical background, 306–8, 313–14, 319–20; politics, 323; pollution, 305–6; Pope John Paul II's visit, 304; races/ethnicity, 307; sandalwood trees, 307; Secretariat of the Pacific Community (SPC), 310; since independence (1970), 308; South Pacific Regional Environmental Programme (SPREP), 310–11; Sustainable Development Bill, 305; Suva (capital), 306; tourism, 308; traditional factors, 314–19; urbanization, 319–20; Vanua Balavu, 303; Vanuabalavu, 313; Vanua Levu, 321; Vatoa Island, 307; villager's perspective, 312–13; Viti Levu, 321; Wolfgramm, Robert, 16, 303–23; *Yavusa Qalitu* tribe, 312–13

Fiji Broadcasting Commission, 303

Fiji Development Bank, 303

Fiji Times, 303

"First-Past-the-Post" electoral system, New Zealand, 340, 347

Fishery resources, Sri Lanka, 191–92

Fishing tradition, Fiji, 306, 316–19

FoE. *See* Friends of the Earth

FON. *See* Friends of Nature

Food and Agriculture Organization, France, 98

Forest Amendment Act, New Zealand, 345

Forests/deforestation: China, 209, 220; India, 166–67, 169–72; New Zealand, 343, 345, 354; protected forests and affected forest communities in South (developing countries), 166–67; reforestation in Turkey, 147–48; Russia, 135; Sri Lanka, 191, 198; Turkey, 147–48; unsustainable deforestation, 166, *See also* Odisha (Orissa), India's protest against protected areas through social networking and external actors

Forest Stewardship Council, Russia, 135

Fossil fuels, 5

Foucault's concept of discourse, 243

"Four Little Dragons" (South Korea, Hong Kong, Singapore, and Taiwan), 236

"Four Pests Campaign," China, 210

"Fourth world societies," 360

Foxconn, 14

Fox, Matthew, 305

Framing process, 9

France, 10, 45, 95–116, 370; from the 1960s to the birth of the Green Party, 97–98; anti-globalization, 107–8; anti-GMO movement, 107–8; antinuclear protest, 95, 97, 109–11; *Associations pour le maintien d'une agriculture paysanne*, 113; Bové, José, 106–9; Cap 21, 104; challenges to the political system, 95–96; civil disobedience, 107–8, 111–12; *Conféderation paysanne* (Peasant

Farmers' Union), 107–8, 109, 114; courts and trials, 112, 113; degrowth theory, 112; Dumont, Rene: first green candidate for president, 98–99; *Ecologie bleue*, 104; *Ecologie et survie* Party, 98; ecologist parties: to the right and left of the Green Party, 104–5; elections and protest: Green Party and green movement, 100–101; environmental networks, 105–6; European Pressurized Reactor, 110–11; Europe Ecologie—Les Verts, 96, 101–2, 111; *Faucheurs volontaires*, 107; *Federation Nature et Environment (FNE)*, 105–6; Food and Agriculture Organization, 98; France Nature Environment, 105–6; French green ideology, 112–14; Friends of the Earth, 108–9; Gaullists, 95, 97; *Génération écologie*, 104; genetically modified (GM) crops, 95; green movement issues, 95–96; Green Party, 96–99, 104–5; Greenpeace, 102, 107, 109; green performances in elections, *103–4*; interregional national council (CNIR), 99, 101; Larzac, 106–7; *Les Amis de la Terre* (Friends of the Earth), 98–99; *Les Verts*: organization and activists, 99–100; *Mouvément écologiste independant*, 104; official creation of the Green Party (1984), 98; Ollitrault, Sylvie, 10, 95–114, 367; "ordinary citizens" activism, 111–12; overview, 10, 95–96; pacifism: from Larzac to antiglobalization—the rebellion of local farmers, 106–7; Parisian Friends of the Earth, 98; Socialist Party (PS), 96, 99, 100–102, 105, 108, 111; World Wildlife Fund (WWF), 108

France Nature Environment, 105–6

Frankfurter Rundschau (German newspaper used for protest data), 29

Friends of Bodrum, Turkey, 151

Friends of Nature (FON), China, 216, 217, 219, 220, 221, 225, 227

Friends of the Earth (FoE): Australia, 289, 293; France, 108–9; Germany, 26; New Zealand, 337–39; United Kingdom, 49–64

Fukushima, Japan, nuclear accident in, 28, 36–37, 111, 254

Fundamentalists (left (or red) greens), 340–41

Fund-raising, Russia, 123

Game Protection Society, Sri Lanka, 187

Gandhian tradition of protests, 12, 173

García-Guadilla, Maria-Pilar, 363–64

Gaullists, 95, 97

Gender roles, 361–62, 367, *See also* Women's movements

Génération écologie, France, 104

Genetically modified (GM) crops: France, 95, 106, 107–9; New Zealand, 354

Geo-culture, environmentalism and, 359–62

Geographical borders, 6

Germany, 7–8, 23–44, 370; activities and issues of the movement, 29–31, *29*; "Activities and Issues of the Movement," 32; air pollution and *Waldsterben* (death of all forests), 30; alternative milieu subculture, 35; anti-nuclear energy movement/

protests, 27–31, 33–34, 37; *Bundesnetzwerk Bürgerinitiativen Umweltschutz* (BBU) Federal Network of Environmental Citizen Initiatives), 26; *Bund für Umwelt und Naturschutz Deutschland* (BUND) Federation for Environment and Nature Protection Germany), 26; Die Grünen (West German Green Party), 7; Die Grünen's four pillars, 18; distrust of research/study findings, 33–34; Eastern GDR, 28; ecologic protests, *29*, 30–31; environmentalism in East Germany, 28; environmentally friendly behavior/products, 35–37; everyday behavior environmentalism domain, 25, 35–37; fall of the Berlin Wall, 28, 30, 32; *Frankfurter Rundschau* (German newspaper used for protest data), 29; Friends of the Earth International, 26; German unification, 32; *Gesellschaft für Natur und Umweltschutz* (Society for Nature and Environment Protection), 28; Green Party as part of the national government, 30, 34; Greenpeace, 26; historical background of the environmental movement, 23–24, 25–28; independent research institutes, 34; institutionalizing environmentalism as a concept, 24–25; knowledge and science environmentalism domain, 25, 33–35; *Naturschutzbund Deutschland* NABU (German Nature Protection Federation), 27; Öko Institut, 34; overview, 7–8; phasing out of nuclear energy (by 2022), 28, 36–37; Politbarometer survey, 32; political environmentalism domain, 25, 31–33; PRODAT time series project, 29, *29*, 31; regulative, normative, and cultural-cognitive institutions, 24, 34, 37; Roose, Jochen, 23–38, 370; *Süddeutsche Zeitung* (German newspaper used for protest data), 29; *Umweltbibliothek* (Environmental Library in East Berlin), 28; *Umweltbundesamt* (Federal Environment Agency), 31; West German Parliament protest, 18; World Wildlife Fund, 27

Gesellschaft für Natur und Umweltschutz (Society for Nature and Environment Protection, Germany), 28

Girmit system, Fiji, 307

Global civil society, 2, *See also* Civil society

Global financial-economic crisis, 350–52, 354

Global North-South dichotomy, 2–6, 360–61

Global Village of Beijing (GVB), China, 217, 227

Global warming, 309, *See also* Climate change

GMO. *See* Genetically modified (GM) crops

Goh Chok Tong, 261

Gonghae (pollution or public nuisance), 237

Goodpaster, K.A., *142*

Gopalpur, Odisha, India steel plant protest, 174

Grassroots environmentalism, 369; Australia, 288; China, 217–18; examples of, 337; historical background, 336–37; Ireland, 83; Kenya, 367–68; New Zealand,

335, 336–37; triggers for, 337; United Kingdom, 8, *See also* Localism
Great Britain: geographical definition, 45–46, *See also* United Kingdom (UK)
Green Beagle, China, 218, 225
Green Belt Movement, Kenya, 361, 367–68
Green Building Product Certification, Singapore, 267–68
Green Camp, China, 215
Green Choice Alliance, China, 225
"Green Choice" campaign, China, 224–25
Green Corridor campaign, Singapore, 268–69
Green Drinks, Singapore, 262, 267–68
Green Earth Volunteers, China, 217, 223, 227
Green imperialism, 361
"Green kaleidoscope," 1
Green Labeling Scheme, Singapore, 268
Green Mark Scheme, Singapore, 267–68
Green Parties: Australia, 289, 290; France, 96–99, 104–5; Germany, 30, 34; New Zealand, 340–41, 348–49, 352, 353, 355; South Korea, 254
Greenpeace: Australia, 288, 289, 293; blowing up of flagship in Auckland Harbour, New Zealand, 17; France, 102, 107, 109; Germany, 26; New Zealand, 337–38; Russia, 136; Turkey, 151; United Kingdom, 51–57, 59–60, 63–64
Green politics, 1, 18; Australia, 297–99; France, 112–14, *See also* Green Parties
Greer, Linda, 225–26
Ground-Up Initiative, Singapore, 274–75
Guanting Reservoir DDT poisoning, China, 208–9
Gumanitarnyi Ekologicheskii Zhurnal (Humanitarian Environmental Magazine), Russia, 129

Habermas, Jurgen, 239
Hambantota Harbour, Sri Lanka, 202
Hansalim organic cooperatives in South Korea, 241, 248–52; aims and core values, 250; beginning of, 250, 251; meaning of *Hansalim* (saving every life), 250; women and, 251–52
Hawke, Robert, 287, 290, 293, 295, 298, 299
Hayvan Haklari Federasyonu (HAYTAP) (Animal Rights Federation), 154
Heritage-based protests, Ireland, 69–74
Hexyl hydride, 225
Hindu campaigns, in Odisha, India, 178, 368
Hindu party (BJP), 178
Hirakud Dam, India, 179
Holism, 141, *142*, 143, 145, 156
Hollande, François, 102
Howard, John, 287, 292–94, 297, 298, 299
Huang Wanli, 221
Hulot, Nicolas, 102
Human Environment Conference, 185
"Humans Do Not Have to Have Reverence for Nature" (*Global* magazine), China, 227
Hunger as a Factor: An Impact of Hunger on Human Behaviour, Social Organization and Public Life (Sorokin), 117
Hydroelectric power plants, Turkey, 153
Hydropower development, China, 221–24

Ice Age, 45
Ico-op, South Korea, 241
Incinerator protest campaigns: China, 217; Ireland, 79, 85, 88
India, 12–13; Appiko forest movement, 171; Chipko (Hug the Trees) movement in Uttar Pradesh, 170–71; colonial period, 170; democracy in ("Indian wonder"), 169; displaced indigenous populations by protected areas, 166–67, 172–73, 176; forest movements, 166–67, 169–72; influence of external factors and actors, 176–79; National Forest Policies, 170–72; postcolonial, 170; study of protest movements, 169–75; Swain, Ashok, 12, 165–80, 368, *See also* Odisha (Orissa), India's protest against protected areas through social networking and external actors
Indigenous populations, 360, 371; displaced by protected areas in India, 166–67, 172–73, 176; Environment Court, 332; Fiji (*iTaukei*), 303–4, 307–8, 311; Maori of New Zealand, 332, 334–36, *335*, *See also* Fiji; Odisha (Orissa), India's protest against protected areas through social networking and external actors
Industrial Revolution, 2, 4; birthplace of (Britain), 46
Institute of Environment and Development (IED), China, 216
Institute of Public and Environmental Affairs (IPE): China, 218, 224–25; China Water Pollution Map, 224
Institutionalization: defined, 38; environmentalism in Germany, 23–25, 31–38; regulative, normative, and cultural-cognitive institutions, 24, 37, *See also* Germany
International Co-operative Alliance (ICA), 249
International Crane Foundation (ICF), 215, 219
International Fund for Animal Welfare (IFAW), 221
Internationalization, of Russian environmental movement, 134–36
International Monetary Fund (IMF), 346; bailout of Ireland, 9, 90
International Socio-Ecological Union (IsoEU), Russia, 125, 127, 134
Interregional national council (CNIR), France, 99, 101
Intrinsic values, 141
iPhones, 225
Iraq, 7, 10, 368, 372
Ireland, 9, 69–94, 367; anti-incinerator campaigns, 79, 85, 88; anti-multinational campaigns, 86–87; architectural heritage, 71–72; bailout of the Irish economy by the IMF, 9, 90; bog lands, 75; Burren Action Group and the Mullaghmore tourism dispute, 72–73, 82, 85, 87–88; campaign/activism outcomes, *80–81*, 82–83; Carnsore Point anti-nuclear protests, 82, 84, 86, 87; categories of protest events and outcomes, *76–77*, 82–83; Celtic Tiger boom, 9, 75, 79, 87, 90; civil society, 83; ecopopulism/rural populism and conservation campaigns, 74–75, *76–77*, 78–79, *80–81*, 82–83, 89; framing environmental conservation in, 69–74; framing of environmental campaigns, 84–86; grassroots

movements, 83; heritage-based protests, 69–74; housing construction, 79; IMF bailout of, 90; Leonard, Liam, 9, 69–90; neoliberalism governments, 90; overview, 9, 69–74; populist political makeup, 83; recycling, 88; road building protests, 73, 82, 85–86, 89–90; Rossport 5 farmers' campaign, 75, 85; Shell to Sea gas pipeline protest campaign, 73–74, 78, 85, 87, 89–90; solid waste disposal crisis, 78–79; tourism, 72–73; trade unions, 87; Tynagh mines dispute, 84–85; Unionist heritage, 71; water contamination crisis, 79
Irish Georgian Society, 71
Iron Curtain, fall of, 10, 117
"Island Ecosystems," Fiji, 311
Israel, 366–67
iTaukei Fijian indigenous people, 303–4, 307–8, 311

Jabiluka uranium mine campaign, Australia, 291
Jadot, Yannick, 102
James, H., *142*, 143–44
Jamil, Sofiah, 15, 259–76, 365
Ji-ha Kim, 246–48
Ji-yul, 242
Joly, Eva, 102

Kaitiakitanga (guardianship), New Zealand, 334
Kaleidoscope, defined, 1
Kalinga Nagar protest deaths, India, 176
Kanaky liberation movements, 306
Kandalama Hotel construction, Sri Lanka, 200
Kant, I., *142*, 144
Katsumi, Yokota, 249
Kenya, 372; Green Belt Movement, 361, 367–68; women's movements, 361–62, 367–68
Kewra flower growers, India, 174
Key, John, 350
Kim Dae-jung, 241
Knowledge and science, environmentalism in Germany and, 25, 33–35
Knowledge production, 372
Koçan, Gürcan, 11, 139–57, 369
Korea Federation of Environmental Movement (KFEM), 241
Korean Anti-Pollution Movement Association (KAPMA), 239–40
Koroi, Mesake, 16, 303–23, 366
Kotagama, Sarath Wimalabadara, 13, 185–205, 370
Krushchev, Nikita, 118
Kuala Lumpur, 260
Kyoto Protocol, 322, 349

Labeling, 35
Labour governments: Australia, 16; New Zealand, 344–50
Lake Pedder campaign, Australia, 289–90
Land degradation, Sri Lanka, 189
Lapita pottery people, Fiji, 308, 313, 321
Larzac, France, 106–7
Latin America, 370
Latouche, Serge, 112
Lau group of islands, Fiji, 313
Leading Group for Environmental Protection, China, 211
Lee Hsien Loong, 270
Lee Kuan Yew, 261
Left-Right spectrum of ideology, 2–4
Legal position, Fiji, 322–23
Leonard, Liam, 9, 69–90, 367; *Environmental Movement in Ireland*, 82

Lepage, Corinne, 102
Les Amis de la Terre (Friends of the Earth), France, 98–99
Les Verts, France, 99–100
Liang Congjie, 227
Liao Xiaoyi, 227
Life, defined (*Oxford English Dictionary*), 247
Lifestyle issues, vs. livelihood issues, 360
Li Lailai, 216
Liu Detian, 216
Localism: France, 99–100; Ireland, 83; Russia, 128, 135–36, *See also* Grassroots environmentalism
Loggerhead sea turtles, Turkey, 149
Lunugamwehera Reservoir, Sri Lanka, 201–2

Maathai, Wangari, 367
MacGregor, Sherilyn, 1–18, 359–72
Mahaweli Project, Sri Lanka, 196–97
Ma Jun, 224–26
Malayan Nature Society, 263
Mana Party, New Zealand, 336
Manifesto of Solidarity against the Reclamation Project in Saemanguem, South Korea, 235–36
Mao Da, 14, 207–29
Maoist movement, in Odisha, India, 176–77, 368
Maori of New Zealand, 332, 334–36, *335*
Mao Zedong, 221
Market economy: environment and Russia and, 132–33; reproduction in a "market economy" context (Russia, after 1993), 122–24
Marsh Arabs, Iraq, 368
Marsh, George Perkins, 339
Marxism, 3–4, 97
Mataqali indigenous clans, Fiji, 312
Mature liberal democracies, 365
Mauri (life force), New Zealand, 334
Menik Ganga, Sri Lanka, 201–2
Merchant, Carolyn, *Death of Nature, The*, 248
Mexico, 370
Military coups, Fiji, 305
Military dictatorships, South Korea, 236, 237–39
Mill, J.S., *142*, 143
Mining protests, New Zealand, 350–51, 354
Ministry for the Environment, New Zealand, 341, 344
Mitterand, François, 107
Mixed economy, Turkey, 146
Mixed Member Proportional (MMP) system, New Zealand, 340, 347–48
Monsanto corporation, 108
Moral particularism, 141
Morals, environmental positions and, 140–45, *142*
Mouvement écologiste indépendant, France, 104
Mukhachev, Sergei, 132
Muldoon, Robert, 343
Muthuki, Janet, 361, 367
Muthurajawela ecological profile and development plan, Sri Lanka, 197–98

Naess, A., *142*, 145, 248
Nakdong River phenol-contamination and anti-developmentalism, South Korea, 243–46
National Biodiversity Centre, Singapore, 265
National Conference on Environmental Protection, China, 210, 211
National Environmental Agency, Singapore, 272
National Forest Policies, India, 170–72
National Parks Board, Singapore, 265

National Party governments, New Zealand, 339, 347, 350–52
National Socialist movement, 26
National Sustainable Development Councils (NSDCs), New Zealand, 341
National Trust for Places of Historic Interest or Natural Beauty, UK, 47, 52, 57
Native Forest Action Council, New Zealand, 343
Native Lands Commission Act (1905), Fiji, 312
Native Land Trust Act (1940), Fiji, 311–12
Natural Heritage Trust, Australia, 293
Natural Resource Defence Council, China, 225–26
Nature Conservation Council, New Zealand, 341
Nature Society of Singapore, Singapore, 262–65, 268, 274
Naturschutzbund Deutschland (German Nature Protection Federation), 27
Neoliberalism, 3–4; environmentalisms and, 359–63; Ireland, 90; New Zealand, 17–18, 344–52
Network governance, New Zealand, 349–51
Networking from below, 370
Neuralgia, in South Korea, 238
"New Left" government, of Hugo Chávez, 364
"New world" societies, 5
New Zealand, 5, 17–18, 331–58, 365; "Blue-Greens," 339; Bührs, Ton, 17, 331–55, 363; "Buy Kiwi-Made" campaign, 348; carbon tax and emissions trading scheme, 349; Clark, Helen, 348; climate change, 348, 349, 354; Commission for the Environment, 341, 344; conclusion and summary, 352–55; Dairy Accord, 349; dairy farming pollution, 349, 351; Department of Conservation, 341, 344–45; destruction of nature and decline of resources, 333; development projects, 339–40; ECO (Environment and Conservation Organisation), 338–39; Ecologic Foundation, 337–39; Environmental Council, 341; Environmental Defence Society (EDS), 338, 344; environmental organizations, *335*, 337–39; Environment Court, 332; "Envirotown" (variation on the "Transition Town") movement, 353; era 1. early 1960s-to the mid-1980s: state-led development, 343–44; era 2. mid-1980s–2000: radical neoliberalism, 344–47; era 3. 2000 to 2008: "mitigated neoliberalism" era, 347–50; era 4. 2008 to present: "obstinate neoliberalism": return of environmental confrontation, 350–52; fifth Labour government, 348–50; "First-Past-the-Post" electoral system, 340, 347; Forest Amendment Act, 345; fourth Labour government, 344–47; Friends of the Earth, 337–39; global financial-economic crisis, 350–52, 354; grassroots environmentalism, *335*, 336–37; Green Party, 340–41, 348–49, 352, 353, 355; Greenpeace, 337–38; institutional environmental advocates, *335*, 341–42; international environmental reputation of ("clean and green"), 331, 345; Key, John, 350; the

Maori indigenous group, 332, 334–36, *335*; Marsh, George Perkins, 339; mining protests, 350–51, 354; Ministry for the Environment, 341, 344; Mixed Member Proportional (MMP) system, 340, 347–48; Muldoon, Robert, 343; National Party governments, 339, 347, 350–52; National Sustainable Development Councils (NSDCs), 341; Native Forest Action Council, 343; native forest protection, 354; Nature Conservation Council, 341; neoliberalism's impact on environmentalisms, 363; network governance, 349–51; New Zealand Forest and Bird Protection Society, 343; nuclear issues, 345; Office of the Parliamentary Commission for the Environment, 344; oil issues, 351; overview, 17–18, 331–33; Parliamentary for the Environment, 341; political (Parliamentary) wing, *335*, 340–41; political structure, 332; population, the people, and geography, 332; professional/expert-based environmentalism, *335*, 339–40; quasi nongovernmental organizations (Quangos), 341; Regional Councils, 344; Resource Management Act (1991), 332, 335–36, 344, 346, 353; Royal Forest and Bird Protection Society (RFSB), 337–38; Save the Rivers Campaign, 343; summary of environmental movements, *335*; "Think Big" projects, 343; Third Way philosophy, 349; Treaty of Waitangi, 335–36; Values Party (green party), 340; Waitangi Tribunal, 335; water pollution, 349, 351, 354; West Coast Forest Accord, 345

New Zealand Forest and Bird Protection Society, 343

Nigeria, 360, 365–66, 372; Ogoni peoples' struggles with Shell, 366

Noeksaek Gagae (The Green Store), South Korea, 240

Non-anthropocentrism, *142*, 144–45

Nongovernmental organizations. *See* Environmental nongovernmental organizations (ENGOs); headings under specific countries

Normative institutions, 24, 34, 37

Norms of reciprocity and trust, 168

North (affluent, minority world), 2, 370, 371, *See also* North-South dichotomy

Northern Ireland, 45–46, *See also* Ireland

North-South dichotomy, 2–6

"Not in my back yard" (NIMBY), 75, 218

Nuclear accidents: Chernobyl, Russia, 27, 28, 37; Fukushima, Japan, 28, 36–37, 254

Nuclear energy: phasing out of nuclear energy (by 2022) in Germany, 28, 36–37, *See also* Antinuclear energy movements/protests

Nuclear Free and Independent Pacific lobby, 306

Nuclear-free status, New Zealand, 345

Nuclear testing: New Zealand, 343, 345; Pacific Islands, 309

Nujiang River hydropower project, China, 222

Nunn, Patrick, 321–22

Objectivism, *142*
Occitan movement, France, 106
Occupy Movement, 352
Occupy Wall Street, 352
Oceania: overview, 15–18, *See also* Australia; Fiji; New Zealand
Odisha (Orissa), India's protest against protected areas through social networking and external actors, 12–13, 165, 172–80, 368; Chilika Lake protest, 173–74; Christian missionaries, 178; class analysis, 173–74; coastal and inland comparisons in Odisha, 173–75; demographics/geography of the state of Odisha, 172; displacement and dispossession in Odisha, 167, 172–73, 176, 179; diversity in tribal groups, 175; external factors/actors, 168–69, 176–79; Gandhian protests, 173; Hindu groups, 178; indigenous tribes in Odisha, 173–74; Maoist movement, 176–77; Odisha Freedom of Religion Act, 178; overview, 165; protected forest areas and affected forest community, 166–67; protest deaths in Kalinga Nagar, 176; social movement theory, 167–69; steel plant protest at Gopalpur, 174; subgroup interactions, 176–77; successful movement strategies, 174–75; Swain, Ashok, 165–80, 368; *Visthapan Virodhi Janmanch* (People's Committee against Displacement), 176, *See also* India
Odisha Freedom of Religion Act, 178
Office of the Parliamentary Commission for the Environment, New Zealand, 344
Ogoni peoples of Nigeria, struggles with Shell, 366
"*Oikos*," 248
Oil exploration, New Zealand, 351
Oil fracking, New Zealand, 351
Öko Institut, Germany, 34
"Old world" societies, 5
Ollitrault, Sylvie, 10, 95–114, 367
Öncü, Ahmet, 11, 139-57, 369
"Ordinary citizens" activism, 111–12
Organic products: Germany, 35; Turkey, 148
Original equipment manufacturers (OEMs), 225
Orissa. *See* Odisha
Other, The (Said), 72
"Out-of-bounds" markers, 272–73

Pacific Concerns Resources Centre (PCRC), Fiji, 305–6
"Pacific Futures," Fiji, 311
Pacific Islanders/Islands: environment issues, 306, 309–11; natural disasters, 312; nuclear testing, 309, 345; size and population, 309, *See also* Fiji
Pacifism, in France, 106–7
Pakeha (Maori for white people), 336
Paraxylene (PX) chemical plants protests, China, 217, 218
Parc de la Vanoise, France, 97
Parisian Friends of the Earth, 98
Park Chung hee dictatorship, South Korea, 236, 237, 245
Parliamentary for the Environment, New Zealand, 341
People's Solidarity for Participatory Democracy, South Korea, 239
Pesticides: in China, 209; women's movements and, 362
Pinochet regime, Chile, 369

Plane Stupid, UK, 62
Poh Ern Shih Buddhist temple, Singapore, 275
Politbarometer survey, Germany, 32
Political (Parliamentary) wing, New Zealand, *335*, 340–41
Political opportunity structure (POS), Russia, 119, 121, 125
Politics: environmentalism in Germany and, 25, 31–33; in Fiji, 323
Politics of the Earth: Environmental Discourses (Dryzek), 243
Pollution: Fiji, 305–6; New Zealand, 349, 351; South Korea, 237–38; Turkey, 150–52, *See also* Air pollution; Waste problems; Water pollution
Pollution Research Group, South Korea, 237
Pollution Research Institute, South Korea, 237–38
Pompidou, Georges, 97
Pope Benedict XVI, 23
Pope John Paul II's visit to Fiji, 304
Positivism, 6
Postcolonial Ireland, 71–72
Post-colonialism, 4–6
Post-industrialism, 4–6
Post-materialism, 4–6
Post-politics, 3–6
Practices, for measuring environmentalism, 25, 32–33
Pragmatism, *142*, 143–44; Turkey, 11–12, 155–57, *See also* Turkey
PRODAT time series project, 29, *29*, 31, *See also* Germany
Professional/expert-based environmentalism, New Zealand, *335*, 339–40
Project ME: Muslims and the Environment, Singapore, 275
Quasi-eco-politicians, Russia, 131
Quasi nongovernmental organizations (Quangos), 341
Qu Geping, 211–12, 215

Race/ethnicity: environmental justice and, 271–72; in Fiji, 307; in New Zealand, 332
Radical etatists, in Russia, 132
Radical Left, 4
Rail Corridor, Singapore, 268
Rainbow Warrior (ship), 309
Rajapaksa, Mahinda, 188
Ramblers' Association, UK, 47
Ranawaka, Champika, 188
Rand, A., *142*
Reagan, Ronald, 346
Realists (right (or blue) greens), 340–41
Real World Coalition, 55
Recycling: Ireland, 88; Singapore, 271
Red-crown crane, China, 212, 219
"Red-Greens," New Zealand, 341
Reductionism/individualism, 141–43, *142*
Reforestation. *See* Forests/deforestation
"Reform and Opening" policy, China, 212, 214
Regan, T., *142*, 144–45
Regional Councils, New Zealand, 344
Regulative institutions, 24, 34, 37
Religious leaders, in South Korea, 242
Re-Live event, Singapore, 266–67
Remember Chek Jawa, Singapore, 265
Repertoires, in environmentalisms, 369–71
Research: attitude, 24–25; distrust of findings in Germany, 33–34; independent research institutes in Germany, 34
Resorts World, 266

Resource Assessment Commission, Australia, 293
Resource Management Act (1991), New Zealand, 332, 335–36, 344, 346, 353
Resources: allocation of, 360–61; exploitation in Fiji, 304; global gap of usage, 360
Reverence for nature debate, China, 226–28
Rio Earth Summit (UNCED), 8, 55
Rising Tide, 60
River activism: China, 212, 221–24; New Zealand, 343; South Korea, 243–45, *See also* Dams
Road building protests, Ireland, 73, 82, 85–86, 89–90
Roissy Charles de Gaulle airport, 97
Roose, Jochen, 7, 23–38, 370
Rootes, Christopher, 8, 45–64, 371
Rossport 5 farmers' campaign, Ireland, 75, 85
Royal Forest and Bird Protection Society (RFSB), New Zealand, 337–38
Royal Society for the Protection of Birds (RSPB), UK, 47, 49, 52, 56–57
Royal Society for Wildlife Trusts (formerly the Society for the Promotion of Nature Reserves), UK, 47
Russia, 10–11, 117–38, 364–65; advocacy science/scientists and citizens as experts, 133–34; alternativists, 128–29; antiglobalist movement, 128–29; basic survival issues ("well-being first, ecology later"), 136; civil initiatives, 126, 130; civil society, 118, 121–22; Communist Party of the USSR, 117–20; conservationists, 125–28, 134–35; deep ecologists (new pastoralists), 129; "democratic façade" in, 364; *Druzhina* movement (students' Nature Protection Movement, est. late 1960s), 117–18, 119–20; ecological "actors," 136; eco-patriots, 127; eco-politicians, 126–27; eco-technocrats, 127, 132–33; ENGOs and networking, 123; etatists, 129, 132; Forest Stewardship Council, 135; fund-raising, 123; Greenpeace, 136; *Gumanitarnyi Ekologicheskii Zhurnal* (Humanitarian Environmental Magazine), 129; the internationalization of the movement, 134–36; International Socio-Ecological Union (IsoEU), 125, 127, 134; localism, 128, 135–36; overview, 10–11, 117–18; political opportunity structure (POS), 119, 121, 125; six major types of milieus, 118; social movement organizations' (SMOs) reproduction, 122; Socio-Ecological Union (SoEU), 119, 125; stage 1: the late bureaucratic socialist society (before 1985), 118–20; stage 2: the beginning of liberalization (1985–1991), 120–21; stage 3: the turning point (1991–1993), 121–22; stage 4: reproduction in a "market economy" context (after 1993), 122–24; stage 5: building a political party, 124–25; structure of social opportunity (SOS), 119; subgroups in, 128; sub-politics in: communitarians (moderate red-green eco-patriots), 132; sub-politics in: developers, 132; sub-politics in: eco-politicians, 125, 131;

sub-politics in: quasi-eco-politicians, 131; sub-politics in: radical etatists, 132; totalitarian period, 117, 122; traditionalists, 126, 129; transnationalists, 135–36; United Party of Russian Greens, 124–25; value basis/types of the movement's differentiation, 125–27; value shifts, 127–33; World Wildlife Fund for Nature—Russia (WWF-Russia), 119, 135, 136; Yanitsky, Oleg, 10, 117–36

Saengmyung (life) discourse, South Korea, 235–36, 246–48; compared to anti-developmentalism, 252, *253*; main ideas, 248
Said, Edward, *Other, The*, 72
"Salamanders' Suit," South Korea, 242
Salih, Mohamed, 10, 368
Samanalawewa Reservoir, Sri Lanka, 195–96
Sampoong department store collapse, South Korea, 247
Sandalwood trees, Fiji, 307
Sanmenxia dam on the Yellow River, China, 221
Sarkozy, Nicolas, 95, 96
Satyagraha (firmness in truth), 12
Saunders Gull Conservation Society of Panjin City (SGCSPJ), China, 216, 219
Save the Rivers Campaign, New Zealand, 343
Scales, 2
Schell, Jonathon, *Fate of the Earth, The*, 305
Scientific neutrality, 33
Scotland, 45–46, *See also* United Kingdom (UK)
Secretariat of the Pacific Community (SPC), 310
Seeto, Johnson, 305
Self-reflexivity, 372
Sentiocentrism, 142–45, *142*
Sepulveda, Claudia, 368
Shell, Ogoni peoples struggle with, Nigeria, 366
Shell to Sea gas pipeline protest campaign, Ireland, 73–74, 78, 85, 87, 89–90
Should the Three Gorges Dam Be Constructed: A Collection of Scholars Debate (Dai), 222
Silent Spring (Carson), 49, 305
Singapore, 15–16, 259–83, 365; Agri-Food and Veterinary Authority, 264; Ang, Wilson, 266; Animal Concerns Research and Education Society (ACRES), 262–66; breaking down barriers between environmentalist currents, 268–69; Bring-Your-Own Bag/Receptacles, 271; Bukit Brown Cemetery, 269; Chek Jawa wetlands campaign, 263, 264, 265; conservation and urbanization, 259–66; Conservation of Nature in Singapore, 264; eco-efficiency, 259, 260, 266–67; ECO-Singapore, 262, 266–67, 268; Endangered Species (Import and Export) Act, 264; environmental justice, 259, 260, 271–75; faith-based initiatives, 275; Federation of Malaysia, 260; GDP and growth rate in, 260; Goh Chok Tong, 261; Green Building Product Certification, 267–68; Green Corridor campaign, 268–69; Green Drinks, 262, 267–68; Green Labeling Scheme, 268; Green Mark Scheme, 267–68; Ground-Up Initiative, 274–75; independence, 260, 261; Jamil, Sofiah, 15, 259–76, 365;

Lee Hsien Loong, 270; Lee Kuan Yew, 261; National Biodiversity Centre, 265; National Environmental Agency, 272; National Parks Board, 265; Nature Society of Singapore (NSS), 262–65, 268, 274; Olivia Choong, 267; "out-of-bounds" markers, 272–73; overview and historical background, 15–16, 259–62; poverty, 273–74; Project ME: Muslims and the Environment, 275; Rail Corridor, 268; recycling, 271; Re-Live event, 266–67; *Remember Chek Jawa*, 265; Singapore Environmental Council, 268; Soup Kitchen Project, 274; Sungei Buloh wetlands campaign, 263, 264, 265; transport sector, 267; tree planting, 261–62; unemployment, 260–61; urban development vs. environmentalism, 269–71; urbanization and economic development strategies, 260–62; waste problems, 270–71; "When sharks die, the oceans die" campaign, 265–66; Wildlife Crime Hotline, 265

Singapore Environmental Council, 268

Singer, P., *142*, 144

Sinharaja Rainforest/World Heritage Reserves, Sri Lanka, 13, 185

Sinharaja World Heritage Reserves, Sri Lanka, 194–95

Snub-nose monkey, China, 219–20

Social capital, 168

Social environmental changes, Sri Lanka, 192

"Socialism of the 21st Century," Chávez government (Venezuela), 363–64

Socialist Party, France, 96, 99, 100–102, 105, 108, 111

Social media, overthrow movements of authoritarian governments and, 352

Social movement organizations' (SMOs) reproduction, Russia, 13, 26, 122

Social movement theory, protest against protected areas and, 167–69

Social networks, 165–69; importance, 168; structure, 168; types, 168, *See also* Odisha (Orissa), India's protest against protected areas through social networking and external actors

Social Trends, 55

Society, domains of, 239

Socio-Ecological Union (SoEU), Russia, 119, 125

Soil preservation, Turkey, 148

Solid and hazardous waste: China, 217, 218; Sri Lanka, 190

Solid waste disposal crisis, Ireland, 78–79

Sorokin, Pitirim, *Hunger as a Factor: An Impact of Hunger on Human Behaviour, Social Organization and Public Life*, 117

Soup Kitchen Project, Singapore, 274

South (poorer, majority world), 2; democratization in, 167; protected forests and affected forest communities in South (developing countries), 166–67, *See also* North-South dichotomy

South Korea, 235–58, 360; Anti-Dong River Dam Campaign, 241; antinuclear movement, 240; Anti-Saemangeum Reclamation Project Campaign, 241, 242; Chang, Woojoo, 15, 235–54,

361, 366; children's diseases, 242; Chun Do Hwan, 236, 238; Citizen's Coalition for Economic Justice (CCEJ), 239; civil society, 239–40, 244–46; comparison of anti-developmentalism and *Saengmyung* (life), *253*; Doorae organic cooperatives, 241; environmentalism and developmentalism, 245; "Four Little Dragons" (South Korea, Hong Kong, Singapore, and Taiwan), 236; *gonghae* (pollution or public nuisance), 237; Green Political Party, 254; *Hansalim* organic cooperatives, 241, 248–52, 361; Ico-op, 241; industrialization and urbanization, 236; Ji-ha Kim, 246–48; Ji-yul, 242; Kim Dae-jung's government, 241; Korea Federation of Environmental Movement (KFEM), 241; Korean Anti-Pollution Movement Association (KAPMA), 239–40; *Manifesto of Solidarity against the Reclamation Project in Saemanguem*, 235–36; military dictatorships, 236, 237–39; Nakdong River phenol-contamination and anti-developmentalism discourse, 243–46; neuralgia, 238; *Noeksaek Gagae* (The Green Store), 240; overview, 15, 235–37; Park Chung hee's government, 236, 237, 238, 245; peaceful protests, 242–43; People's Solidarity for Participatory Democracy (PSPD), 239; Pollution Research Group, 237; Pollution Research Institute, 237–38; religious leaders in, 242; *Saengmyung* (life) discourse, 235–36, 246–48, 366; "Salamanders' Suit," 242; Sampoong department store collapse, 247; stage 1: environmental movement prior to 1987, 237–38; stage 2: environmental movements between 1987 and the mid-1990s, 238–40; stage 3: mid-1990s to present, 240–43; student movements, 238–39; Sungsoo Bridge collapse, 247; women's movements, 240, 241–42, 244, 361–62

South Pacific Regional Environmental Programme (SPREP), 310–11

Sri Lanka, 13, 185–205, 370; air pollution, 13, 190, 370; British capture of, 186; Caring for the Environment, 188; Central Environment Authority, 203; coal power plant program, 199–200; diversion of Menik Ganga to augment Lunugamwehera Reservoir, 201–2; ENGOs in, 203–5; environmental health, 192; environmental impact assessments (EIA), 195, 199–202; environmental movement of, 194; Environment Protection Society, 185; Eppawala phosphate mining issue, 198–99; expressway development and the environment, 200–201; Fauna and Flora Protection Ordinance, 191; fishery resources, 191–92; Forest Conservation Act, 191; forestry sector, 191; Forestry Sector Master Plan, 198; forestry sector master plan and its environmental concerns, 198; future trends, developments, and requirements, 192–93; Game Protection Society, 187; Greater Colombo Economic Commission, 197; Hambantota Harbour,

202; institutional changes, 202–5; Kandalama Hotel construction, 200; Kotagama, Sarath Wimalabadara, 13, 185–205, 370; land degradation, 189; loss of biological diversity, 190–91; Mahaweli Project, 196–97; March for Conservation, 204; Muthurajawela ecological profile and development plan, 197–98; National Action Plan for Haritha Lanka Programme, 188–89; National Environment Action Plans, 188; overview and background, 13, 185–89; Samanalawewa Reservoir, 195–96; Sinharaja World Heritage Reserves, 194–95; social environmental changes, 192; solid and hazardous waste, 190; Strategy for Sustainable Development, 188; water pollution, 189–90; Wilderness Heritage Act, 195; Wildlife and Nature Protection Society, 194; wildlife conservation, 187
Star, Cassandra, 16, 287–99, 367
Starhawk, 305
State cooperation issues, 362–65
State Environmental Protection Administration (SEPA), China, 211–12, 214
State-led development, New Zealand, 343–44
Stockholm summit (1972), 97
Stop Climate Chaos, UK, 59–61
Structuralism, 6
Structure of social opportunity (SOS), Russia, 119
Studies, varying standards for, 2
Subjectivism, *142*, 156
Süddeutsche Zeitung (German newspaper used for protest data), 29
Sungei Buloh wetlands campaign, Singapore, 263, 264, 265
Sungsoo Bridge collapse, South Korea, 247
Sung-tae Hong, 245
Sustainable Development Bill, Fiji, 305
Sustrans, UK, 52
Suva, Fiji (capital), 306
Swain, Ashok, 12, 165–80, 368

Taiga Rescue Network, 128
Tal, Alon, 366
Tang Xiyang, 212–13, 220
Tasman, Abel, 307
Tasmania's hydro-industrialization, 290
Taxes, on environmentally friendly products in Germany, 35–36
Taxonomies, 2
Taxonomy of environmental ethical positions, 140–45, *142*
Thatcher, Margaret, 8, 54, 294, 346
Thermal power plants, Turkey, 150
"Think Big" projects, New Zealand, 343
Third Way philosophy, New Zealand, 349
Thomas, Gubboo Ted, 305
Thoreau, Henry David, 97
Three Gorges dam on the Yangtze, China, 221–22
"Three Parallel Rivers of Yunnan Protected Areas" (World Natural Heritage), 222
Tibetan Antelope protection, China, 217
Tickner, Arlene, 372
Tourism: Burren Action Group and the Mullaghmore tourism dispute in Ireland, 72–73, 82, 85, 87–88; Fiji, 308; Turkey, 149
Town and Country Planning Acts (1932), UK, 48

Toyne, Phillip, 292
Trade unions, Ireland, 87
Transnational corporations, 3–4; anti-multinational campaigns in Ireland, 86–87
Transnationalists, in Russia, 135–36
Treaty of Waitangi, New Zealand, 335–36
Tree Planting Day, Singapore, 261–62
Tribals, 167, *See also* Indigenous populations; Odisha (Orissa), India's protest against protected areas through social networking and external actors
Turkey, 11–12, 139–61, 369; Animal Party, 154–55; animal rights, 149, 153–55; architectural preservation, 149; Chamber of Forest Engineers, 147; Chamber of Mechanical Engineers, 152; the commons and commoning, 147–50; energy, 152–53; etatist model, 146; farming, 148; forest conservation movements, 147–48; founding of the Republic of Turkey (1923), 145–46; Friends of Bodrum, 151; Greenpeace, 151; *Hayvan Haklari Federasyonu* (HAYTAP, Animal Rights Federation), 154; historical background, 145–47; hydroelectric power plants, 153; Koçan, Gücan, 11, 139–57; as link between global West and global East, 7; military intervention (1980), 139, 146; mixed economy period, 146; overview, 11–12, 139–40; pollution, 150–52; pragmatic environmental ethics in, 149, 155–57; return to parliamentary system (1982), 139; soil preservation, 148; taxonomy of environmental ethical positions, 140–45, *142*; thermal power plants, 150; Turkish Foundation for Combating Soil Erosion, for Reforestation and the Protection of Natural Habitats (TEMA), 147–48; Zafer Park, 148–49
Turkish Foundation for Combating Soil Erosion, for Reforestation and the Protection of Natural Habitats (TEMA), 147–48
Tynagh mines dispute, Ireland, 84–85

UK. *See* United Kingdom
Uli drano (fishing ritual), Fiji, 306
Umweltbibliothek (Environmental Library in East Berlin), 28
Umweltbundesamt (Federal Environment Agency, Germany), 31
Unification, German, 32
Unified global movement, 6
Union pour la majorité populaire, France, 96
United Kingdom (UK), 8–9, 45–67, 369; the 1990s: consolidation and challenge, 52–55; British Trust for Conservation Volunteers, 48; Campaign against Climate Change (CCC), 61; Campaign to Protect Rural England, 58–59; Camp for Climate Action, 62; climate change issues, 8, 60–63; Commons Preservation Society, 47; Council for Environmental Conservation (CoEnCo), 48, 57; Council for the Preservation of Rural England (CPRE), 47–48, 56–57; decline of protests, 57–60; Earth First!, 53–54, 60; England, 46; environmental nongovernmental organizations

(ENGOs), 47–49; Friends of the Earth (FoE), 49–64; geographical features, 45–46; Great Britain, 46–47; Greenpeace, 51–57, 59–60, 63–64; National Trust for Places of Historic Interest or Natural Beauty, 47, 52, 57; Northern Ireland, 45–46; origins and development of environmentalism, 46–49; overview, 8–9, 45; Plane Stupid, 62; poll tax, 54; protests, 52–60; Ramblers' Association, 47; rise of the modern movement, 49–52; Rootes, Christopher, 8, 45–64, 371; Royal Society for the Protection of Birds (RSPB), 47, 49, 56–57; Royal Society for Wildlife Trusts (formerly the Society for the Promotion of Nature Reserves), 47; Scotland, 45–46; Stop Climate Chaos, 59–60, 61; Sustrans, 52; Town and Country Planning Acts (1932), 48; at the turn of the millennium, 55–57; urbanization, 45–46; Wales, 45–46; Wildlife Link, 52, 56; Wildlife Trusts, 57, 61; Woodland Trust, 52; World Wildlife Fund (WWF), 48, 52, 56

United Nations, climate conference (COP6), 60

United Nations Conference on Human Environment, 210

United Nations Conference on the Environment and Development (Earth Summit, 1992), 166, 240, 290

United Party of Russian Greens, Russia, 124–25

United States, 5, 363, 365, 369

"Universal We," 360

UN Earth Summit Rio+20, 102

Urbanization: United Kingdom, 45–46, *See also* Singapore

Utilitarianism, *142*, 143, 156

Utopian socialism, cooperative movement and, 249

Values Party (green party), New Zealand, 340, 353

Value types of civil initiatives, in Russia, 126

Vanua Balavu, Fiji, 303

Vanuabalavu, Fiji, 313

Vanua Levu, Fiji, 321

Vatoa, Fiji, 307

Veddah people, Sri Lanka, 196

Vegetarians, 113

Venezuela, 363–64

Viking Dublin, 71

Villarroel, Pablo, 368

Visthapan Virodhi Janmanch (People's Committee against Displacement), India, 176

Viti Levu, Fiji, 321

Waever, Ole, 372

Waitangi Tribunal, New Zealand, 335

Waldsterben (death of all forests in Germany due to air pollution), 30

Wales, 45–46, *See also* United Kingdom (UK)

Wang Yongchen, 223, 227

Waste problems: China (solid, water, gas), 209, 218; Singapore, 270–71; solid waste disposal crisis in Ireland, 78–79

Water contamination crisis in Ireland, 79

Water pollution: Guanting Reservoir DDT poisoning in China, 208–9; Ireland, 79; Nakdong River spills in South Korea, 243–46; New Zealand, 349, 351, 354; Sri Lanka, 189–90

Watling, Dick and Stuart Chape, *Environment Fiji: The National State of the Environment Report*, 305
Weltanschauung: rural agricultural cycles/subsistence economies, Fiji, 303; Turkey, 149
Wen Jiabao, 222
West Coast Forest Accord, New Zealand, 345
Western-centric field of scholarship, 1
Western/Northern framing of environmentalism, 5, *See also* North-South dichotomy
"When sharks die, the oceans die" campaign, Singapore, 265–66
Whistleblowers, 106, 111
Who's Who in the Environment?, 56
Wilderness protection, 5
Wildlife conservation movements: China, 218–21; Singapore, 262–66; Sri Lanka, 187, 194
Wildlife Crime Hotline, Singapore, 265
Wildlife Link, UK, 52, 56
Wildlife Trusts, UK, 57, 61
Wild Yak Troupe, 220–21
Wolfgramm, Robert, 16, 303–23
Women's movements: *Hansalim* cooperatives, South Korea, 251–52; Kenya, 361–62, 367–68; South Korea, 240, 241–42, 244, 361–62, *See also* Gender roles
Woodland Trust, UK, 52
World Bank, 346
World Environment Conference, 185
"Worlding process," 1, 372
World Natural Heritage, "Three Parallel Rivers of Yunnan Protected Areas," 222
World Trade Organization (WTO), 249, 250
World War II, destruction of civil society and, 26
World Wide Fund for Nature, 119, 215, *See also* World Wildlife Fund (WWF)
World Wildlife Fund (WWF): Australia, 289; China, 215, 219, 221; France, 108; Germany, 27; Russia, 119, 127, 128, 135, 136; United Kingdom, 48, 52, 56
WTO. *See* World Trade Organization
Wu Dengming, 212, 215
WWF. *See* World Wildlife Fund

Xiaonanhai Hydropower Station, Yangtze River, China, 223
Xi Zhinong, 220
Xu Xiujuan, 212

Yablokov, Alexi, 124–25
Yancheng Nature Reserve, Jiangsu Province, China, 212
Yangtze River, China, 212, 223
Yang Xin, 220–21
Yanitsky, Oleg, 10, 117–36
Yarlung Zangbo River, China, 222
Yavusa Qalitu tribe, Fiji, 312–13
Youth groups, in Australia, 288–89

Zabelin, Sviatoslav, 127
Zafer Park, Turkey, 148–49
Zhalong Crane Nature Reserve, Heilongjiang Province, China, 212
Zhou Enlai, 208, 210

About the Editors and Contributors

Editors

TIMOTHY DOYLE, BA Hons. (Melb), MA (Adel), PhD (Griffith), is professor of politics and international studies at the University of Adelaide in Australia, and founding chair of management, for the Indo-Pacific Governance Research Centre (IPGRC), also at Adelaide. Professor Doyle is also chair of politics and international relations at Keele University in the United Kingdom. He has published widely in a diverse range of journals. His most recent works include the books: *Environmental Politics and Policy-Making in Australia*; *Environment and Politics* with D. McEachern; *Green Power: the Environment Movement in Australia*; *Environmental Movements in Minority and Majority Worlds: A Global Perspective*; *Beyond Borders: Environmental Movements and Transnational Politics*, edited with B. Doherty; and *Crucible for Survival: Environmental Security and Justice in the Indian Ocean Region*, edited with M. Risely. He has taught and contributed to university courses in the United Kingdom, the United States, Malaysia, India, and Australia and been a dedicated environmental and human rights activist since the 1980s. He serves on a variety of editorial boards and has a long and extensive background in editing and co-editing.

SHERILYN MACGREGOR is senior lecturer in environmental politics at Keele University in the U.K. Her published works include *Beyond Mothering Earth: Ecological Citizenship and the Politics of Care* (University of British Columbia Press, 2006) and numerous journal articles and book chapters on gender, feminist theory, and environmental justice, including "A Stranger Silence Still: The Need for Feminist Social Research on

Climate Change" (*Sociological Review* 2009). MacGregor is joint editor of *Environmental Politics* journal and a convener of the Environmental Politics Standing Group of the European Consortium for Political Research. She holds a PhD in environmental studies from York University in Toronto, Canada.

Contributors

ADELA ALFONSI holds a doctorate in history from the University of Adelaide, Australia. She teaches history and politics, and is a researcher at the Indo-Pacific Governance Research Centre (IPGRC), both at the University of Adelaide, Australia. Adela is the commissioning editor of the *Journal of the Indian Ocean Region* (Routledge).

TON BÜHRS, PhD, is associate professor of environmental policy at Lincoln University, New Zealand. His published works include *Environmental Policy in New Zealand: The Politics of Clean and Green?* (Auckland: Oxford University Press, 1993) [co-authored with Robert V. Bartlett], and *Environmental Integration: Our Common Challenge* (New York: SUNY Press, 2009). He holds a doctorate in political science from the University of Auckland, New Zealand.

WOOJOO CHANG studied philosophy at Ewha Womans University in South Korea. She was involved in the Korea Dialogue Academy and the Korea Women's Environmental Network, focusing on feminism, environment and activism in South Korea. Later she got her PhD from Lancaster University in the U.K. The title of her thesis is "Women's Collective Caring Practices in the South Korean Environmental Activism."

SOFIAH JAMIL is an adjunct research associate at the Centre for Non-Traditional Security (NTS) Studies, S. Rajaratnam School of International Studies (RSIS), Nanyang Technological University (NTU), Singapore. Her published works include two co-authored chapters in an edited volume on Energy and Non-Traditional Security in East Asia; and a chapter on "Islam and Environmentalism: Greening Our Youth" in *Igniting Thought, Unleashing Youth: Perspectives on Muslim Youth and Activism in Singapore.* She holds a master's in international relations from NTU and is currently pursuing her doctorate in international, political, and strategic studies at The Australian National University.

GÜRCAN KOÇAN, PhD, is professor of political philosophy at Istanbul Technical University, Istanbul, Turkey. His published works include "From Morality of Living to Morality of Dying: Hunger Strikes in Turkish Prisons" (*Citizenship Studies* Vol. 10, No. 3 pp. 349–70) and "Democracy and Politics of Parliamentary Immunity in Turkey" (*New Perspectives on Turkey* No. 33, pp. 121–43). Koçan holds a doctorate in political science from Bilkent University.

MESAKE KOROI was born in 1947. He began his schooling at Mavana District Village School and completed it at Ratu Kadavulevu School and the Fiji College of Agriculture. He has been in print and radio journalism, and media–public relations in Fiji all his working life—as a journalist with the Fijian language *Nai Lalakai* and chief-of-staff with the *Fiji Times* newspaper, then as a radio announcer and news editor for the Fiji Broadcasting Corporation, and finally, as general manager and publisher of the *Fiji Daily Post.* In between these positions, Mesake has worked in various public relations positions for the Fiji Development Bank, and the Fiji police force. A survivor of four coups (two in 1987, and one in 2000 and 2006), Mesake contributed a chapter to the recent book *1987, Fiji Twenty Years On.* Mesake last worked as training officer of staff of the *Fiji Sun* newspaper, but is now retired and spends his days farming and reading.

SARATH W. KOTAGAMA is professor of environmental science at the University of Colombo. He is the author of *The National Wildlife Policy for Sri Lanka* (1990), the National Wetland Conservation Policy for Sri Lanka (1997), the sector author on Environmental Education for the National Conservation Strategy (1985), a member of the National Science Policy Committee (1984), and the author of the strategy for the preparation of the National Biodiversity Action Plan for Sri Lanka (1977). He is also a member of the Ministerial Task Force on Wildlife (1999–2000) and of the Presidential Task Force on Wildlife (2000). His published works include over 75 peer-researched publications, and books on birds in Sinhala and English. Professor Kotagama holds a doctorate in ornithology from the University of Colombo.

LIAM LEONARD is an award-winning author and lecturer who specializes in the areas of environmental justice, sociology, criminology, and human rights. He has won the Research Excellence Award from Sage Publishing and the NAIRTL Award for research-based innovation in teaching.

Dr. Leonard is the senior editor of the *Advances in Ecopolitics* book series. He is author or editor of over 10 books, including *Sustainable Politics and Crisis of the Peripheries: Ireland and Greece, Global Ecological Politics, The Environmental Movement in Ireland,* and *Community Campaigns for Sustainable Living: Health Waste and Protest in Civil Society.*

DA MAO is postdoctoral fellow at the School of Chemistry, Beijing Normal University, China. His doctoral dissertation has been published under the title, "The Sea Hath Bounds: Waste Disposal in the Ocean in the United States, 1870s–1930s." He holds a doctorate in world history from Beijing Normal University.

SYLVIE OLLITRAULT is research fellow at Centre National de la Recherche Scientifique (France), at University of Rennes 1, France. Her published works are mostly in French, including: *Militer pour la planète, la désobéissance civile* (civic disobedience). She has also edited many articles on environmental topics such as social movements, pressure groups, and the history of French Green parties. She was senior associate member of St Antony's College (Oxford University) in 1997–1998 and she is involved in the Green Politics Standing Group of the European Consortium for Political Research. She holds a PhD from the University of Rennes 1, France.

AHMET ÖNCÜ, PhD, is professor of sociology at Sabancı University, Istanbul, Turkey. His published works include "Wither Business Ideology: Revisiting Veblen's Theory of Engineers as Revolutionary Actors," in *Review of Radical Political Economics* 41, no. 2 (2009): 196–215; "Imbalances in the World Economy and Congestion in the Periphery," in *Neoliberal Globalization as New Imperialism: Case Studies on Reconstruction of the Periphery,* New York, NY: Nova Science, 2007, 55–76; "Dictatorship Plus Hegemony: A Gramscian Analysis of the Turkish State," in *Science and Society* 67, no. 3 (2003): 303–28. Öncü holds a doctorate in sociology from University of Alberta.

CASSANDRA STAR, PhD, is a senior lecturer in public policy, and director of studies of the graduate program in public administration, at Flinders University, South Australia. She is author of more than 40 book chapters and papers on climate change and other public policy issues. Her current research is a multi-country comparative study of climate activism in the developed world. Cassandra holds a doctorate in environmental politics from Monash University, Victoria, Australia.

JOCHEN ROOSE is assistant professor of sociology at the Freie Universität Berlin, Germany. Dr. Roose's published works include *Vergesellschaftung an Europas Binnengrenzen* (*Societization at European Inner Borders*); *How European is European Identity* (JCMS 2013); *Europäisierung von Umwelt-Organisationen* (*Europeanization of Environmental Organizations*). Roose holds a doctoral degree from the Freie Universität Berlin.

CHRISTOPHER ROOTES is professor of environmental politics and political sociology, and director of the Centre for the Study of Social and Political Movements, at the University of Kent, Canterbury, U.K. As well as publishing many articles and chapters on social movement theory, student and environmental movements, protest and Green parties, he has edited: *The Green Challenge: The Development of Green parties in Europe* (with D. Richardson) (Routledge 1995); *Environmental Movements: Local, National and Global* (Cass/Routledge 1999); *Environmental Protest in Western Europe* (Oxford 2003); *Acting Locally: Local Environmental Mobilizations and Campaigns* (Routledge 2008); and *Environmental Movements and Waste Infrastructure* (with L. Leonard) (Routledge 2010).

ASHOK SWAIN is professor of peace and conflict research at the Uppsala University, Sweden, and he also serves as the director of Uppsala Centre for Sustainable Development. His published works include *The Environmental Trap: The Ganges River Diversion, Bangladeshi Migration and Conflicts in India*; *Managing Water Conflict: Asia, Africa and the Middle East*; *Struggle against the State: Social Network and Protest Mobilisation in India*; and *Understanding Emerging Security Challenges: Threats and Opportunities.* Swain holds a doctorate in international politics from Jawaharlal Nehru University, New Delhi.

ROBERT WOLFGRAMM was born in Fiji in 1952 and arrived in Australia in 1963 for his secondary and subsequent education. After failing theology, he turned to sociology and completed his BA in 1980, his MA in 1983 and his PhD in 1994. After 24 years of tertiary teaching he retired to the country of his birth in 2004 as editor-in-chief of the *Fiji Daily Post* newspaper. Following the military coup of 2006 and the censorship issues it brought in its train, the newspaper suspended publication in 2009. Robert now works voluntarily as a Bible editor, and Australian Rules football promoter.

OLEG YANITSKY, PhD, is a professor of environmental sociology at the Institute of Sociology of the Russian Academy of Science. Professor

Yanitsky has published extensively on environmental movements and environmental policy in Russia, as well as on the risks of post-socialist transition. His main books are: *Cities of Europe: The Public's Role in Shaping the Urban Environment* (1991, co-editor); *Russian Environmentalism: Leading Figures, Facts, Opinions* (1993); *Ecological Movement in Russia: Critical Analysis* (1996, in Russian); *Russia: Risks and Threats of 'Transition' Society,* editor (1998, in Russian); *Russian Greens in A Risk Society: A Structural Analysis* (Helsinki, 2000); *Sociology of Risk* (2003, in Russian); *Russian Environmentalism: The Yanitsky Reader* (2010).